MOLECULAR IONS:
SPECTROSCOPY, STRUCTURE AND CHEMISTRY

MOLECULAR IONS: SPECTROSCOPY, STRUCTURE AND CHEMISTRY

Edited by

TERRY A. MILLER

and

V. E. BONDYBEY

Bell Laboratories
Murray Hill
New Jersey
U.S.A.

1983

NORTH-HOLLAND PUBLISHING COMPANY
AMSTERDAM · NEW YORK · OXFORD

© Bell Telephone Laboratories, Incorporated, and North-Holland Publishing Company - 1983

ISBN: 0 444 86717 1

Published by:
NORTH-HOLLAND PUBLISHING COMPANY
AMSTERDAM · NEW YORK · OXFORD

Sole distributors for the U.S.A. and Canada:
ELSEVIER SCIENCE PUBLISHING COMPANY, INC.
52 VANDERBILT AVENUE
NEW YORK, N.Y. 10017

Library of Congress Cataloging in Publication Data
Main entry under title:

Molecular ions.

    Includes bibliographical references and index.
    1. Ions. 2. Molecular structure. I. Miller,
Terry A. (Terry Alan), 1943-   . II. Bondybey,
V. E. (Vladimir E.), 1942-   .
QD561.M6876 1983   541.2'2   83-12100
ISBN 0-444-86717-1 (U.S.)

Printed in The Netherlands

PREFACE

For many years the rich chemistry and physics of molecular ions were the almost exclusive domain of mass spectroscopists. While mass spectroscopy and related techniques can give very useful data about the existence and stability of molecular ions, they provide little information about their structures and usually cannot even differentiate between isomeric species. Recently a variety of other, structurally more sensitive experimental methods, have been applied to the study of molecular ions. Most of these techniques involve absorption or emission of photons and fall into the general area of spectroscopy. Application of these techniques has yielded a wealth of new information about molecular ions, their geometry, and their vibrational and electronic structures and their chemistry. It is this rather explosive growth of available information that has prompted us to edit this book.

We are very fortunate indeed that one of the pioneers and leading scientists in this field, Dr. Gerhard Herzberg, has agreed to contribute an introductory chapter and provide us with a historical overview of the subject. He has appropriately divided the spectroscopic study of ions into three periods - early, middle, and late. The early period was characterized by the first observations of the optical emission spectra of ions and highlighted by the interplay between the experimental ion work and the development of molecular orbital theory by Hund, Mulliken, and others. The second period, starting in about 1950, saw the observation of the optical absorption spectra of ions, the spectra of negative and polyatomic molecular ions, as well as the development of photoelectron spectroscopy. The final period, in which we find ourselves presently, can probably best be dated from the first application of lasers to the study of ions. However, the laser is but one of the recent developments which has led to remarkable advances in our knowledge about ions.

The rest of the chapters in this volume can best be described as belonging to the late (present) period of ion study. We have tried to arrange the remainder of this book in a somewhat logical fashion. The three chapters following Herzberg's historical overview deal primarily with high and very high resolution studies of some of the simplest molecular ions. Woods describes in considerable detail his efforts to obtain the spectra of di- and tri-atomic molecular ions in the microwave spectral region, and comments lucidly on the astrophysical implications of his work. Carrington and Softley move the frequency up a bit, but for the most part maintain the resolution, with their survey of the infrared spectroscopy of molecular ions. Particular attention is given to the fundamental ions HD^+, HeH^+, CH^+, and H_3^+. The latter ion, H_3^+, is so important that our next chapter by Oka is entirely devoted to it. It is clear that the motivation for much of the high resolution work described in Chapters 2-4 is two-fold. Because of the extreme simplicity of the ions involved, a detailed experimental knowledge of their spectra and structure provides the ideal benchmark for the testing of our detailed theoretical understanding of molecular bonding. On the other hand, because of the species' very simplicity, they constitute some of the most abundant and important molecules in the universe. Thus a significant portion of the laboratory work is in support of the observation of these same ions in interstellar space. Ions like H_3^+, HCO^+, etc. likely play key roles in the basic underlying chemistry upon which the universe is based. All of our first four chapters comment upon the astrophysical significance of molecular ions.

Chapters 5 and 6 are the only chapters in this book which do not involve ions in the gas phase. Rather they detail how molecular ions can be isolated in cryogenic inert gas matrices for study. Chapter 5 by Andrews centers upon the study of the vibrational structure and interactions of matrix isolated ions by infrared spectroscopy. In the chapter by Bondybey and Miller, a great deal of information about the photophysics, vibrational, and electronic structure of matrix isolated ions is presented. In this case, the principal probe is electronic spectroscopy, particularly laser induced fluorescence.

In Chapter 7 by Klapstein, Maier, and Misev our steady progress towards higher frequencies and larger molecular ions is maintained. However, now the subject is the decay of excited electronic states of organic ions in the gas phase. In particular, comparison is made between electron impact excited emission and laser excited fluorescence of gaseous organic ions.

Chapter 8 by Miller and Bondybey continues the interest in laser induced fluorescence studies of organic cations, but the interest is now restricted to a number of highly symmetric benzenoid cations, e.g. $C_6F_6^+$, sym $C_6F_3H_3^+$, etc., which have electronically doubly degenerate ground states and thus are subject to the Jahn-Teller effect. The basic theory of the Jahn-Teller effect is developed and stabilization energies and distorted geometries given for several of these ions.

In the final chapter by Dunbar, the interest remains with large gas-phase molecular ions. However, unlike the previous eight chapters, the goal is now to understand the photophysics of molecular ions when they are exposed to radiation which, once absorbed, leads to their destruction (fragmentation). This is an area of much current interest for both neutral and ionic species. However, in many cases the photofragmentation of ions may provide more information than can be obtained from neutrals because of the ease of detecting ionic species.

We believe that overall the nine chapters in this book provide a good perspective upon the exciting recent developments in the understanding of the spectroscopy, structure, and chemistry of ions. We have tried to provide a balanced account ranging from the elegant high resolution spectroscopic experiments on very simple ions to a variety of lower resolution work on much more complex molecular ions. We particularly wish to thank the authors of the various chapters. Each has made outstanding contributions to the work described, and has in this present volume presented his work in a manner easily assimilable.

CONTENTS

5.   Spectroscopy of Molecular Ions in Noble Gas Matrices
Lester Andrews

6.   Vibronic Spectroscopy and Photophysics of Molecular Ions in Low Temperature Matrices
V. E. Bondybey and Terry A. Miller

7.   Emission and Excitation Spectroscopy of Open-shell Organic Cations
Dieter Klapstein, John P. Maier, and Liubomir Misev

List of Contributors

LESTER ANDREWS, Department of Chemistry, University of Virginia, Charlottesville, Virginia, U.S.A. 22901.

VLADIMIR E. BONDYBEY, Bell Laboratories, Murray Hill, New Jersey, U.S.A. 07974.

ALAN CARRINGTON, Department of Chemistry, University of Southampton, Southampton, England S09 5NH.

ROBERT C. DUNBAR, Department of Chemistry, Case Western Reserve University, Cleveland, Ohio, U.S.A. 44106.

GERHARD HERZBERG, National Research Council of Canada, Ottawa, Ontario, Canada K1A OR6.

DIETER KLAPSTEIN, Physikalisch-Chemisches Institut, Universität Basel, CH-4056 Basel, Switzerland.

JOHN P. MAIER, Physikalisch-Chemisches Institut, Universität Basel, CH-4056 Basel, Switzerland.

TERRY A. MILLER, Bell Laboratories, Murray Hill, New Jersey, U.S.A. 07974.

LIUBOMIR MISEV, Physikalisch-Chemisches Institut, Universität Basel, CH-4056 Basel, Switzerland.

TAKESHI OKA, Department of Chemistry and Department of Astronomy and Astrophysics, The University of Chicago, Chicago, Illinois, U.S.A. 60637.

T. P. SOFTLEY, Department of Chemistry, University of Southampton, Southampton, England S09 5NH.

R. CLAUDE WOODS, Department of Chemistry, University of Wisconsin, Madison, Wisconsin, U.S.A. 53706.

Molecular Ions: Spectroscopy, Structure and Chemistry
Terry A. Miller and V.E. Bondybey (editors)
© North-Holland Publishing Company, 1983

INTRODUCTION TO MOLECULAR ION SPECTROSCOPY

G. Herzberg

National Research Council of Canada
Ottawa, Ontario, K1A 0R6
Canada

The field of molecular ion spectroscopy has developed in recent years into a very active subject. A detailed review of this subject as presented in this monograph seems therefore useful for those now involved in it as well as those who want to begin studies in this field or try to apply the understanding gained to other fields.

During the early period of the history of the subject it was of course by no means obvious which band spectra observed in electric discharges were due to neutral molecules and which to molecular ions. In 1922 Wien [1] showed that in a positive ray deflected by an electric or magnetic field the emission of the so-called "negative nitrogen bands" unlike the "second positive group" goes with the deflected beam; that is, the carrier of this spectrum must be the positive ion of the nitrogen molecule, N_2^+. The designation "negative bands" of nitrogen arose from the observation that these bands occur near the cathode in the negative glow. It was then realized that bands of other molecules that are predominant in the negative glow must also be due to the corresponding positive ions. The first to be so identified were the negative bands of oxygen and carbon monoxide. These identifications were later confirmed when the doublet or quartet character of the bands assigned to the ions N_2^+, O_2^+ and CO^+ was recognized, since for an odd number of electrons the multiplicity must be even. Other ion spectra identified in this way were those of HCl^+, HBr^+, BH^+, AlH^+ and much later HF^+. Also in the early period the spectra of the ions BeH^+, MgH^+, ZnH^+, CdH^+ and HgH^+ as well as CH^+ and OH^+ were identified mainly on the basis of the observation that all of these are singlet (or triplet) spectra while the corresponding neutral molecules have even multiplicities.

The early work on molecular ion spectra just described was closely tied to the development of molecular orbital theory by Hund and Mulliken. For example, the comparison of the spectra of N_2^+ and CO^+ with that of CN made it very clear that the most loosely bound orbital in N_2 and CO must be a σ orbital while the next lowest orbitals must be π and σ since the first and second excited states of the ions are $^2\Pi$ and $^2\Sigma$. To be sure, at the time the $^2\Pi$ state of N_2^+ had not yet been observed but the similarity to CO^+ and CN left no doubt that it existed. It was actually observed first in the spectrum of the aurora by Meinel [2] and soon after in the laboratory by Dalby and Douglas [3,4].

Another interesting group of diatomic molecular ions which were studied by the same conventional methods as those mentioned in the preceding paragraphs but much more recently are the ions consisting of two inert gas atoms. Druyvesteyn [5] in 1931 was the first to notice two band groups near 4250 and 4100 Å in discharges through mixtures of He and Ne, but it was only in 1975 that it was recognized

by Tanaka, Yoshino and Freeman [6] that these band groups were due
to HeNe$^+$ and that similar band groups corresponding to XY$^+$ occur in
all binary mixtures of inert gases X and Y. Tanaka, Yoshino and
Freeman found that these band groups occur approximately at a wavelength corresponding to the difference of the ionization potentials
of X and Y, strongly suggesting that they correspond to charge
exchange between the two ends of the molecule, i.e. XY$^+$ → X$^+$Y. The
dissociation energies are generally very small corresponding to van
der Waals binding. There is one exception, the HeNe$^+$ ion. Its
spectrum (both that of ^{4}HeNe$^+$ and ^{3}HeNe$^+$) has been analysed in detail
[7] and shows that the ground state has a dissociation energy of
0.69 eV and an equilibrium internuclear distance r_e = 1.30 Å
indicating a much stronger bond than for any of the excited states
of this molecule and all states of the other XY$^+$ molecules. Figure 1

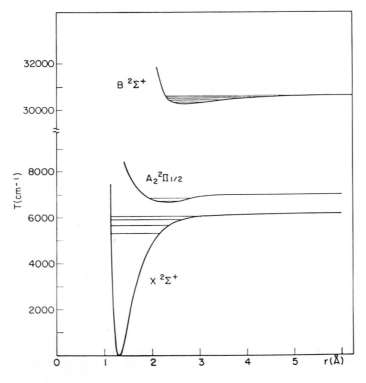

Figure 1
RKR Potential Functions of the
Three Observed States of HeNe$^+$

shows the potential functions of the observed states of HeNe$^+$. In
contrast, in HeAr$^+$ the dissociation energy of the ground state is
only 0.026 eV [8]. The reason for this difference is clearly the
interaction between B $^2\Sigma^+$ and X $^2\Sigma^+$ which because of the much
greater proximity is much larger for HeNe$^+$ than for HeAr$^+$.

The second (middle) period of the history of our subject starting in
about 1950 but overlapping the first period just described included

the following developments:

    (a) the observation of <u>absorption</u> spectra of molecular ions;

    (b) the observation of spectra of negative ions;

    (c) the observation of spectra of polyatomic molecular ions;

    (d) the study of electronic states of molecular ions by means of photo-electron spectra.

A few remarks about each of these steps may be in order.

The N_2^+ ion was the first to be detected in absorption [9]. This was done by means of the flash discharge method illustrated in Figure 2. The ions produced by a mild flash discharge in the long

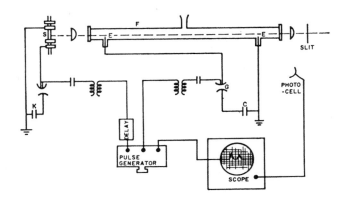

Figure 2
Flash Discharge Method for the Study of
Absorption Spectra of Molecular Ions

and fairly wide tube F were studied in absorption using the continuous spectrum produced by a second narrow flash discharge tube S as a continuous background. This second tube was triggered with a short delay after the flash in the absorption tube. Figure 3 shows

3914 4 Å

Figure 3
Absorption Spectrum of N_2^+ (B $^2\Sigma_u^+$ - X $^2\Sigma_g^+$)

as an example the 3914 Å band of N_2^+ in absorption. By the same method the corresponding spectrum of CO^+ at 2189.8 Å was obtained in absorption. When the absorption spectrum of a flash discharge through methane was studied (with the aim of obtaining the absorption spectrum of CH_4^+) a very simple spectrum similar to that of N_2^+ was

found near 5400 Å (both in emission and absorption) which
was shown [10,11] to be due to the negative ion C_2^-. It represents
the same electronic transition in this negative ion as the 3914 Å
system in the positive ion N_2^+ (having the same number of electrons).
In Figure 4 the electronic states of C_2^-, CN, N_2^+ and CO^+ are
compared.

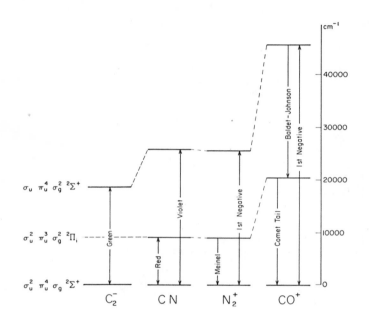

Figure 4
Observed Electronic States of 13-electron Systems

The first polyatomic ion spectrum, that of CO_2^+, was already
observed in emission in the early period of the subject [12,13,14].
In 1967 when the flash discharge method was applied to CO_2 both band
systems of CO_2^+, that is, $\tilde{A}\,^2\Pi$ - $\tilde{X}\,^2\Pi$ and $\tilde{B}\,^2\Sigma$ - $\tilde{X}\,^2\Pi$, were observed
in absorption [9,15]. Other polyatomic molecular ions observed in
this intermediate period (before the advent of laser techniques) are
$C_4H_2^+$ (the diacetylene ion) studied in emission [16], N_2O^+ studied
both in emission and absorption [15,17], CS_2^+ [18], COS^+ [19,20],
H_2S^+ [21] and H_2O^+ [22] studied in emission only. Extensive
rotational and vibrational analyses of these ions were carried out
by the authors mentioned.

Toward the end of the intermediate period the method of photoelectron spectroscopy was developed by several groups,under
A.N. Terenin, D.W. Turner, K. Siegbahn and W.C. Price. Here the
energy levels of the ions are determined from the energies of the
peaks of the photo-electrons produced by photons of short (fixed)
wave lengths, usually the He line at 584 Å. At low resolution,peaks
are observed which correspond to the various electronic states of
the ion, at higher resolution these peaks are resolved into bands
corresponding to different vibrational transitions and in a few
cases at very high resolution the rotational structure is resolved.

This method has supplied data about a large number of molecular ions. A few ions of free radicals have also been studied in this way.

The first break in the rather slow development of the field of molecular ion spectroscopy came when low voltage electron beam excitation of various parent molecules was used to produce ion emission spectra. This method was pursued by a number of independent groups. In our laboratory Lew [22] discovered in this way an extensive spectrum of H_2O^+ and D_2O^+. I tried to find in this way a spectrum of the NH_3^+ ion. Although I did observe interesting spectra in the region 2200-2400 Å which I considered as strong candidates for NH_3^+ and NH_2^+ a confirmation of this assignment is still lacking. At about the same time Maier and his collaborators in Basel, Miller and his collaborators at the Bell Laboratories, and Leach and his associates at Orsay started their detailed and comprehensive work on a large number of organic ions, especially ions of derivatives of benzene such as $C_6F_6^+$, $C_6HF_5^+$, etc. Of special interest is the detailed study made possible by these experiments of the Jahn-Teller effect in the degenerate electronic ground states in ions like $C_6F_6^+$ and $C_6H_3F_3^+$. The use of laser excitation makes possible the study of the absorption spectra of these ions. The work of Maier and Miller is fully described in the chapters contributed by these authors to this volume.

The power of laser techniques is beautifully illustrated by the study of N_2^+ by Rosner, Gailey and Holt [23] who, using the laser beam-ion beam technique, were able to resolve the hyperfine structure of the N_2^+ lines. Figure 5 shows an example from their paper.

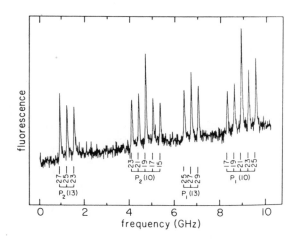

Figure 5
Fluorescence vs Doppler-shifted Laser Frequency
(with Arbitrary Origin) for a Part of the 0-1
Band of the B $^2\Sigma_u^+$ - X $^2\Sigma_g^+$ System of N_2^+

During the same period R.C. Woods and his students obtained microwave spectra of a number of triatomic molecular ions like HCO^+, HOC^+, N_2H^+, HCS^+ and their isotopes, obtaining precise frequencies and rotational constants. This work will be described in the Chapter contributed by Woods.

The addition of laser techniques has greatly aided in the further development of the subject. The use of the combination of a laser beam with a molecular ion beam of variable velocity allowed Wing and his associates to obtain for the first time an infrared spectrum of the HD^+ ion which, as was first pointed out by E. Teller, has a strong oscillating dipole moment. Here also ab initio calculations have been possible which show a remarkable agreement with the observed spectra. Carrington and his students have observed by a similar method many of the higher vibrational levels of HD^+ which show almost as good agreement with ab initio calculations as the lower observed levels (see the Chapter contributed by Carrington and Softley).

The simplest stable polyatomic molecular ion is H_3^+ which is the most prominent ion in the mass spectrum of hydrogen unless the pressure in the source is very low. The infrared spectrum of this ion could in principle be observed by orthodox methods. However the stationary concentration of H_3^+ (like that of many ions) in electric discharges is very low and its detection requires therefore more sophisticated methods. These were first applied by Oka [24] as described more fully in his Chapter. He used a difference laser and with it reached a detection limit of about 4×10^{-6} cm^{-1}. In this way a fairly complete infrared band of H_3^+ was obtained and several molecular constants were determined. Almost simultaneously Wing and his associates [25] obtained the spectrum of D_3^+ by the use of the molecular ion-beam laser-beam technique. The same two methods were applied to the discovery of a spectrum of HeH^+, first by Tolliver, Kyrala and Wing [26] by the beam method and later by Bernath and Amano [27] using Oka's method.

Carrington, Buttenshaw and Kennedy [28] using the ion-beam laser-beam method have recently obtained extensive spectra of H_3^+ corresponding to transitions between levels near the dissociation limit. This spectrum is still awaiting a detailed analysis.

J.D. Morrison and his students [29] have studied the spectra of the ions CH_3I^+ and CD_3I^+ using a triple quadrupole mass spectrometer with a tunable dye laser. The region studied is one in which predissociation occurs and thus a study of the absorption spectrum by detecting the photodissociation products was possible under fairly high resolution. A full analysis of this spectrum has not yet been achieved. A similar method was used by the same group for the study of the spectra of O_2^+ [30] and O_3^+ [31]. Ion-beam laser-beam studies of O_2^+ have also been carried out by Carrington's group.

Still another method, using a drift tube mass spectrometer combined with a tunable dye laser, was applied by Mosely, Cosby and their associates to the study of the absorption spectra of a number of negative ions especially O_3^- [32], [33] and CO_3^- [34]. As in the previous methods absorption is detected by way of predissociation which leads to the occurrence of product ions observed by the mass spectometer. The observed vibrational structure allows determination of vibrational frequencies in the excited states. From the long wave length limits of the absorptions the electron affinities can be determined.

There has been an ever increasing interplay in the last 50 years between laboratory spectroscopy and astronomy [35]. This interplay has been very active with regard to molecular ion spectra. The

first incidence of this interplay was the laboratory observation in
1909 by Fowler [36] of the prominent doublet bands observed in the
spectra of comet tails by Deslandres. While at the time it was not
yet established that these bands are due to CO^+ it did become clear
somewhat later. Spectra of comet tails of reasonable resolution
became available only in the 1940's when bands of N_2^+, CH^+, CO_2^+,
and more recently OH^+ were identified. No neutral molecules were
found in the tails. With the advent of Comet Kohoutek in 1973, for
the first time comet tail spectra of reasonably high resolution in
the red and near infrared became available. At this time the
laboratory spectrum of H_2O^+ had just been discovered and analysed by
H. Lew and turned out to be the key to the identification of a
considerable number of comet tail features [37][38]. Only low
rotational lines of H_2O^+ are observed but these are quite strong.
The identification of strong H_2O^+ bands in comet tails supplied a
welcome confirmation of Whipple's comet model, i.e. "dirty ice".

Interestingly the same H_2O^+ bands were recently identified [39] in
spectra of the twilight glow of the earth's atmosphere obtained by
Krassovsky and Viniukov [40]. It should be noted however that up to
now no mass-spectrometric observation of the H_2O^+ ion has been made
in the upper atmosphere.

The spectrum of the CH^+ ion in the visible region was observed in
the interstellar medium before it was observed in the laboratory.
Rather the astronomical observation stimulated the laboratory
studies which established the identification [41]. The next three
ions, HCO^+, N_2H^+ and HCS^+, were also first observed in the interstellar medium, in the microwave region, before they were found in
the laboratory. It was first suggested by Klemperer [42] that a
strong unidentified microwave line at 89190 MHz observed by Buhl
and Snyder [43] might be due to the HCO^+ ion but it took several
years before this identification was definitely established, first
by the observation of corresponding lines of DCO^+ and $H^{13}CO^+$ in
interstellar clouds and finally by the observation in the laboratory
(see Woods' contribution to this volume). For N_2H^+ the identification was easier because of the observation in the interstellar
feature of hyperfine structure both corresponding to the inner and
the outer nitrogen nucleus. Soon after, N_2H^+ was observed in the
laboratory by Woods and his associates. The last ion so far
observed in the interstellar medium, viz. HCS^+, was first identified
by Thaddeus, Guélin and Linke [44] by comparison with theoretical
rotational constants and was finally observed in the laboratory
again by Woods and his associates [45].

Although the number of molecular ions observed in the interstellar
medium is still small it is generally agreed, on the basis of the
work by Herbst and Klemperer [46], Watson [47] and others, that ions
play a major role in the chemical reactions that take place in the
interstellar medium.   This is so largely because ion-molecule
reactions are fast reactions; they have no activation energy.

Molecular hydrogen is by far the most abundant molecule in the interstellar medium; it may be ionized by UV photons or cosmic rays but
the H_2^+ ion immediately reacts (on the first collision) with H_2
according to the well-known reaction

$$H_2^+ + H_2 \rightarrow H_3^+ + H \quad .$$

The H_3^+ so formed is very stable and has a relatively long life
which is ended only by the recombination with an electron and

subsequent dissociation. This dissociative recombination of H_3^+ in
all probability proceeds via the Rydberg states of (neutral) H_3
whose ground state of course is unstable. The same process may be
assumed to be responsible for the emission of the Rydberg spectrum
observed in hollow cathode discharges [48][49]; it thus seems
likely that such an emission (at 5600 and 7100 Å) may be observed
from suitable interstellar clouds. It is possible (even though by
no means certain) that NH_3 in interstellar clouds is formed in a
similar way by dissociative recombination of NH_4^+ with electrons.
In this process again the Rydberg spectrum of neutral NH_4 recently
identified in the laboratory [50] might be emitted.

Interstellar H_3^+ is generally held responsible for the formation of
HCO^+ by the reaction

$$H_3^+ + CO \rightarrow HCO^+ + H_2 \quad .$$

Many other subsequent reactions lead to the formation of the many
observed interstellar molecules. To put all these considerations
on firm ground it will of course be necessary to detect H_3^+
directly.

In view of the known fairly high abundance of H, He, Ne and Ar it
would not be surprising if evidence for the presence of the
previously discussed ions HeH^+. $HeNe^+$ and $NeAr^+$ in interstellar
clouds could be found. The mechanisms by which these ions could be
formed are however rather uncertain and may have a low yield.
Further work on the spectra of these ions is clearly desirable.

REFERENCES

[1] W. Wien, Ann. de Phys. (4) 69, 325 (1922), 81, 994 (1926).
[2] A. Meinel, Ap. J. 114, 431 (1951)
[3] P. Dalby and A.E. Douglas, Phys. Rev. 84, 843 (1951).
[4] A.E. Douglas, Ap. J. 117, 380 (1953).
[5] M.J. Druyvesteyn, Nature 128, 1076 (1931).
[6] Y. Tanaka, K. Yoshino and D.E. Freeman, J. Chem. Phys. 62, 4484
    (1975).
[7] I. Dabrowski and G. Herzberg, J. Mol. Spectrosc. 73, 183 (1978).
[8] I. Dabrowski, G. Herzberg and K. Yoshino, J. Mol. Spectrosc.
    89, 491 (1981).
[9] G. Herzberg, Pontif. Acad. Sci. Commentarii 2, N.15, 1 (1968).
[10] G. Herzberg and A. Lagerqvist, Can. J. Phys. 46, 2363 (1968).
[11] W.C. Lineberger and T.A. Patterson, Chem. Phys. Lett. 13, 40
     (1972).
[12] R.S. Mulliken, J. Chem. Phys. 3, 720 (1935).
[13] F. Bueso-Sanllehi, Phys. Rev. 60, 556 (1941).
[14] S. Mrozowski, Phys. Rev. 60, 730 (1941), 62, 270 (1942), 72,
     682, 691 (1947).
[15] G. Herzberg, Quarterly Rev. Chem. Soc. 25, 201 (1971).
[16] J.H. Callomon, Can. J. Phys. 34, 1046 (1956).
[17] J.H. Callomon, Proc. Chem. Soc. p. 313 (1959); J.H. Callomon
     and F. Creutzberg, Phil. Trans. 277, 157 (1974).
[18] J.H. Callomon, Proc. Roy. Soc. 244A, 220 (1958).
[19] S. Leach, J. chim. phys. 61, 1493 (1964).
[20] M. Horani and S. Leach, C.R. Paris 248, 2196 (1959).
[21] G. Duxbury, M. Horani and J. Rostas, Proc. Roy. Soc. 331A, 109
     (1972).
[22] H. Lew, Can. J. Phys. 54, 2028 (1976).
[23] S.D. Rosner, T.D. Gailey and A. Holt, Phys. Rev. 26A, 697 (1982).

[24] T. Oka, Phys. Rev. Lett. 45, 531 (1980).
[25] J.T. Shy, J.W. Farley, W.E. Lamb, Jr. and W.H. Wing, Phys. Rev. Lett. 45, 535 (1980).
[26] D.E. Tolliver, G.A. Kyrala and W.H. Wing, Phys. Rev. Lett. 43, 1719 (1979).
[27] P. Bernath and T. Amano, Phys. Rev. Lett. 48, 20 (1982).
[28] A. Carrington, J. Buttenshaw and R. Kennedy, Mol. Phys. 45, 753 (1982).
[29] S.P. Goss, J.D. Morrison and D.L. Smith, J. Chem. Phys. 75, 757 (1981).
[30] D.C. McGilvery, J.D. Morrison and D.L. Smith, J. Chem. Phys. 70, 4761 (1979).
[31] S.P. Goss and J.D. Morrison, J. Chem. Phys. 76, 5175 (1982).
[32] P.C. Cosby, J.T. Moseley, J.R. Peterson and J.H. Ling, J. Chem. Phys. 69, 2771 (1978).
[33] G.P. Smith and L.C. Lee, J. Chem. Phys. 71, 2323 (1979).
[34] J.T. Moseley, P.C. Cosby and J.R. Peterson, J. Chem. Phys. 65, 2512 (1976).
[35] G. Herzberg, Highlights of Astronomy, Vol. 5, 3 (1980).
[36] A. Fowler, M. N. Roy. Astr. Soc. 70, 176 (1909).
[37] G. Herzberg and H. Lew, Astron. & Astrophys. 31, 123 (1974).
[38] W.A. Wehinger, S. Wyckoff, G.H. Herbig, G. Herzberg and H. Lew, Ap. J. 190, L43 (1974).
[39] G. Herzberg, Ann. de Géophys. 36, 605 (1980).
[40] V.I. Krassovsky and K.I. Viniukov, Ann. de Géophys. 35, 109 (1979).
[41] A.E. Douglas and G. Herzberg, Ap. J. 94, 381 (1941), Can. J. Res. 20A, 71 (1942).
[42] W. Klemperer, Nature 227, 1230 (1970).
[43] D. Buhl and L.E. Snyder, Nature 228, 267 (1970)
[44] P. Thaddeus, M. Guélin and R.A. Linke, Ap. J. 246, L41 (1981).
[45] C.S. Gudeman, N.N. Haese, N.D. Piltch and R.C. Woods, Ap. J. 246, L47 (1981).
[46] E. Herbst and W. Klemperer, Ap. J. 185, 505 (1973).
[47] W.D. Watson, Ap. J. 183, L17, 188, 35 (1973).
[48] G. Herzberg, J. Chem. Phys. 70, 4806 (1979).
[49] I. Dabrowski, G. Herzberg, J.T. Hougen, H. Lew, J.J. Sloan and J.K.G. Watson, Can. J. Phys. 58, 1238, 1240 (1980), 59, 428 (1981), 60, 1261 (1982).
[50] G. Herzberg, Faraday Disc. No. 71, 165 (1981).

Molecular Ions: Spectroscopy, Structure and Chemistry
Terry A. Miller and V.E. Bondybey (editors)
© North-Holland Publishing Company, 1983

SPECTROSCOPY OF MOLECULAR IONS
IN THE MICROWAVE REGION

R. Claude Woods

Department of Chemistry
University of Wisconsin
Madison, Wisconsin
U.S.A.

In the microwave portion of the electromagnetic spectrum
absorption due to a half dozen molecular ions has been
detected in laboratory low pressure discharges, and
emission from most of these same species has been observed
by radioastronomy. The astrophysical work has shown that
these ions are widely distributed in the interstellar
medium and has provided a great deal of information on the
latter's chemistry and dynamical conditions. The more than
fifty neutral molecules that have been observed in interstellar gas clouds are generally thought to be formed in
sequences of ion-molecule reactions, and the microwave
observations of the ions themselves has provided critical
evidence in support of this mechanism. The laboratory studies,
in addition to providing unambiguous identifications and
accurate frequencies for astronomical work, yield very detailed
and precise information on molecular structure and bonding.
Relative intensity data can reveal the extent of an ion's
vibrational excitation, Doppler shifts can determine velocities,
and pressure broadened line widths can define the dominant
intermolecular forces in collisions between ions and neutrals.

INTRODUCTION

Spectroscopic transitions occuring in the microwave region, which we shall take to
include frequencies between roughly 4 GHz and perhaps 600 GHz, are generally
those between the various rotational levels of a given vibronic state of a molecule. For certain paramagnetic species transitions between various spin or fine-
structure components of a given rotational state may also occur at these frequencies. Since the leading terms in the model Hamiltonian used to predict the
rotational energy levels are the rigid rotor ones, the relation of the measured
transition frequencies to the species' moments of inertia, and thus its geometrical structure parameters (bond distances and angles) is a very direct one.
Furthermore, great precision is routinely obtained in these measurements, so
that the possibility exists to obtain very accurate and reliable molecular structures. In fact, a very large number of neutral molecules have been studied by
microwave spectroscopy, and the technique has yielded the most accurate information available on vapor phase molecules with up to a dozen atoms — a data base
upon which much of the modern understanding of structural chemistry is founded.
Microwave spectroscopy always employs coherent and highly monochromatic sources,
and at suitably low pressures, 100 mTorr or less, the linewidths of the transitions are very small. The resulting intrinsically very high resolution makes
the method ideal for the observation of small splittings due to hyperfine interactions,Zeeman or Stark effects, internal rotation or inversion, etc. Absolute
intensities of microwave transitions provide a measure of species concentration,
and relative intensities in a gas of known and definite temperature can yield
otherwise undetermined vibrational spacings and isomerization energies. In non-

equilibrium environments, e.g., electric discharges, relative intensities yield
dynamical information, i.e., the extent of vibrational, rotational or other excitation. In the late 1960's the advent of molecular radioastronomy highlighted
yet another strength of this spectroscopic tool, its extreme specificity in the
qualitative identification of an observed molecule. With sufficiently precise
frequency measurements, one or two or three observed transitions can constitute
an absolutely unambiguous fingerprint.

In spite of the obvious and compelling desirability of extending the application
of this powerful and versatile technique to the class of molecular ions, the
prospects for doing so had hardly even been seriously considered or discussed
prior to 1970 — for a variety of reasons. Because of the low photon energy the
sensitivity of microwave spectroscopy was generally less than that of optical
spectroscopy, and certainly much less than that of mass spectrometry, which had
been used for most studies of ion chemistry. (It will be seen below that in
favorable cases and with modern instrumental improvements, microwave sensitivity
can become quite high.) Large steady state densities of ions were notoriously
difficult to produce, and the plasma electrons which would inevitably accompany
them were expected to interfere drastically with the propagation of the radiation and to contribute excessive noise. The chemistry of production and destruction of specific ions in suitable low pressure electric discharges was
usually ill-characterized, and the dynamical situation of ions, their vibrational,
rotational, and translational temperatures, in such media were even more enig-
matic. There was concern that pressure-broadening might be unworkably large.
High resolution optical spectroscopic data was available for only a few molecular
ions, many of which were not particularly suitable for microwave work, and use-
fully accurate ab initio structure calculations on molecular ions were practically
non-existent, so that predictions of microwave transition frequencies was quite
problematical. Notwithstanding these real and imagined barriers, microwave spectroscopy of molecular ions has indeed become a reality in the intervening years,
and much of the techniques's promise alluded to earlier, has been demonstrated in
actual studies of specific ions.

At the present time six molecular ions, all light diatomics or triatomics, have
been studied by microwave techniques. In the following section a brief historical
outline of the observations of these six will be given to provide the reader with
general time perspective for the more detailed discussion to follow and also to
emphasize the close and complementary relationship that has existed between the
laboratory and astronomical work throughout the course of these studies. Five of
the six ions have been seen in the interstellar medium, in some cases before the
laboratory work and in some cases after it. Radio observations of these ions,
especially of HCO^+, have become a well developed area of astronomical research
and have contributed much to our understanding of the chemical processes and
physical conditions of the interstellar medium. Although the initial detection
of the laboratory spectrum of a molecular ion was rendered quite difficult by
the low available signal to noise ratio, steady and considerable improvements in
sensitivity have been made since then. Several different research groups are
now active in the laboratory studies, using various ion production schemes and
spectroscopic instrumentation, and are making important contributions to these
improvements. A separate section will summarize these various experimental
approaches and the progress that has been made. For most of the ions that have
been studied spectra of several isotopic species were obtained, and accurate
molecular structures were obtained for three of the triatomics. In a later
section of this chapter these experimental results will be shown to agree to a
very gratifying extent with molecular structures obtained from ab initio quantum
chemical calculations, which have also exhibited dramatic progress in the last
dozen years. In two cases (in two different laboratories), vibrational satellite
spectra have been obtained. The latter contribute significantly to a more
quantitative picture of the ions' structures and molecular force fields and also
provide a concrete measure of the extent of vibrational excitation of the ions
in their respective discharge types. Other types of dynamical information have

come from Doppler shifts, where frequency measurement precision has become
sufficient for meaningful measurements of ion velocity and mobility to be made,
and from pressure broadening linewidth studies, where the Langevin type interaction, i.e., the monopole-induced dipole force, has been demonstrated to be the
dominant consideration.

A previous review article [1], has briefly summarized microwave spectroscopy
of molecular ions in the larger context of all their high resolution spectroscopy.
The discussion of the present chapter provides a more in-depth treatment and
also covers a period of two additional full years, during which a great deal of
additional progress has been made.

AN HISTORICAL SKETCH

The first major step in the study of rotational spectra of molecular ions came
unexpectedly, and to a large extent accidentally, in 1970, when radioastronomers
Buhl and Snyder [2] discovered the first interstellar unidentified line. Molecular radioastronomy had only really begun in the late 1960's,and prior to this
observation all the spectral lines that had been seen had been readily assigned
to a handful of familiar molecules, whose microwave spectra were already well
characterized from laboratory work [3]. (A great many other unidentified astronomical lines (U-lines) have been discovered since that time, and most of them
remain unassigned even now.) In the absence of a concrete identification of the
molecular species responsible for their newly discovered transition, they provided it with the colorful name X-ogen. Soon afterwards Klemperer [4] made the
suggestion, which in retrospect seems a very farsighted one, that X-ogen was in
fact the molecular ion HCO^+. Using a simple rigid rotor model with $r(CH) = 1.06$ Å
and $r(CO) = 1.115$ Å (taken from HCN and CO^+, respectively), he predicted the
$J = 0-1$ transition of HCO^+ to fall at 89.246 GHz, in very good agreement with the
observed value for X-ogen, 89.190 GHz. These bond length guesses are not particularly close to the now accurately known values (see below), so that such close
frequency agreement was almost entirely fortuitous. Nevertheless, the correct
conclusion had been reached, even though almost five years were to elapse before
its validity could be proven. Very fortunately for the development of this subject the element of mystery surrounding the identification of X-ogen quickly
provided a major stimulus to further astronomical, theoretical, and laboratory
work.

The first ab initio theoretical calculations on protonated carbon monoxide
actually predate slightly the X-ogen controversy [5,6]. These Gaussian basis set
SCF computations had demonstrated clearly that HCO^+ was the most stable protonated form, although the other linear isomer, HOC^+, was seen to exist as a metastable form in a deep local minimum of the ground state potential surface, and
they had also given preliminary structural parameters. In 1973 Wahlgren et. al.
[7] published a much more accurate calculation (large Gaussian basis set with
extensive configuration interaction (CI)) and obtained $r(CH) = 1.095$ Å, $r(CO) =$
1.1045 Å, and a predicted X-ogen transition frequency of 89.36 GHz, which provided strong support for Klemperer's hypothesis. Bruna et al. [8] did further
SCF level calculations of the HCO^+ and HOC^+ potential surfaces (ground and excited) but did not attempt to improve upon the structure. The most comprehensive
theoretical calculations on the problem were those of Kraemer and Diercksen [9],
who used a very large basis set CI potential surface for HCO^+ to predict not only
the equilibrium structure (r_e's), but also the vibration-rotation interaction
corrections (α's) and the B_0 value. Their ab initio transition frequency of
89.61 GHz and their empirically corrected (by comparison to HCN results) one of
89.19 GHz were very strongly supportive too, but unfortunately they only appeared
after the X-ogen problem had been otherwise resolved.

Observational astronomical efforts to confirm the identity of X-ogen were intense,
and even negative results were published. A 1971 search for neutral formyl

radical [10] by chance covered the region around Klemperer's prediction [4] of
86.708 GHz for the $H^{13}CO^+$ species, but with the low sensitivity then available
nothing was found. By 1972 Buhl and Snyder [11], had found X-ogen in a total of
nine interstellar sources. Herbst and Klemperer [12] attempted an elaborate
empirical estimation of the $H^{13}CO^+$ frequency, but their result (86.718 ± .009 GHz),
was not much closer to the (now well known) true value (86.754 GHz) than
Klemperer's previous estimate [4]. By early 1975 Hollis et al. [13] had observed
X-ogen in twenty eight emission sources, obtained a better rest frequency, and
shown that the U89.189 line was not part of a multiplet. Other improved rest
frequency values [14,15] were also observationally deduced. While all of this
work was going on the question of whether X-ogen was HCO^+ had become a more
urgent one because of the kinetic models that were evolving to explain the formation the many molecules that had been and were being found in the interstellar
clouds. Herbst and Klemperer [16] and Watson [17-19], had independently concluded that a mechanism dominated by gas phase ion-molecule reactions could best
explain the great diversity of interstellar chemistry. Their models placed HCO^+
in a preeminent role and predicted an abundance for it that was large and not
unlike what would be deduced from the intensity of the X-ogen transition. Since
no other suitably nearby and intense candidate lines could be found, if U89.189
had not been due to HCO^+, the viability of these gas phase models would have been
very seriously compromised.

In the late 1960's we had become convinced that a microwave absorption spectrometer in which the sample medium was the plasma of a DC glow discharge was
feasible and very promising as a tool for the study of transient molecular
species and had begun to build one. Previous designs of transient spectrometers
had yielded spectra of just a few free radicals, e.g., OH [20], CS [21], or SO
[22,23], but they were always produced outside the microwave radiation path and
then flowed into it. An internal discharge, even though it promised to provide
a higher transient concentration, had been avoided because of concern about the
microwave attenuation and noise generation by the plasma electrons. We predicted
and then demonstrated that these effects could be kept manageably small and in
1973 published the design of the spectrometer and some preliminary results on
previously known radicals [24,25]. As development of this new instrument had progressed it had become more and more evident (from measured ion densities, observed
spectrometer sensitivity, and calculated absorption coefficients) that detection
of microwave spectra of molecular ions was an entirely realistic possibility.
The first proof of that assertion came at the successful conclusion of a long
and tedious search for the N = 0-1 transitions of CO^+ in late 1974 [26]. A 300 MHz
search region, chosen with the aid of the best optical spectroscopic constants
[27,28], was covered, but success came only on the second time through, after the
walls of the cell had been cooled with liquid nitrogen. Although the attainable
signal to noise ratio for the two lines (J = .5 – .5 and J = .5 –1.5) was rather
low at that time, they were unambiguously assignable to CO^+ on the basis of relative intensities, Zeeman effects, chemical behavior, and derived spectroscopic
constants. The electric dipole moment of CO^+ had already been shown to be quite
favorable for microwave detection by an ab initio calculation [29]. The feasibility of laboratory microwave spectroscopy of molecular ions having thus been
demonstrated, the choice of the second ion for study was assuredly an easy one –
HCO^+.

An unequivocal resolution of the X-ogen dilemma came in 1975 from two independent
and nearly simultaneous sources: astronomical detection of the long sought after
$H^{13}CO^+$ line [30,31], and laboratory detection of the original 89.189 GHz transition in a hydrogen-carbon monoxide discharge [32]. Snyder, et al. [30,31], had
finally achieved sufficient radioastronomical sensitivity to see the ^{13}C line,
and they showed that its location in space correlated well with that of X-ogen.
The laboratory observation required no frequency search and yielded a strong
HCO^+ signal as soon as availability of a microwave source made the attempt
possible. The close agreement between the best astronomical X-ogen frequency
(89,188.55 ± 0.10 MHz [14]) and the new laboratory one (89,188.545 ± 0.020 MHz [32])

left no serious doubt that the same species, HCO^+, was observed in each case. Further indisputable proof that X-ogen was HCO^+ followed almost immediately as a laboratory measurement of the $H^{13}CO^+$ frequency [33] was compared very successfully to the newest astronomical result [31]. A vast amount of additional work has been done on HCO^+ since that time and will be discussed later in this chapter, but now we turn our attention to the isoelectronic, and in many respects very similar, HNN^+ ion.

In 1974 a triplet of new U-lines at 93.174 GHz was discovered accidentally by Turner [34] and found by him to occur in a large number of interstellar sources. Identification of the carrier proved much more straightforward than it had been for X-ogen because of U93.174's distinctive triplet structure. The 1-5-3 intensity ratio and 1/1.5 frequency splitting ratio were characteristic of a $J = 0-1$ transition of a linear molecule with quadrupole hyperfine structure due to a nucleus with spin 1. Recognizing this, Green et al. [35], carried out SCF ab initio calculations on HNN^+, which gave a moment of inertia in reasonable agreement with the observed frequency, a quadrupole coupling constant eQq = -5.3 ± .5 MHz for the outer nitrogen in good agreement with the observed value (- 5.7 MHz), an eQq for the inner nitrogen too small to have produced resolvable splittings (-1.2 ± .5 MHz), and a large dipole moment (3.4 ± .5 D). All this left little doubt that U93.174 was HNN^+ as claimed by Green et al. [35]. Somewhat later Thaddeus and Turner [36] obtained a higher resolution astrophysical spectrum of the same transitions which showed clear evidence of the hyperfine structure of the inner, as well as the outer, ^{14}N nucleus, providing additional strong confirmation. Just a few weeks after the HCO^+ laboratory microwave spectrum was obtained the triplet at 93.174 GHz was seen in the same apparatus using a hydrogen-nitrogen discharge by Saykally et al. [37]. Laboratory rest frequencies were obtained and the assignment of the spectrum to HNN^+ could not be questioned. The ion-molecule reaction theories of interstellar chemistry had now been greatly strengthened by definitive observational proof that both HCO^+ and HNN^+ were widely distributed throughout our Galaxy as those theories predicted [16-19].

In astronomical observations spanning several years Thaddeus et al. [38], found two series of transitions that clearly belonged to (two different) linear molecules or near-linear molecules. They assigned one set to the HCS^+ ion on the basis of the observed spectroscopic constants (B_0 and D_0) and astronomical distribution. The B_0 value agreed well with that from an ab initio structure calculation [39] and the centrifugal distortion constant D_0 was almost identically the same as that of the isoelectronic neutral molecule HCP [40]. In 1980 Gudeman et al. [41], were able to see the $J = 1-2$ transition of this molecule in a $He - CO - H_2S$ discharge, confirming that it was HCS^+. They also provided additional evidence from ab initio calculations of the structure in support of that conclusion. Thaddeus et al. [38], suggested that their second spectrum belonged to $HOCO^+$ or HOCN. The former choice is somewhat more compatible with available theoretical calculations [42], but the identity of the species giving rise to this second spectrum must be considered very much open to question until additional information can be obtained. Attempts to observe it in the laboratory have been made, but have so far been unsuccessful.

In 1980 Erickson et al. [43] finally were able to observe CO^+ in the interstellar medium, although they only saw it in a single source, OMC-1 (Orion molecular cloud). They were compelled to look for the $N = 1-2$ transition, since absorption by molecular oxygen makes the earth's atmosphere almost opaque near 118 GHz, where the $N = 0-1$ transitions occur. They detected both the $J = 1.5 - 2.5$ transition at 236,063 MHz (repeatedly) and the weaker $J = .5 - 1.5$ line at 235,790 MHz. These frequencies had been computed from the earlier laboratory microwave frequencies for the $N = 0-1$ transitions [26] and the optical centrifugal distortion constant [27]. Using these as assumed rest frequencies with the observed (Doppler shifted) spectrum gave a CO^+ velocity of 7.9 ± .2 km/sec compared to a typical value for other molecules in OMC-1 of 8.5 km/sec, corresponding to frequency agreement of better than 1 MHz. Since then actual laboratory measurements of these two transitions (and several other high N lines of CO^+) have been reported by Sastry et al. [44], and the frequencies are a few hundred kilohertz lower than the above prediction and accurate to about 30 kHz. The uncertainty of the astronomical CO^+ frequency data and the scatter of molecular velocities seen in OMC-1, however, still limit the level of agreement between laboratory and astrophysical frequencies to roughly the 1 MHz level. The surprising thing about the detection of CO^+ in the interstellar medium is not that five years elapsed after the frequencies could be predicted accurately [26] before it occurred (mainly a matter of receiver availability at 236 GHz), but rather that it was possible at all. The CO^+ ion in space is subject to fast destruction by the reaction

$$CO^+ + H_2 \rightarrow HCO^+ + H \,. \tag{1}$$

Recent model calculations [45,46] of the interstellar abundance of CO^+ give results one to three orders of magnitude below the (roughly determined) observed value. Erickson et al. [43] suggest that the CO^+ in OMC-1 may reside in a local interfacial zone, where the abundance of C^+ (a precursor of CO^+) is particularly high, but much additional work appears necessary to provide a complete and satisfactory explanation of the CO^+ concentration discrepancy.

The history of the early work on the HNC molecule parallels that of HCO^+. Its $J = 0-1$ transition was first discovered by Snyder and Buhl in 1971 [47] as U90.663. They suggested the carrier, often called Y-ogen, was probably HNC, and eventually that identification was confirmed by laboratory microwave spectroscopy [48-50]. The fact that HNC appeared as abundant in many sources as HCN has long been attributed to their formation in the dissociative recombination of a common ionic parent

$$HCNH^+ + e^- \rightarrow \begin{cases} HCN + H \\ HNC + H \end{cases} \text{(branching ratio near unity)} \tag{2}$$

and considered as strong support of the ion-molecule reaction mechanisms [19]. (Goldsmith et al. [51] have recently published careful measurements of [HNC]/[HCN] is several giant molecular clouds, and their paper provides further discussion of the chemistry and references to other pertinent work.) The longstanding prominence of HNC as an interstellar molecule and the previously mentioned ab initio calculations on protonated carbon monoxide [5,6,8], naturally prompted speculation and further work on the analogous ionic species HOC^+. In 1976 Herbst et al. [52] published the first CI level structure calculation on HOC^+ and predicted the $J = 0-1$ transition to lie at 88.83 GHz, slightly below the X-ogen line. Additional correlated electron calculations by Hennig et al. [53] and Nobes and Radom [54] (discussed later) followed and suggested an extended microwave search. In late 1981 a slow search over the entire range from 88,200 MHz to 89,800 MHz was carried out by Gudeman and Woods [55], and the HOC^+ absorption was indeed detected, at $89,487.414 \pm 0.015$ MHz, along with several vibrational satellites of HCO^+ [56] (see below). The assignment of the 89,487 MHz line as the $J = 0-1$ of HOC^+ was unambiguously confirmed by the observation of the corresponding $HO^{13}C^+$ and $H^{18}OC^+$ lines, line width measurements, and considerable additional evidence [55]. This represented the first certain detection of this species by any experimental technique. Once the above accurate transition frequency became known, Woods et al. [57] were able to initiate radioastronomical searching for HOC^+. In 16 sources no signal was observed and the upper limit of HOC^+ concentration was established to be at least two orders of magnitude below the HCO^+ abundance in each of them. In one source, the galactic center cloud Sagittarius B2, a weak but clearly reproducible emission at the expected frequency was seen and provisionally assigned to the HOC^+ ion, at an intensity corresponding to $[HCO^+]/[HOC^+] = \sim 330$ [57]. Further observational work aimed at confirming this identification, especially an attempt to detect the $J = 2-3$ transition in Sgr B2 is in progress.

The last of the six molecular ions that have been detected by microwave spectroscopy is NO^+, an important constituent of the earth's ionosphere. This species
is readily produced in discharges, but gives much weaker spectra than ions like
HCO^+ because of its much smaller dipole moment. The later has been computed to
be 0.66 ± 0.38 D by Jungen and Lefebvre-Brion [58], and .31 D by Billingsley [59].
Bowman et al. [60] have surmounted this difficulty by a novel technique for
increasing the total ion density in their glow discharge, namely the application
of an axial magnetic field of about 200 G. They have observed the $J = 1-2$
transition near 238.4 GHz (split by ^{14}N hyperfine) and the unsplit $J = 2-3$ transition. The assignment to NO^+ is unequivocal and is based on the agreement of
B_0 with the optical value, the agreement of D_0 with that calculated from
$D_e = 4 B_e^3 / \omega_e^2$, the reasonable eQq for ^{14}N (-6.76 ± 1.0 MHz), etc. NO^+ is the only
one of the six ions that has not been observed in the interstellar medium.
Radioastronomical detection is a distinct possibility for the future, but it is
expected to be difficult.

EXPERIMENTAL CONSIDERATIONS

With a single exception to be described later all the laboratory spectroscopy
described in this chapter fits into the common conceptual mold of a classic
absorption experiment, with a source, a cell, and a detector. Radiation from a
tunable, monochromatic source is passed through an extended bulk plasma sample
containing a relatively small fractional concentration of the ion to be studied
and detected in a sensitive detector. Absorption is monitored and displayed as
the radiation frequency is scanned over a narrow range. The original spectrometer [24,25] used for the detection of CO^+ [26], HCO^+ [32], HNN^+ [37], and
HNC [48] will be described first. The sample cell is a DC glow discharge tube
15 cm in diameter and 3.5 m long with hollow cylinder electrodes of the same
diameter near the ends. An appropriate gas mixture flows in at the anode end and
is pumped away at the cathode end by a 500 ℓ/min mechanical pump or a throttled
15 cm oil diffusion pump. Operating pressures are typically 1-100 mTorr,
discharge voltages are generally 1-2 kV, and in the original version discharge
current ranged up to 400 mA. This corresponds roughly to plasma densities
(electron or total positive ion concentrations) of 10^9-10^{10}/cm^3 and a fractional
ionization of 1-10 ppm. Microwave horns and teflon lenses at each end of the
tube collimate radiation from the source, a reflex klystron, for free-space
propagation through the discharge region and then refocus it into a silicon
point-contact diode detector. The klystron was free-running, i.e., not phase-
locked to a more stable oscillator, and a repetitive analog electronic sweep
over perhaps a 15 MHz range was used for a signal averaging, typically with
hundreds of scans of a few seconds each being averaged. The computer would
digitize and store absorption data at regular time intervals, e.g., 1000 points
at 5 msec/point, with the start of data taking in a given scan synchronized to
an accurate frequency marker to compensate for long term drift of the klystron.
Double square wave modulation [61] of the klystron frequency and lock-in
amplification were used to separate the weak ionic signals from noise and background absorption. As previously mentioned the final addition to this apparatus
that rendered the 1974 detection of CO^+ possible was the provision for liquid
nitrogen cooling of the walls of the discharge tube, a modification which has
also proved very helpful in most subsequent work. The system that has just been
outlined is the configuration that was used for work in this laboratory through
1979. Before describing the later improvements that have been made to it or
the other systems that have been built, we pause to consider some of the
fundamental concepts that influence spectrometer design and motivate such
development.

The intensities of microwave absorption transitions are normally expressed in
terms of the peak absorption coefficient [62]

$$\gamma = \frac{8\pi^2 N f |\mu_{ij}|^2 \nu_o^2}{3 c k T \Delta\nu} \tag{3}$$

when N is the density of the observed species, f is the fraction of them in the
initial quantum state of the transition, and ν_o is the center frequency. The
dipole moment matrix element is denoted by $|\mu_{ij}|^2$ and the line width (half width
at half maximum), by $\Delta\nu$. The temperature factor T arises from the partial cancellation of absorption and stimulated emission, and in a non-equilibrium environment like a discharge it is the rotational temperature. A previous paper by
the author [63] has discussed the estimation and measurement of the various
quantities that enter into this formula when it is applied to transient species
in electric discharges. Here we focus on the frequency and linewidth factors
and their relationship to the selection of the operating pressure and the dimensions of the discharge tube. At microwave frequencies below 100 GHz pressure
broadening is normally dominant over Doppler broadening, and for best microwave
resolution and convenient modulation of the signal pressures of 10-50 mTorr are
desirable. The large tube diameters used are dictated both by microwave optics
and kinetic theory considerations. Geometrical optics applies when the aperture
of a lens is very large compared to the wave length of radiation, and the larger
the tube diameter the lower one can go in frequency before diffraction losses
become intolerable. In order for a glow discharge to be maintained electron
motion to the walls must be a diffusion process, i.e., the electron mean free
path must be small compared to the tube diameter. For a pressure as low as
10 mTorr, diameters of 10 cm or so are required, with the exact relation dependent on temperature, gas composition, etc. Since the actual spectrometer signal
goes as the product of γ and the path length, a long discharge tube is obviously
advantageous, but there are practical limitations. In addition to the obvious
problems of laboratory space and handling difficulties one must consider standing
waves. With the coherent microwaves reflecting off irregularities in the transmission path, e.g., the horns at the two ends of the cell, an interference
pattern is set up, and the microwave power reaching the detector varies more or
less periodically with frequency, with maxima separated by c/2L, where L is
usually somewhat more than the cell length. This standing wave pattern is detected and amplified by the source modulation scheme, and the true molecular
signal is normally distinguished from it on the basis of a faster frequency
variation of the true signal. This discrimination may be visual, electronic,
or computer based, but in any case the interference fringes become more difficult
to separate from the desired signal when the cell becomes too long. The γ
formula indicates a strong sensitivity advantage in going to higher frequencies
due to the presence of the ν_o^2 term. Actually for a linear molecule γ goes up
as ν_o^3 rather than ν_o^2, since the rotational factor in the fraction f is proportional to ν_o, so the high frequency advantage becomes even more dramatic in that
important case. Again there are limits. Whether and to what extent this theoretical advantage is realized in practice depends on the characteristics of
available sources and detectors at various frequencies. Non-linear molecules
exhibit more complicated dependence of γ on the frequency than linear ones. At
too high rotational states the Boltzmann factor for the lower rotation state can
no longer be neglected and decreases γ. As frequency goes up Doppler broadening
begins to dominate, resolution is degraded, and $\Delta\nu$ becomes proportional to ν_o,
canceling one of the ν_o's in the numerator of γ. Despite these limitations
increasing frequency is normally quite advantageous, as will be seen below. The
cell diameter choice is also influenced by operating frequency ν_o. As wave
length decreases, lens aperture may also shrink for constant diffraction losses.
Furthermore, as Doppler broadening becomes higher at high frequencies, higher
pressure broadening, and consequently higher gas pressures and lower tube
diameters, become acceptable.

The plasma electrons do cause attenuation of the microwave beam, create a non-
unity refractive index, and emit a microwave noise spectrum [64]. For the high
radiation frequencies and fairly low plasma densities (N_e) considered here the

refractive index effect is the most important one. The refractive index (μ) of the plasma at (high) frequency ν_0 is given by [64]

$$\mu = (1 - \frac{N_e e^2}{m \varepsilon_0 \, 4\pi^2 \, \nu_0^2})^{\frac{1}{2}} \qquad \text{(MKS units),} \qquad (4)$$

where m,e, and ε_0 are the usual fundamental constants. The plasma frequency ν_p for a given density is defined by

$$\nu_p^2 = \frac{N_e e^2}{4\pi^2 \, m \, \varepsilon_0} . \qquad (5)$$

In the low density, high frequency case ($\nu_0 \gg \nu_p$) we may further approximate μ as

$$\mu = 1 - \frac{N_e e^2}{8\pi^2 \, m \, \varepsilon_0 \, \nu_0^2} . \qquad (6)$$

Random fluctuations of the plasma density in a discharge that is at all unstable thus create a modulation of the refractive index of the plasma and the phase of the microwave radiation at the detector. Because of the previously mentioned standing waves this sporadic phase modulation is converted by the detection process into amplitude excursions, i.e., spectrometer noise. From equation (6) it can be seen that this effect is inversely proportional to the square of ν_0, and that dependence constitutes a second major advantage of operation at higher frequency. As frequency increases higher plasma densities or less stable discharges become tolerable. Using equation (6) a simple proportionality between the microwave phase shift (phase with plasma present minus phase with plasma absent) of radiation traversing a plasma sample of length L_p and the latter's average plasma density may be derived [64]:

$$\overline{N}_e[cm^{-3}] = 118.4 \; \nu_0[Hz] \; \Delta\phi(\text{radians}) \; /L_p[cm] \qquad (7)$$

Since $\Delta\phi$ is fairly readily measureable with the instrumentation available to the microwave spectroscopist, the average electron density is usually easy to obtain.

The modulation and computer control schemes of the previously described spectrometer were completely changed in early 1980. The tone-burst modulation scheme of Pickett [65], in which frequency modulation of the microwave carrier (at 1.0 MHz for example) is turned on and off at an audio frequency, e.g., 50 kHz, was substituted for the earlier double square wave method. This new method permits the klystron to be phase-locked (using the digital phase detector scheme of Pickett [66]) and modulated simultaneously and gives an improved demodulated line shape, in which the various lobes are separated by exactly 1.0 MHz. The klystron is locked to a harmonic of a computer controlled frequency synthesizer and the sweep is purely digital. The klystron becomes extremely monochromatic, the sweep is absolutely linear, and the frequency is known to nine significant figures at each sweep point. Long term signal averaging without frequency drift or any source broadening is possible, and both the line shape and the digital frequency data format are ideal for computer suppression of the spurious background signal due to the standing waves and for non-linear least squares fitting to a model line shape. The latter is always done to extract accurate center frequencies and linewidths from the spectral data. These changes substantially enhanced the sensitivity of the spectrometer and also were essential for the studies of Doppler shifts [67] and pressure broadening [68] that will be

described later in the chapter. At about the same time the Si point contact
diode detector was replaced by a more sensitive Schottky barrier diode.

In 1981 a second glow discharge spectrometer was completed in our laboratory,
and it has been employed in all our work on HOC$^+$ [55] and the vibrational
satellites of HCO$^+$ [56] and for the most recent measurements of pressure
broadening [68] and Doppler shifts of ions [67]. The absorption cell is 3 m long
and just 10 cm in diameter, and the water cooled electrodes are in long side arms
of larger diameter for greater power handling capacity. The discharge current
maximum is now 1.1 A. The microwave radiation samples only the positive column
of the glow, whereas in the older system it traversed all the different zones
of the discharge. Thus measured dynamical properties, e.g., relative intensities
or line widths, apply to a more homogeneous and well defined sample in the newer
spectrometer. The two electrodes are identical, so that the discharge polarity
can be easily reversed for convenient Doppler shift and rest frequency
measurements. A fast flow system with pumping by a 15 cm diffusion pump is
normally employed. Liquid nitrogen cooling, digital frequency sweep, and other
features are as previously described. Measurements of the average plasma
density in the new system (in the positive column) by the microwave phase shift
method gives values of 10^{10}-10^{11}/cm^3 for various gases at full discharge current.
A major improvement of the new system is the provision for mass spectrometric
sampling of the species in the discharge. Ions escaping through a pinhole at
the wall of the discharge tube are focused through a differentially pumped
intermediate chamber into a quadrupole mass spectrometer. (Major neutral species
can also be sampled by deflecting away the initial ion beam and ionizing the
neutral molecular beam by electron impact in the mass spectrometer chamber.)
The production of a given ion can be optimized with respect to gas composition,
flow rate, current, etc., while its concentration is monitored with the quadrupole.
Actually the mass spectral relative intensities and the microwave phase shift
measurements of N_e can be combined to give an approximate absolute ion density for
the calculation of γ in equation (3), so that the feasibility of detecting a
given ion may be judged in advance. Measurements with the mass spectrometer have
already been very helpful in illucidating discharge ion chemistry. They showed
that HCO$^+$ was produced very copiously in a discharge mixture that was primarily
argon with only a few percent each of hydrogen and carbon monoxide, and it was
this mixture that actually permitted the detection of HCO$^+$ in vibrationally
excited states [56] and that of the isomer HOC$^+$ [55]. In another application the
chemistry that had been used to obtain the laboratory microwave observation of
HCS$^+$ [41] was shown by mass spectrometry to indeed be an efficient means of
producing that ion. A final feature of the new system is a high resolution 1.5 m
double pass monochromator with cooled photomultiplier detection for monitoring
ultraviolet-visible emission from species within the discharge plasma.

While work with the previously described spectrometers has for the most part been
restricted to the spectral range below 125 GHz, two other groups have used
harmonic generation techniques to extend the observations of molecular ions into
the region of roughly 150-500 GHz, which includes the frequencies of the
J = 1-2 through J = 4-5 transitions of most of the ions that have been discussed
here. Independently in 1981 Bogey et. al. [69] at Lille University and Sastry
et. al. [70] at Duke University extended the measurements on several isotopic
forms of HCO$^+$ into this region, and subsequently the latter group has also done
similar extensions on HN$_2^+$ [71], CO$^+$ [44], and HOC$^+$ [72], as well as carrying
out the previously discussed initial detection of NO$^+$ [60]. Bogey et. al. [69]
used a commercial InSb detector operating at 1.7°K and a commercial harmonic
generator. Sinusoidal source modulation at 12.5 kHz with lock-in detection at
25 kHZ was employed. For ion production they used an electrodeless RF discharge
at 50 MHz in a pyrex cell 1 m long and 5 cm in diameter at 15-50 mTorr pressure.
They had previously used this same apparatus for studies of highly excited
vibrational states of neutral unstable molecules like CS [73]. Bogey et. al.
[74] have also shown that the propensity of the RF discharge to create high

vibrational excitation extends to the ionic case by observing CO^+ in v = 1 and
v = 2 (with a klystron fundamental oscillator at the N=0-1 transitions). The
Duke group has used custom built InSb detectors (at 1.7°K) and harmonic
generators of the King-Gordy type [75], which were developed there. Their
studies of HCO^+ [70], CO^+ [44], and HN_2^+ [71], employed a DC discharge in a 10 cm
by 1 m long tube with discharge currents up to 500 mA. Their work on HOC^+ [72]
and NO^+ [60] has utilized a novel alternative method for ion production. A DC
discharge at 5 kV and 10 ma in a 4 cm diameter by 1.5 m long tube with a 200 G
axial magnetic field was employed, and the latter was reported to increase the
ion signal by as much as two orders of magnitude. All the cells mentioned in
this paragraph have been operated with liquid nitrogen cooling. In all this work
very good signal to noise ratios have been obtained, clearly demonstrating that
the theoretical sensitivity advantages of these higher frequencies can be
effectively exploited with current multiplier and detector technology.

The pure rotational spectroscopy of ions recently reported by van der Heuvel and
Dymanus [76], at frequencies near 1000 GHz, while it falls somewhat outside of
our definition of microwave spectroscopy, is very closely related to the foregoing discussion. They have obtained tunable submillimeter wave radiation by
mixing the output of klystrons in the 50-100 GHz range with that of an HCN laser
in an open diode mixer and have detected it after a single pass in a cryogenic
bolometer. They have observed strong absorption signals for the J = 8-9, 10-11,
and 11-12 transitions of HCO^+ and also one similarly high J transition each for
CO^+ and HNN^+. The ions in these experiments are produced in a hollow cathode
discharge at 300 mA. The absorption cell is the 5 cm diameter by .7 m long
copper hollow cathode, which is cooled with liquid nitrogen. As in all the
previously mentioned experiments that cooling was observed to be very beneficial.
The observed absorption coefficient was used to estimate an HCO^+ density of
$3 \times 10^{10}/cm^3$ in the hollow cathode discharge.

The last laboratory microwave experiment to be included here constitutes the
exception alluded to in the first sentence of this section. Brown et. al. [77]
have carried out microwave-optical double resonance (MODR) on CO^+, using the
two N = 0-1 microwave transitions near 118 GHz. A mass selected ion beam is
illuminated by a pulsed dye laser and the laser induced fluorescence (LIF) is
detected with a photomultiplier. Changes to the LIF signal effected by double
resonant microwave radiation (delivered from a klystron into a microwave cavity
containing the beam) are observed in a complicated time sequenced photon
counting procedure. Initial experiments required long experiment times to
achieve a modest signal to noise ratio, but future technical developments may
improve this situation considerably. Interestingly one of the two CO^+ lines
showed some asymmetry, which was attributed to the Stark effect (see below
for discussion of dipole moments).

Finally we turn to the instrumental or observational aspects of radioastronomy.
Since radioastronomical observation of molecular ions is carried out in exactly
the same way and with exactly the same equipment as that of neutral molecules,
we will not attempt to give a comprehensive review of the techniques or their
development. Instead a few major points will be outlined for the non-astronomer,
especially points of contrast with laboratory methods and concepts. Most radioastronomical studies of ions have detected emission, while the laboratory studies
always involved absorption. In a typical millimeter wave radiotelescope
radiation from the source is collected and focused onto a mixer, which may or may
not be cryogenically cooled, by a large parabolic antenna (typically 5-20 m in
diameter). The surface accuracy requirements on the dish become high at short
wavelengths and provide an upper frequency limit for a given telescope. Angular
resolution is typically 1-3' of arc. Superheterodyne detection, in which the
incoming radiation is mixed with a local oscillator (LO) to give an intermediate
frequency (IF), is always used. The IF is amplified in a low-noise, broadband
amplifier and then, after possible further down-conversion with additional local

oscillator-mixer stages, applied to a filter bank, e.g., 256 channels separated by 1 MHz. Thus an entire segment of the spectrum (256 MHz in the example) is accumulated simultaneously, saving a great deal of time relative to a swept LO scheme. Receiver design, mixer performance, etc. have been steadily improved since 1970, and the minimum detectable signal level has decreased accordingly. Work at 200-300 GHz has become feasible in the last few years, and in that region the LO frequency is derived from harmonic multiplication of the output of a lower frequency klystron [78]. A recent volume [79] has been devoted entirely to the design, construction, and operation of radiotelescopes and provides much detail on all the foregoing matters.

Spectral intensities in radioastronomy are expressed as a brightness or line temperature (T_L) in °K and refer to the temperature of a black body radiator that would give that same intensity level at the frequency in question. Actually the intensity measured is the excess above that at an off-source reference point nearby in the sky. Lines with temperatures of 2-3°K are considered strong, while lines at .1°K are quite weak but detectable. The general radiative transfer problem in radioastronomy, i.e. the calculation of the observed T_L from the profile of concentration, velocity, and rotational excitation of molecules along the line of sight, is exceedingly complex. A good summary of the basic ideas and equations and the various possible limiting cases has been given recently by Winnewisser et. al. [80]. The relative population of the upper and lower state of the transition observed is normally expressed as an excitation temperature T_{ex}

$$\frac{n_2}{n_1} = \frac{g_2}{g_1} \exp\left(-\frac{h\nu}{kT_{ex}}\right) , \tag{8}$$

where the g_i's are the usual statistical weights. Since the time between collisions may be long compared to the spontaneous emission lifetime of the upper state, T_{ex} may be distinctly lower than T_K, the local kinetic temperature. Population of the upper state is mainly accomplished by collisions with hydrogen (density n = [H] + 2[H_2]), while depopulation is primarily due to spontaneous emission, so at high n^2 one expects $T_{ex} \sim T_K$. The optical depth is often not small compared to one, so that radiation trapping or self-absorption effects must be explicitly accounted for. The optical depth can influence the radiative cooling rate and thus T_{ex}, so the latter quantity may be quite different even for different isotopic forms of the same ion or molecule. At the millimeter wavelengths used for molecular ion studies the cosmic 3°K blackbody radiation from the big bang and any galactic continuum radiation from behind the observed object are normally negligible. In the simplest possible case, where one further assumes the cloud is uniform (in T_{ex}, concentration, and velocity) along the line of sight, one obtains [80]

$$T_L = \frac{h\nu}{k} \frac{1 - e^{-\tau_\nu}}{\exp(h\nu/kT_{ex}) - 1} , \tag{9}$$

where the optical depth τ_ν is

$$\tau_\nu = \frac{c^2 g_2 A_{21}}{8\pi g_1 \nu^2} N_1 [1 - \exp(-h\nu/kT_{ex})] \Phi(\nu) . \tag{10}$$

Here A_{21} is the Einstein coefficient, $\Phi(\nu)$ is a lineshape function, and N_1 is the column density of molecules in the lower state ($N_1 = n_1$ d, where d is the cloud thickness). In one extreme of high optical depth ($\tau_\nu \gg 1$) and excitation ($T_{ex} \gg h\nu/k$) one has simply $T_L = T_{ex}$. For low optical depth ($\tau_\nu \ll 1$) (still assuming $T_{ex} \gg h\nu/k$ and integrating over the lineshape) one obtains [80]

$$\int T_L d\nu = \frac{8\pi^3 \nu^2}{3kc} \frac{g_2}{g_1} |\mu_{21}|^2 N_1 , \tag{11}$$

which is seen to be independent of T_{ex}. In this case molecular abundances may be deduced from the observed integrated intensity, but only after one makes some assumptions about the cloud thickness d and the fraction of all the molecules that are in the lower state of the observed transition (N_1/N). Furthermore, one must consider effects of finite beam width, antenna efficiency, atmospheric absorption, etc., and one does not expect the uniform cloud assumption to be entirely realistic, so the determination of accurate molecular abundances by radioastronomy is generally a very difficult problem. In order to resolve the various ambiguities relating to density (n), geometry, homogeneity, excitation, and optical depth and obtain reliable results for abundances one must normally combine all available kinds of observational information: velocity profiles, spatial contour maps of intensities, observations of several different molecules, or especially, observations of several different isotopic forms or rotational transitions of the molecule of interest.

MOLECULAR IONS IN THE INTERSTELLAR MEDIUM

The regions of the interstellar medium (ISM) in which molecular ions like HCO^+ or HNN^+ are usually observed and studied are the so-called dense clouds, defined as those whose molecular density exceeds approximately $10^4/cm^3$, and these same dense clouds are also the home of most of the complicated neutral molecules that have been seen in the ISM. There also exist an intercloud medium (density $< 1/cm^3$), diffuse clouds ($10^1 \lesssim$ density $\leq 10^3/cm^3$), where ions (HCO^+) and a few important molecules are found, and circumstellar shells, which contain fairly complicated neutral molecules but generally no observable quantities of molecular ions. The density cutoff (near $10^4/cm^3$) between the diffuse and dense regimes is a fairly distinct one. In the diffuse clouds the hydrogen is atomic and penetrating starlight destroys most molecules through photodissociation. In dense clouds, however, the hydrogen is almost completely in molecular form, and the opacity to starlight, due to light scattering off dust grains, becomes high enough that photo processes are essentially negligible. In this paragraph we shall provide enough information about the dense clouds and their chemistry to make the following review of ion studies intelligible. Much more detailed descriptions and many additional references are contained in the reviews by Watson [19] and by Winnewisser et. al. [80]. Most of the mass (~70%) is H_2, and most of the rest is He (~28%). The elements C, N, and O are down by 3-4 orders of magnitude from H_2, and all others are even less common. About 1% of the mass is in dust grains, and these probably contain most of the metals and other heavy elements, although their exact composition and nature is highly uncertain. A large fraction of the gas phase carbon is in the form of CO, which is the most abundant molecule other than H_2. Energy for the chemistry of molecule formation comes from cosmic ray ionization of H_2 and He. The initially formed H_2^+ is immediately converted to H_3^+ by reaction with an H_2 molecule. The He^+ and H_3^+ then initiate complex chains of ion-molecule reactions leading to all the other ions and neutrals according to the ion-molecule theories [16,18]. In particular, HCO^+ is formed in the reaction

$$H_3^+ + CO \rightarrow HCO^+ + H_2 , \tag{12}$$

which is known to be fast. It is destroyed by dissociative recombination

$$HCO^+ + e^- \rightarrow H + CO \tag{13}$$

and to a lesser extent by proton transfer to species with greater proton affinities than CO. Similarly HNN^+ is formed from N_2, but it possesses an additional loss mechanism, proton transfer to CO. Thus HCO^+ and HN_2^+ are of extreme interest as astronomical probe molecules not only because they give strong signals in a wide variety of sources, but also because they enter into the chemical model in a very simple, predictable, and fundamental way. Their abundances are straightforwardly related to that of CO, which is directly observable, and to those of N_2 and e^- and to the cosmic ray flux, which are all of great interest but not directly measurable. Dense clouds are conveniently further subdivided into dark clouds and giant molecular clouds, both of which are widely distributed in the Galaxy. Dark clouds are typically characterized by densities of 10^4-10^5/cm^3, kinetic temperature of 10-20°K, Doppler linewidths of .1-1 km/s, total masses of 10^2-10^4 solar masses, and linear dimensions of up to 30 light-years. The corresponding ranges of these parameters for giant molecular clouds are 10^4-10^7/cm^3, 20-80°K, 3-30 km/s, 10^5-10^7 solar masses, and 3-200 light-years, respectively. The linewidths are primarily due to either turbulence or collective motion (expansion, contraction, or rotation) rather than thermal velocity, and pressure broadening is of course utterly negligible. Giant molecular clouds are usually associated with imbedded HII regions (ionized zones where the hydrogen is in the form of H$^+$). The HII regions in turn contain and are caused by a star or a small group of stars, which presumably formed because that part of the cloud became unstable to gravitational collapse. The stars themselves are normally visible only in the infrared, or not at all, because of the optical opacity of the surrounding cold cloud material. Thus the giant molecular clouds are of intense astronomical interest as the birthplace of stars, and radioastronomy provides a new and incisive tool for studying the intriguing process of star formation. The two most famous giant molecular clouds are Orion (OMC-1) and Saggitarius B2. Orion is near (1500 light-years) the sun, which is fairly far out one of the spiral arms of the Milky Way Galaxy, and this proximity permits it to be mapped with greater spatial resolution than any other giant molecular cloud. The molecular cloud OMC-1 surrounds the Kleinmann-Low (KL) nebula, which contains a star called the Becklin-Neugebauer (BN) object, visible in the infrared. Sgr B2 on the other hand is very near the center of our Galaxy (30,000 light-years away) and appears to be the largest single object in it (10^7 solar masses). Most of the molecules that have been seen anywhere in the ISM have been seen in Sgr B2. The Sgr B2 cloud is known to have complex internal structure and is often described in terms of a core region (density 10^5-10^6/cm^3) and a surrounding halo (density ~ 10^4/cm^3) for discussion purposes [80].

The clear demonstration that HCO$^+$ and HNN$^+$ were widely distributed in dense clouds in itself provided strong support for the ion-molecule theory of molecule formation in the ISM, and the detailed further studies we will now describe have collectively constituted an even more decisive confirmation of its essential correctness. In addition, they have provided a great deal of specific information about the chemistry, physical conditions, and structure of dense interstellar clouds. In early 1976 Woods et al. [33,81] were able to extend the laboratory observations of HCO$^+$ to a total of six isotopic forms including DCO$^+$ and HC^{18}O$^+$. Using these laboratory frequencies Hollis et al. [82] were soon able to locate and identify the spectrum of DCO$^+$ in the molecular clouds NGC 2264 and DR 21(OH) and the cool dark cloud L134, and Guélin and Thaddeus [83] were able to observe that of HC^{18}O$^+$ in Sgr B2. Remarkably in L134 the DCO$^+$ emission was almost as intense as that of ordinary HCO$^+$, and this was attributed to extreme chemical fractionation of deuterium into H$_2$D$^+$ in the reaction

$$H_3^+ + HD \rightarrow H_2D^+ + H_2 \ . \tag{14}$$

The reverse reaction is made practically negligible by zero-point vibrational effects at the very low temperature of L134. An upper bound to the fractional abundance of the electrons in a cloud ($x_e = [e^-]/[H_2]$) can be estimated from an analysis of Watson [84] and the observed ratio R = [DCO$^+$]/[HCO$^+$] = x(DCO$^+$)/ x(HCO$^+$), where x represents the abundance of a species relative to molecular hydrogen. Comparing formation of H$_2$D$^+$ via equation (14) to destruction by dissociative recombination (rate constant k_e) leads to x(H$_2$D$^+$)/x(H$_3^+$) < k_{14} x(HD)/k_e x_e. Further assuming that HCO$^+$ comes only from H$_3^+$ and DCO$^+$ comes only from H$_2$D$^+$ (in one third of the collisions with CO) yields R $\cong$ x(H$_2$D$^+$)/3x(H$_3^+$). The upper bound on x_e is then

$$x_e < \frac{k_{14} \ x(HD)}{3 \ k_e \ R} \ . \tag{15}$$

Guélin et al. [85] detected the J=1-2 transition of DCO^+ in 5 sources including three cool dark clouds (L63, L134, and L134 N), and deduced R's of between .3 and .8 for the dark clouds. With numerical estimates of the rate constants in equation (15) and assuming $x(HD) \leqslant 4 \times 10^{-5}$, i.e., twice the cosmic atomic D/H ratio [86], they obtained upper bounds of $x_e < 10^{-8}$ from these R's, much lower than any previously available upper bounds for the fractional ionization in the ISM. They further concluded that gas phase metals at the known cosmic abundances would have led to much higher ionization, and thus that the metals were very heavily depleted by condensation onto dust grains. In 1977 Anderson et al. [87] obtained a laboratory frequency for the J=0-1 transition of DNN^+, and very soon afterwards Snyder et al. [88] were able to observe DNN^+ in L134 N. They obtained an abundance ratio $x(DNN^+)/x(HNN^+) \cong .45$, which with essentially the same analysis just described led to $x_e \lesssim 10^{-8}$ again. No significant abnormalities in the ^{14}N hyperfine intensities were seen, suggesting that even normal HNN^+ was in the optically thin limit and thus (seemingly) eliminating one possible uncertainty in the earlier DCO^+ work. The HNN^+ spectrum in L134 N showed even higher resolution than the previous one of Thaddeus and Turner [36] in OMC-2; all seven quadrupole components were clearly visible. This is the highest resolution obtained for ions in the microwave region ($\Delta\nu \sim$ 70kHz compared to ~200kHz in the best laboratory spectra [68]), and clearly shows that radio-astronomy is a powerful tool for obtaining high quality spectroscopic constants, as well as for studying the ISM. Both Guélin et al. [85] and Snyder et al. [88] pointed out that their very low values for the fractional ionization in dark clouds suggested the prevailing small interstellar magnetic field would not couple strongly to the mass motion of the cloud and would thus have less influence on the dynamics of gravitational collapse than previously suspected. In early 1978 Watson et al. [89] reported new observations of the J=0-1 transitions of HCO^+, DCO^+, and $H^{13}CO^+$ in L134 and concluded from the relative brightness temperatures of HCO^+ and $H^{13}CO^+$ that the former was subject to serious optical depth effects. They revised the estimate of R to .12 and that of the upper bound to $x_e \leqslant 4 \times 10^{-8}$ accordingly, but they felt the qualitative conclusions of the previous studies were still valid. Using similar arguments they also gave estimates of the upper bounds to the CO and N_2 abundances ($x(CO) \lesssim 10^{-5}$ and $x(N_2) \lesssim 10^{-5}$). Langer et al. [90] looked at the HCO^+ isotopes in two more dust clouds (L134 N and TMC-1) and found surprisingly that the $H^{13}CO^+$ line was about twice as intense as that of normal HCO^+ at some points in these sources. They argued that no uniform cloud model could be consistent with their observations and that instead a dense background region was responsible for emission and a less dense foreground region of much lower excitation (T_{ex}) was responsible for serious self absorption in the normal HCO^+ species. Their upper bounds for x_e were in general agreement with that of Watson et al. [89]. Turner and Zuckerman [91] at about the same time reported studies of the ratios of DCO^+, DNN^+, DCN, and DNC to their normal isotopic forms in 13 interstellar clouds, including both giant molecular clouds and dark clouds. They also concluded that self absorption effects in HCO^+ in dark clouds were very significant.

Before concluding our discussion of the deuterium enhancement issue and the estimates of electron abundance limits we divert our attention to some of the other kinds of studies that have been done, since they all tend to be inter-related to a certain extent. In 1977 Turner and Thaddeus [92] published a survey of HCO^+ and HNN^+ emission in 73 sources, including detailed maps of four giant molecular clouds, and compared the distribution of the two ions with that of HCN and CN. HCO^+ was found to correlate well spatially with HCN, and HNN^+, with CN. HNN^+ appeared to avoid regions of high (H_2) density, while HCO^+ (or HCN) did not. This was particularly true in the Orion (OMC-1) region, which was one of those sources mapped in detail. They felt their observations were generally consistent with the ion-molecule theory, but would be in somewhat better agreement if most of the interstellar carbon were depleted onto grains or tied up in molecules other than CO, contrary to the assumption of Herbst and Klemperer [16]. At about the same time Snyder et al. [93] reported a detailed study of the OMC-1 region, in which they compared the distributions of HCO^+, HNN^+, and SO_2. They also observed the anticorrelation of HNN^+ abundance with that of HCO^+ and with density; the HNN^+ abundance was seen to be low near the KL center position and to increase in all directions away from that point. They argued that this behavior was due to the destruction of HNN^+ by proton transfer to CO (and thus supportive of the ion-molecule description of the chemistry) and they also used a simple kinetic model along with their data to estimate that H_2O (unobserved) was six times less abundant than CO in Orion. The spectral lines of HCO^+ and several other molecules show in OMC-1 a complicated two component velocity profile attributed to a "ridge" source and a "plateau" source [78,92, 94-97]. The former is spatially spread out over a relatively long North-South ridge and has a fairly narrow Doppler profile, while the plateau source has a broad line profile due to very high velocity gas but is spatially tightly confined to a region within about 1' of the KL position. Rydbeck et al. [94] have mapped the J=0-1 transitions of HCN and HCO^+ with 20" spacing in OMC-1 and shown that the plateau source is very near the KL/BN position in both, as have Kuiper et al. [95]. Rydbeck et al. [94] also conclude that $x(HCN)/x(HCO^+)$ is about 3 in the ridge cloud, but that the HCN is about one order of magnitude more abundant than HCO^+ in the plateau source. CO observations of Scoville [96] indicate the plateau source contains about 10 solar masses of gas expanding about the position of the BN object, and he suggests the original kinetic energy of expansion may have been ~10^{50} ergs, or comparable to that of a supernova. Huggins et al. [97] were the first to observe the J=2-3 transition of HCO^+, and they were also looking at OMC-1. Weak plateau emission of HCO^+ is seen in this transition too, while the plateau in the J=2-3 of HCN is much more prominent. The work of Huggins et al. [97] came before the previously mentioned laboratory observations of the J=2-3 of HCO^+ [69,70] and constituted the first measurement of that ion's centrifugal distortion coefficient (D_0). Erickson et al. [78] have obtained spectra of the J=2-3 transitions of HCO^+ and HNN^+ in 27 and 10 sources, respectively, including better signal to noise ratio observations of the plateau and ridge features of HCO^+ in OMC-1.

Wootten et al. [98] have determined densities, temperatures, and abundances of HCO^+, H_2CO, HNC, HCN, and ^{13}CO in 13 dense cloud regions from model calculations and intensity measurements, and Wootten et al. [99] have described the results for HCO^+ and H_2CO in much greater detail. Observations of CO, for which $T_L \cong T_{ex} \cong T_K$, were used to establish the cloud temperature, and simultaneous observations of the $2_{12}-1_{11}$ and $2_{11}-2_{12}$ transitions of formaldehyde were used to obtain its excitation temperature (T_{ex}), and thus the local density ([H_2]), and the abundance of H_2CO. A large velocity gradient (LVG) model was used for the treatment of radiative transfer. The abundances of HCO^+ or the other species could then be obtained from their line temperatures (T_L) and the density, which determines their excitation. The abundance of HCO^+ was found to vary dramatically from region to region as temperature and density changed. In the typical dark dust clouds with densities near $10^4/cm^3$ and temperatures below 20°K they obtained $x(HCO^+)$ ~ 10^{-9}, while in the giant molecular clouds temperatures were 25-50°K, densities ranged up to $10^6/cm^3$, and $x(HCO^+)$ fell off sharply with increasing density to around 10^{-11} at the highest densities. The other molecules sampled behaved similarly, i.e. their absolute concentrations were relatively constant among the various clouds. They concluded that molecules are depleted onto dust grains in the denser, warmer regions [99]. Loren and Wootten [100] have observed self-reversal in the line shapes of the J=0-1 transition of HCO^+ in seven giant molecular clouds. The line profiles for HCO^+ are similar to those of CO in the same sources, and most of them consist of two unequal peaks separated by a central absorption dip. Emission is presumably due to HCO^+ located in the dense core of the cloud, and the absorption is attributed to a surrounding layer of more diffuse gas containing HCO^+ that is abundant enough to be optically thick, but that has very low excitation (T_{ex}) because of the low density. The sense of the asymmetry can be used to distinguish between expansion and collapse of the peripheral material. In a collapsing cloud the foreground absorbing material will exhibit a negative Doppler shift and the high frequency emission peak will be strongest. Four clouds appeared to be contracting, while one appeared to be expanding, and the other two showed more complex profiles. Snell [101] has investigated HCO^+ abundances (using methods similar to Wootten et al. [99]) and other properties of nine dark clouds and concludes that the density in these sources falls off rapidly with distance away from a small dense central region (roughly like r^{-2}). HCO^+ abundances are again seen to decrease greatly as density increases. A survey of HCO^+, HCN, HNC, and CCH emissions towards 9 molecular clouds has been reported by Baudry et al. [102], and Baudry et al. [103] have made an extensive study of HCO^+ and $H^{13}CO^+$ spectra in the Taurus system of dark clouds. A two component model (denser core, surrounding less dense layer) with strong self absorption in normal HCO^+ was used to explain the observed $HCO^+/H^{13}CO^+$ intensity ratios.

In recent years HCO^+ has also been observed in some other types of astronomical objects. Dickinson et al. [104] have seen it in the supernova remnant IC 443, where it exhibits a J=0-1 line profile with great velocity dispersion, extending from 0 to -60 km/sec. The latter is attributed to the passage of the intense shock front associated with the supernova explosion. The ratio $x(HCO^+)/x(CO)$ is said to be nearly 100 times greater than that expected for more normal molecular clouds. Fukui et al. [105] have looked at the HCO^+ J=0-1 spectrum in several points of Sgr B2 and Sgr A, another giant molecular cloud even closer to the exact galactic center than Sgr B2. In the latter the profile extends from -45 to +100 km/sec and shows a core component (from +20 to +80 km/sec) and a wing component. The wing or large velocity dispersion part is interpreted as arising in a less dense surrounding region (just the opposite situation to the plateau source in Orion). The velocity integrated abundance ratio of HCN to HCO^+ was found to be 1.1 in Sgr A (the galactic center), which is similar to that in other giant molecular clouds. Fukui et al. [105] also saw a strong absorption dip (nearly 100% absorption) in the Sgr A emission profile near 0 km/sec due to HCO^+ in a very diffuse region, where it is essentially in equilibrium with the cosmic background radiation (T_{ex} ~3°K), lying between the galactic center and the earth. Linke et al. [106] observed a broad HCO^+ absorption feature towards Sgr B2 between -125 and +5 km/sec, velocities (frequencies) at which there is no HCO^+ emission. Sgr B2, however, is a continuum source with T (continuum) = .21°K in this spectral range, and the HCO^+ absorbs this continuum radiation. Between -116 and -12 km/sec the absorption is essentially complete, i.e., the HCO^+ absorbing zone has quite high optical depth. A similar absorption feature was also seen towards Sgr A, and both were ascribed to HCO^+ in the galactic nuclear disk, the diffuse interstellar gas in the nuclear part of the Galaxy. HCN is also observed, and the column densities of both are estimated to be about $5 \times 10^{14}/cm^2$. Linke et al. [106] assume that this figure is representative of the distribution of these molecules throughout the nuclear disk, whose total mass (of interstellar gas) is consequently 8×10^9 solar masses by their estimate. The reader may note that this implies that several solar masses of HCO^+ are contained in the nuclear disk. In Sgr A Linke et al. [106] also see four sharper absorption dips against the HCO^+ emission profile at -51, -30, -2, and +32 km/sec. The third of these is the same as that seen by Fukui et al. [105] (see above), and all are assigned to HCO^+ in the diffuse gas in the individual spiral arms of the galaxy (with T_{ex} ~ 3°K). Thus HCO^+ appears to be pervasively distributed in the galactic interstellar gas, even though it only emits in the dense regions, where T_{ex} is sufficiently large. Stark and Wolff [107] have detected HCO^+ emission in an external galaxy, M82, and Rickard and Palmer [108] have made a similar HCO^+ detection in another one, NGC 253. In both these galaxies the observed HCN to HCO^+ intensity ratio is near unity, as is typical for giant molecular clouds in our Galaxy (see above). Both sets of authors conclude that this indicates the HCN and HCO^+ emissions from the other galaxies are dominated by giant molecular clouds. Rickard and Palmer [108], however, failed to detect HCO^+ in three other galaxies. They suggest that this failure indicates that the cosmic ray flux density is lower in these three, and in support of that hypothesis they show the synchrotron emission from relativistic electrons is lower in the three galaxes where HCO^+ was not seen than in M82 and NGC 253. They predict that ultimately HCO^+ observations might be a useful probe for determining

cosmic ray fluxes in other galaxies, although many difficulties in doing so can
be envisioned.

The abundances of deuterated versions of interstellar molecules and ions relative
to those of the corresponding non-deuterated forms are potentially of great
cosmological interest, insofar as they reveal the underlying atomic D/H ratio as
a function of location in the Galaxy. Penzias [109] has measured intensity
ratios of DCO^+ to $HC^{18}O^+$ and DCN to $D^{13}CN$ in several sources of widely differing
galactocentric radii, ranging from galactic center sources (Sgr A or B) to some
others far out the spiral arms, and concludes that the deuterium abundance
increases significantly with increasing radius. Of course, variations due to
differing degrees of chemical fractionation are a very serious complicating
factor, but Penzias argues that the sources sampled are chemically similar. In
a later paper [110] he has reanalyzed the same data, especially by using
different ^{18}O abundances in the galactic center and elsewhere, but reached the
same conclusion. As discussed in these two papers such a positive radial
gradient of deuterium abundances is considered as strong support for formation
of deuterium solely in the big bang origin of the universe and for the open
universe. The argument is that deuterium can be consumed but not produced in
the thermonuclear reactions within stars, so the ratio of D/H can only decrease
as matter is processed through stars. The interstellar matter near the galactic
center is expected to have been processed through several generations of stars,
and thus to have lost most of its deuterium, whereas that near the outer edge
of the galactic disc should show nearly the primordial D/H ratio. Furthermore,
the actual value of D/H in the outer part of the Galaxy falls in a range that
is indicative of an open or ever expanding universe according to the standard
big bang model [111]. Penzias [110] also discusses the observations of the
relative isotopic abundances of other elements, e.g., $^{13}C/^{12}C$, $^{15}N/^{14}N$, and
$^{18}O/^{17}O/^{16}O$, and the thermonuclear processes of production and destruction in
stars that are thought to be responsible for the existing ratios. Stark [112]
has measured the ratio $x(HCO^+)/x(H^{13}CO^+)$ and $x(H^{13}CO)/x(HC^{18}O^+)$ in several
sources and considers that his data tend to confirm the earlier $^{12}C/^{13}C$ ratios
of Wannier et al. [113], namely 20 in the galactic center, 40 in Orion, and 60
elsewhere in the galactic disk, i.e., outside the nucleus. Recently Guélin et
al. [114] have observed in Sgr B2 the J=0-1 transition of $HC^{17}O^+$, a rare
isotopic species not yet detected in the laboratory. Its frequency, however,
can be reliably enough predicted from those of the other isotopic forms that
assignment is not a problem. The observed $x(HC^{18}O^+)/x(HC^{17}O^+)$ ratio of 3.1 ±
0.6 is in good agreement with $^{18}O/^{17}O$ ratios obtained from CO or OH. The ratio
$^{18}O/^{17}O$ appears to be fairly constant throughout the Galaxy except in the solar
system, where the value 5.5 applies, but such an invariance is somewhat
unexpected, since the two nuclei are thought to be produced in very different
ways [110].

Considerable further progress on the determination of limits to electron
abundances from HCO^+ observations has been made in recent years. Wootten et al.
[115] have proposed a method for establishing an upper bound to x_e that is
independent of any deuterium observations and is instead contingent upon a
reliable measurement of the parameter $Z \equiv [H_2] x(HCO^+)/ x(CO)$. By considering
the formation of H_3^+ from cosmic ray ionization (with probability 2ζ of
ionizing any H_2 molecule in 1 sec), the formation of HCO^+ in reaction (12),
and the destruction of both H_3^+ and HCO^+ by dissociative recombination (with
rate constants k_e and k_e', respectively) they derive the inequality

$$x_e \leqslant \left(\frac{2\zeta\ K_{12}}{k_e\ k_e'\ Z} \right)^{\frac{1}{2}} \quad , \qquad\qquad (16)$$

whose successful application requires that ζ be known. Conversely, if an upper
bound to electron abundance ($max(x_e)$) is otherwise available, e.g., from the
deuterium ratio method, this relation can be turned around to provide an upper
limit to the cosmic ray ionization rate

$$\zeta \leqslant k_e \ k_e' \ [\max(x_e)]^2 \ Z/ \ (2 \ K_{12}).\tag{17}$$

The previously described method for finding $\max(x_e)$ (as embodied in inequality (15)) depends critically on knowledge of several rate constants, especially k_{14}. Adams and Smith [116] have recently measured k_{14} and the rate constant for the reverse (slightly endothermic) reaction, k_{-14}, at 80°, 200°, and 295°K. Their value of k_{14} is several times larger than the previously employed ICR value [117], and the sum $k_{14} + k_{-14}$ is observed to be roughly equal to k_L, the Langevin rate. At very low cloud temperatures $k_{14} \cong k_L$, and k_{-14} becomes quite small. Adams and Smith [116] attempted to extrapolate their results for k_{-14} to low inter-stellar temperatures, but the method they used, while it represented an improvement over some previous discussions in the astrophysical literature, was still too oversimplified to obtain the correct temperature dependence. Herbst [118] has carried out a proper statistical mechanical treatment to obtain purely theoretical values of k_{14} and k_{-14} as a function of temperature, and his results agree well with the measured values of Adams and Smith [116] at the higher temperatures. In this fairly pathological case nuclear spin statistics, rotational zero-point energy, and exact rotational partition functions must be considered. At temperatures of giant molecular clouds as opposed to cool dust clouds, i.e., for $T \gtrsim 20°K$, the reverse reaction in equation (14) becomes more important than dissociative recombination as a loss process for H_2D^+ ($k_{-14} > k_e \ x_e$). Then instead of inequality (15) one obtains from the deuteration ratio R a limit on k_{-14},

$$k_{-14} \leqslant \frac{k_{14} \ x(HD)}{3 \ R} \ ,\tag{18}$$

and thus an upper limit on temperature. At slightly higher temperatures this inequality becomes essentially an equation, and the deuterium ratio R can serve as an alternative cloud thermometer. Herbst [118] has used this idea, his theoretical rate constants, and published R values to obtain cloud temperatures for several giant molecular clouds that are in reasonable agreement with those obtained in the more usual way from NH_3 or CO line temperatures. Wootten et al. [119] have made an extensive study of many dense clouds using observations of the J=0-1 transitions of HCO^+, $H^{13}CO^+$, and DCO^+ and the J=1-2 transition of DCO^+. Densities and abundances were determined using LVG model calculations that were similar to the previously mentioned ones of Wootten et al. [99] except that the two transitions of DCO^+ were used for the density determination instead of the two formaldehyde lines. The ratio R was obtained from DCO^+ and $H^{13}CO^+$ data to avoid the very serious optical depth effects in the main isotopic form. Upper limits to x_e were obtained from the method of inequality (15), using the new rate constant values [116]. In the coolest dark clouds, e.g., L134 N, the limit was 2×10^{-7}, but larger and less meaningful limits were obtained in warmer sources (giant molecular clouds), where R is more a reflection of temperature than of x_e. A strong dependence of R on T was observed, similar to the trend seen earlier in DNC-HNC comparisons [120] and in general agreement with the predictions of equation (18). The abundance of HCO^+ was seen to fall off very rapidly with increasing density in agreement with the results of Wootten et al. [99]. The method of inequality (16) was also used to establish upper bounds to x_e, and these upper bounds had values near 10^{-7} in almost all cases, including the warmer clouds, where the R method is not too helpful. The value of ζ was taken as 1×10^{-16}/sec [121], and in order to obviate the problems of optical depth the ratio $Z^{13} = [H_2] \ x(H^{13}CO^+)/ \ x(^{13}CO)$ was used instead of Z. Finally Wootten et al. [119] warned that the very strong dependence of deuterium enhancement in HCO^+ on temperature would tend to confuse the determination of the variation of the underlying D/H ratio with galactocentric radius, and they suggested that the conclusion of Penzias [109,110] regarding a positive radial deuterium gradient should be treated with caution. Guélin et al. [122] have made a detail mapping of isotopic HCO^+ in five sources. The HCO^+, $H^{13}CO^+$, $HC^{18}O^+$,

and DCO$^+$ species were all studied, and the work also included the first astronomical detection of D^{13}CO$^+$. The five sources are all cool and mostly consist
of typical dark dust clouds. The ratio of DCO$^+$ to H^{13}CO$^+$ was found to be
remarkably constant from cloud to cloud and within individual clouds. Upper
limits to x_e (using the new rate constant data) were 1-3 x 10^{-7} and were found
to apply rather uniformly to the dense cores of the observed clouds. The normal
HCO$^+$ observations suggested self-absorption in surrounding more diffuse regions,
as had earlier studies [90], and the boundaries of normal HCO$^+$ emission
correlated well with those of the high optical obscuration due to dust.
Abundance determinations for HC^{18}O$^+$ indicated that x(HCO$^+$) $\cong$ 10^{-8}, which is
within one order of magnitude of the total ionization limit, inside these cool
clouds. The depletion of metals onto grains was examined, and the max(x_e)
results indicated that no more than 1 part in 40 of the metals remain in the gas
phase. The basic argument here is that gas phase metals (those of low
ionization potential) will be ionized by charge exchange with molecular ions
much faster than they can be removed by collision with dust grains or neutralized
by radiative recombination, so that x(metals) $\leqslant x_e$. Some consideration is given
to DNN$^+$ and HNN$^+$, and Guélin et al. [122] conclude that optical depth effects
are important even in HNN$^+$ (contrary to [88]) and that x(DNN$^+$)/x(HNN$^+$) is
probably similar to R, although the available data are inadequate to confirm
that similarity. Guélin et al. [122] and Wootten et al. [119] (via inequality
(17)) obtain similar maximum ionization rates (ζ) in cool clouds of 3.5 x 10^{-16}/
sec and 7 x 10^{-16}/sec, respectively. The ratio R in cool dark clouds is .03-
.05 according to both groups of investigators. Thus the limits for ionization
in dark clouds now appear to be fairly well established, while those in typical
giant molecular clouds seem to be numerically similar, and the data and
interpretations of the different groups are all in reasonable harmony.

The chemistry of HCS$^+$, CO$^+$, and HOC$^+$ in the interstellar medium is far less well
characterized and understood than that of HCO$^+$ and HNN$^+$ at the present time.
Erickson et al. [43] only observed CO$^+$ at a single point in OMC-1, where they
estimated a column density of 2 x 10^{12}/cm^2 and a fractional abundance of
x(CO$^+$) ~ 10^{-10} - 10^{-12}. Although this observation was in the direction of the
KL nebula, the velocity distribution was that characteristic of the ridge source
with no evidence for any plateau component. Earlier model calculations [45,46]
had predicted x(CO$^+$) $\leqslant$ 10^{-13} in dense clouds. Huntress et al. [123] have
measured the rate constant k$_1$ over the temperature range 100-390°K using ICR and
have found no variation; their value (1.4 $\pm$ .3 x 10^{-9} cm^3/sec) for the rate of
reaction (1) is essentially the Langevin rate. They also discuss the chemistry
of CO$^+$ in OMC-1 and conclude that only unusual, non-equilibrium conditions can
explain the reported CO$^+$ abundance, since its destruction in reaction (1) is so
fast in relation to any expected formation mechanism. According to Thaddeus
et al. [38] HCS$^+$ is well correlated in intensity with CS in the ten sources in
which they detected it, and the column density estimates for HCS$^+$ are typically
1-3% of those for CS. Model calculations [45,124], however, predict this
proportion to be nearer .1%. The discrepancy between observation and model in
this case is clearly not due to abnormal or specialized physical conditions,
and it is small enough to be reasonably attributable to some combination of
errors in rate constants, abundance estimates of other species, or observationally determined column densities. The other molecule of Thaddeus et al. [38],
which they suggest may be HOCO$^+$, was only detectable in Sgr B2. From the line
temperatures of the several different J transitions, they obtain T$_{ex}$ = 8.7 $\pm$
.8°K for the molecule. Although HOCO$^+$ (or HOCN) would be slightly non-linear
and thus have satellite lines for K $\neq$ 0, their intensities at the low temperature
of Sgr B2 were predicted to be too low for detection, so the observed psuedo-
linear molecule spectrum is not unexpected in this source. HOC$^+$ has also only
been detected in Sgr B2, but upper limits to x(HOC$^+$)/x(HCO$^+$) have been
established in a number of other sources. Woods et al. [57] have described a
simple kinetic model for HOC$^+$ production and destruction which attempts to
account for the observed low abundance of HOC$^+$ relative to HCO$^+$ or HNN$^+$. The
two isomers, HCO$^+$ and HOC$^+$, are assumed to be formed primarily by proton transfer

from H_3^+ to CO with a branching ratio near unity. All exothermic proton transfers (HOC^+ to N_2, CO, or X; HNN^+ to CO or X; HCO^+ to X, where X = H_2O, ...) are presumed to have near Langevin rates, and the destruction of HOC^+, HNN^+, and HCO^+ by dissociative recombination are all expected to be about equally fast. Finally, isomerization by atomic hydrogen

$$HOC^+ + H \rightarrow HCO^+ + H \qquad (19)$$

is assumed to be fast, and the observations are interpreted to indicate that $x(H)$ may be higher than previously thought in dense clouds and that reaction (19) may be a major contributor to the low observed values of the abundance of HOC^+. Direct proton transfer from HOC^+ to CO and indirect proton transfer from HOC^+ to N_2 and then to CO also contribute to the isomer conversion, while the reactions with e^- and with X destroy both isomers and tend to equalize their abundances. The observation that $x(HNN^+)/x(HOC^+) \gg 1$ argues strongly against the isomer conversion being primarily due to the HOC^+ plus CO reaction, since HNN^+ would be expected to be destroyed equally fast by HNN^+ plus CO. Before the inferred predominance of reaction (19) can be considered well established three important assumptions must be confirmed: (1) that k_{19} is indeed large, (2) that HOC^+ does not have a fast reaction with H_2, and (3) that the branching ratio in H_3^+ plus CO is actually close to one. Illies et al. [125] have recently reported mass spectrometry experiments from which they deduce a branching ratio of 16/1 in favor of HCO^+, but such a value makes the observed intensity ratio, $\gamma(HCO^+)/\gamma(HOC^+)$, in the laboratory microwave experiments [55] very difficult to understand. The latter seem most consistent with a branching ratio near unity. Measurements of the rates of the reactions of HOC^+ with H or H_2 have not even been attempted. Because the chemistry of HOC^+ is so intimately connected with that of HCO^+ and HNN^+, astronomical observations of it are potentially a very useful tool, and the clarification of the above three points of uncertainty is of particular astrophysical importance.

MOLECULAR STRUCTURE AND FORCE FIELDS

Probably the most well known application of microwave spectroscopy is the determination of molecular structures of small gas phase molecules. Spectroscopic transition frequencies are used to determine effective rotational constants, B_v, for various vibrational states, using model Hamiltonians that usually also have small correction terms for centrifugal distortion. In a true rigid rotor the reciprocals of the B's would be (within a constant factor) just the moments of inertia of the molecule, and the determination of molecular structure to high precision would be a straightforward matter. In a real moelcule, however, the effective rotational constants always contain additional terms, the vibration-rotation interaction terms or α terms, and the accuracy and meaning of a microwave molecular structure is almost always limited by the nature of the treatment of these α terms, rather than by the precision in the experimental frequency measurements. In the case of a linear triatomic molecule we have

$$B_{v_1 v_2 v_3} = B_e - \alpha_1 \left(v_1 + \frac{1}{2}\right) - \alpha_2 \left(v_2 + 1\right) - \alpha_3 \left(v_3 + \frac{1}{2}\right) + \cdots, \qquad (20)$$

where B_e is reciprocally related to the true Born-Oppenheimer equilibrium moments of inertia and bond distances (r_e's), and higher order terms, e.g., γ_{ij} $(v_i + 1/2)$ $(v_j + 1/2)$, have been indicated only by dots. For larger polyatomic molecules there are even more α terms (one for each normal mode of vibration) and possibly A and C rotational constants, while for a diatomic there is only a single vibrational mode and a single α constant. Ideally one must obtain in a polyatomic molecule the effective rotational constants in the ground vibrational state and one excited vibrational state of each normal mode in order to separately determine B_e and the individual α's, and this must be done for several isotopic species for an equilibrium molecular structure determination (r_e structure). Such extensive data is rarely available even for neutral polyatomic molecules (only for a few triatomics), and the great majority of molecular structure determinations by microwave spectroscopy have relied solely on ground vibrational state data (B_0's) for various isotopic forms. The most useful and widely known structure of this type is the substitution or r_s structure, in which a parent isotopic form and isotopic variants that are singly substituted (relative to the parent) at each atomic position are employed. First the center of mass coordinate of each atom in the parent species is calculated from Kraitchman's equation [126]

$$z_s^2 = \Delta I_0/\mu ,$$ (21)

where ΔI_0 is the change of $I_0 = \hbar^2/2B_0$ on isotopic substitution at that atom and μ is a known function of the atomic masses. Then the z_s's for each atom are combined to give the r_s structure. The underlying premise or hope of the method is that the α constants are at least similar in the various isotopic forms and to a large extent cancel in formation of the shifts, ΔI_0. The r_s and r_e structures have typically been found to be within a few thousanth's of an angstrom of each other in cases where both are known. Further discussion of the r_s structure method has been given by Costain [127]. The laboratory data on the $J = 0$-1 frequencies for six isotopic forms of HCO^+ [81], whose application to astronomical identification of these species has already been discussed, was used to determine an r_s structure for HCO^+. Actually two determinations, with either HCO^+ or DCO^+ as parent species, were made and found to agree closely (within 0.0003 Å). The averaged results were $r_s(CH) = 1.0930(1)$ Å and $r_s(CO) = 1.1071(2)$ Å. Since only $J = 0$-1 data were available, the centrifugal distortion effects were not treated, and this r_s structure of Woods et al. [81] was calculated from values for $B_{eff} = B_0 - 2\ D_0$, rather than from B_0. Bogey et al. [69] have determined B_0 and D_0 separately for all six isotopic forms and recalculated the r_s structure using B_0's, obtaining essentially the same result ($r_s(CH) = 1.0930(1)$ Å and $r_s(CO) = 1.1070(2)$ Å). The measurements of Woods et al. [81] used the older computer control scheme with a free running microwave source, but Gudeman and Woods [56] have obtained measurements of the $J = 0$-1 transitions of HCO^+, $H^{13}CO^+$, $HC^{18}O^+$, and $H^{13}C^{18}O^+$ (observed for the first time) with improved frequency accuracy using the digitally programmed, phase-locked klystron scheme (see Experimental Considerations section). With measured [69] or estimated centrifugal distortion corrections this new data set [56] gave a value $r_s(CO)$ which varied only from 1.10723 Å to 1.10719 Å as the choice of parent species ranged over any of these four species. Thus with sufficiently accurate frequency data the r_s structure seems incredibly stable with respect to choice of parent species. Guélin et al. [122] obtained accurate $J = 0$-1 frequencies of $H^{13}CO^+$ (86,754.294 ± 0.030 MHz) and $HC^{18}O^+$ (85,162.256 ± 0.040 MHz) from high resolution astronomical data in selected dark clouds. The comparison of these to the corresponding values of Gudeman and Woods [56] (86,754.279 ± 0.010 and 85,162.222 ± 0.012 MHz, respectively) indicates the current level of agreement between the best astrophysical and laboratory frequency measurements for ions. An r_s type structure for HNN^+ has been obtained by Szanto et al. [128] based on $J = 0$-1 measurement of a total of 6 isotopic forms: the main species, $HN^{15}N^+$, $H^{15}N^{15}N^+$, and their deuterated analogues. The reported bond lengths are $r_s(HN) = 1.0320(1)$ Å and $r_s(NN) = 1.0947(4)$ Å, using an average of results for either $HN^{15}N^+$ or $DN^{15}N^+$ as parent species. (The variation with parent species is shown in parentheses.) Gudeman and Woods [129] have made more precise frequency measurements for HNN^+, $HN^{15}N^+$, $H^{15}N^{15}N^+$, and $H^{15}NN^+$ and obtained a slightly changed structure based on averaging results for the four hydrogen species as parent: $r_s(HN) = 1.03162(7)$ Å and $r_s(NN) = 1.09505(3)$ Å. (The previously reported $J = 0$-1 frequency for $HN^{15}N^+$, 90,261.06 MHz [128], appears to be in error; the new value is 90,263.833 ± 0.015 MHz [129].) For both HCO^+ and HNN^+ several different groups have carried out theoretical structure predictions using elaborate ab initio calculations including the effects of electron correlation. In Table I these are compared to the microwave r_s structures of these two ions. In general, the agreement is extremely satisfactory and demonstrates the great power of modern theoretical methods for structure predictions on molecules of

TABLE I

A comparison of ab initio theoretical structures and microwave experimental structures of 14-electron ions (all distances in angstroms)[a].

	r(HX)	r(XY)
HCO$^+$		
r_S [56,69,81]	1.093	1.107
CI-SD [7]	1.095	1.105
CI-SD [9]	1.091	1.103
CI-SD/DZ+P [130]	1.095	1.115
MP3/6-311G** [54]	1.093	1.099
HNN$^+$		
r_S [128,129]	1.032	1.095
CI-SD [53]	1.032	1.095
CI-SD [131]	1.031	1.097
SCEP [132]	1.030	1.093
CI-SD/DZ+P [130]	1.035	1.099
HOC$^+$		
r_S [55]	.934	1.159
CI-SD [52]	.976	1.159
CI-SD [53]	.987	1.151
CI-SD/DZ+P [130]	.992	1.167
MP3/6-311G** [54]	.988	1.152

[a]The labels applied to the theoretical calculations are mainly intended to indicate which of several calculations by given sets of authors are being quoted. The original papers should be consulted for a precise definition of the basis set and correlation method used.

this complexity. In assessing the quality of the theoretical values one must remember that they are all predictions of r_e, while they are being compared to spectroscopic values of r_s, which are themselves somewhat different from the true r_e. In the isoelectronic molecule HCN both the r_e and r_s structure have been obtained experimentally [133,134], and in that case the magnitude of the difference between the r_s and r_e values are about 0.002 Å for each of the two bonds. For both HCN and its isomer HNC ab initio calculations [53,54,130,135, 136] similar to those referred to in Table I have been reported, in most instances by the same research groups. The agreement between experimental r_s values for these neutral analogues is not unlike that seen in Table I for the ions. In Table I it can be seen that the calculation of Walgren et al. [7], and Kraemer and Diercksen [9] on HCO$^+$, both of which were intended particularly for structure prediction and done before the experimental structure was known, are in excellent agreement with the r_s results. The perfect agreement of the calculation of Hennig et al. [53] with the r_s structure for HNN$^+$ does not hold for the other isoelectronic species (although they are all in quite satsifactory agreement), and if the experimental r_e were available, instead of r_s, there would probably be small discrepancies for HNN$^+$ too. Although the calculations represented in Table I all involve large basis sets and extensive configuration interaction, they are still not fully converged with respect to the predicted structure. In a number of cases the same authors have done other calculations with different basis sets or different levels of treatment of electron correlation and obtained other structures differing somewhat from those shown in

Table I. In some cases a larger basis set gives a poorer structural result.
Best results are obtained when the basis set and CI treatment are well matched
in the sense that errors from incomplete basis set and those from incomplete
treatment of electron correlation tend to cancel each other, as judged from
results for known reference molecules, e.g., HCN.

As previously mentioned the sensitivity enhancements associated with the improved
glow discharge spectrometer and computer controlled phase-locked microwave source,
along with the utilization of a discharge gas mixture that was mainly argon and
only a few percent each of hydrogen and carbon monoxide, has recently enabled
Gudeman and Woods to observe the microwave spectrum of HCO^+ in excited
vibrational states [56] and also that of the isomer HOC^+ [55]. The ab initio
calculations of Hennig et al. [53] were of particular interest and use in
connection with both of these studies, since they included predictions not only
of the equilibrium structure, but also of the harmonic and anharmonic terms in
the force field, the vibrational frequency constants (ω_e and $\omega_e x_e$), the
vibration-rotation interaction constants (α), and the B_0's and $Bv_1v_2v_3$'s.
Furthermore, all five isoelectronic species (HCO^+, HOC^+, HNN^+,
HCN, and HNC) were treated in a consistent way, and the spectroscopic parameters
were given for all reasonable isotopic combinations. The vibrational satellite
lines of the J = 0-1 transition corresponding to the (100), (02°0), and (001)
states of HCO^+ were detected in the main isotopic species and $H^{13}CO^+$ and $HC^{18}O^+$
[56]. (The fundamental mode of the bending vibration (01'0) has no J = 0-1
transition.) This permitted B_e and each of the three α constants to be
determined from equation (20). The theoretical α values were generally within
25% of the observed ones and proved useful in establishing a search region for
the satellite spectra. The theoretically predicted isotopic variation of the
α's turned out to be in very close agreement with that observed, and this fact
provided a powerful test of the validity of the spectral assignments. Thus
once α_i had been measured for one isotopic form (B) it could be predicted for
another (A) by the scaling relationship

$$\alpha_i(\text{scaled, A}) = \frac{\alpha_i(\text{observed, B})}{\alpha_i(\text{CI-SDQ, B})} \ \alpha_i(\text{CI-SDQ, A}), \qquad (22)$$

and $\alpha_i(\text{scaled, A})$ was usually within 1-2 MHz of $\alpha_i(\text{observed, A})$. Since there are
only two bond distances, in principle B_e's for any two isotopic forms suffice to
determine the r_e structure, and the three available B_e's for HCO^+ allowed three
different determinations of the r_e's, and thus a consistency check.
Unfortunately, the three determinations were not as consistent as one would have
hoped. The values of $r_e(CO)$ were 1.1050, 1.1043, and 1.1063 Å and those of
$r_e(CH)$ were 1.0954, 1.0989 and 1.0885 Å for the HCO^+/$H^{13}CO^+$, HCO^+/$HC^{18}O^+$ and
$H^{13}CO^+$/$HC^{18}O^+$ pairs, respectively [56]. Analogous measurements on HCN, $H^{13}CN$,
and $HC^{15}N$ were carried out [137] to test relation (21) and to check the stability
of the r_e's to choice of isotope pair. In HCN the heavy atom distance variation
over the three pair choices was only 3 x 10^{-5} Å, and the light atom one was only
3 x 10^{-4} Å. The most plausible explanation of the much greater variations in
HCO^+ is a vibrational perturbation of the (100) state in one or more isotopic
forms. The ab initio calculations exclude any such perturbation for either the
(001) or (02°0) states, but the (100) state may be close enough for a non-
negligible interaction with either the (04°0) state (via third order anharmonic
resonance) or the (01'1) state (via coriolis resonance). Some additional support
for this interpretation comes from the fact that the scaling procedure of equation
(22) is less successful for α_1 of HCO^+ than for the other two α's of HCO^+ or any
of those for HCN. Additional microwave or infrared data for HCO^+ will be required
to fully characterize the vibrational perturbations of HCO^+ or otherwise explain
the variation in the r_e structure with isotope pair. Any of the three possible
r_e structures are in reasonable agreement with the r_s one and the theoretical
results shown in Table I, but a final evaluation of the accuracy of various
theoretical structures must await further improvement of the experimental r_e

determination. The α constants and their isotopic dependence contain very useful information on the force field of a triatomic molecule, particularly the cubic force constants. There are, however, six of the latter, and the α's also depend on the quadratic force constants, so it is not convenient to determine force constants solely from the present data set for HCO^+. When infrared data (band origins) become available, they can be combined with the microwave vibration-rotation interaction constants to obtain a very accurate experimental characterization of the potential function.

For HOC^+ [55] isotopic substitution was only done at the carbon and oxygen atoms so the center of mass coordinate of the hydrogen in the r_s structure determination had to be obtained from the so-called first moment condition, $\Sigma mz_s=0$. The result is shown in Table I along with structural predictions from several ab initio calculations on HOC^+. The heavy atom r_s distance is seen to be in excellent agreement with the various ab initio calculations, but $r_s(OH)$ seems much too short. This discrepancy, which is somewhat magnified by the use of the first moment condition to locate the H atom, is simply a manifestation of the large amplitude bending motion in HOC^+. Even the first SCF calculation [5] on this isomer had indicated that it had a very flat bending potential by comparison to that of HCO^+, and the results of Hennig et al. [53] (with some scaling based on the HCN and HNC cases) indicate the bending vibrational frequency is in the range 200-300 cm^{-1}. In such a molecule the r_s procedure tends to yield the projection of the light atom distance on the axis of the heavy atom bond, rather than the true light atom distance; such apparent shortening of the OH bond had already been observed in the CsOH and RbOH molecules [138], which also have low bending frequencies (306 and 309 cm^{-1}). The exact shape of the bending potential of HOC^+, particularly the possible presence of the phenomenon of quasilinearity (a small potential bump at the linear configuration), is now a matter of considerable theoretical and experimental interest. Spectroscopic data on excited states of the bending vibration can eventually yield a reliable potential for this bending motion. Several groups [52,53,54,130] made careful attempts to predict the J = 0-1 transition frequency of HOC^+ before it was observed using ab initio results. Both DeFrees et al. [130] and Nobes and Radom [54] attempted to correct the frequency prediction for HOC^+ based on errors observed for identical calculations on isoelectronic species with known structures and frequencies. Such efforts, however, led to frequencies 400-700 MHz lower than the observed one, largely because they used the unsound assumption that the α corrections for HOC^+ would be similar to those of the reference molecules. Actually α_2 is unusually large for HOC^+ due to the large amplitude bend. By correcting the theoretical B_0 of Hennig et al. [53] for HOC^+ as indicated by comparison of their B_0's and observed ones for reference molecules, however, a very good result (89,400 MHz [55]) was obtained, because these B_0's automatically include α's appropriate to each species.

For HCS^+ isotopic substitution data (and thus an experimental structure) are not yet available, but the spectroscopic B_0 value for the main species [38,41] has been compared to the results of several ab initio calculations [39,41,139,140]. These comparisons strongly support the assignment of the observed spectrum to HCS^+. As previously mentioned extensive ab initio calculations have also been done on $HOCO^+$ (and HOCN) [42], and they show the former is a more likely prospect for the carrier of the psuedo linear molecule spectrum of Thaddeus et al. [38]. High quality ab initio predictions of the structures and rotational constants of many other small molecular ions, particularly those of suspected astrophysical importance or observability, have been reported. No attempt will be made here to review that active area of theoretical research, but a few comments are suggested by the discussion of the last several paragraphs. While reasonably modest level calculations usually suffice to predict microwave transition frequencies of light polyatomic ions to 1-2% accuracy, such uncertainties may be quite large in the contexts of a high sensitivity astrophysical or laboratory search for or identification of an unknown spectrum. Accuracies of a few tenths of a per cent, which are very much more useful to the spectroscopist, are correspondingly much

more difficult. Large basis sets, extensive configuration interaction, and
equivalent calculations on similar known reference molecules are usually
required, and even with all of that unforeseen difficulties may arise. Most of
the extant calculations have only provided equilibrium structure predictions,
but estimates of anharmonic force fields and spectroscopic constants, e.g., the
α's, would be of very great interest. They can help in finding, understanding,
and assigning spectra, and the measured and theoretical values can be compared
very successfully. Hopefully future theoretical studies will provide more of
this valuable additional information.

ELECTRONIC STRUCTURE AND PROPERTIES

The discussion of the preceding section has focused on the measurement of
structural and force field parameters, which may be deduced from the eigenvalues
of a molecule's or ion's electronic Schrödinger equation. There are other
properties, e.g., electric dipole moments, hyperfine coupling constants, or fine
structure parameters for open-shell species, which strongly influence microwave
spectra and which can in principle be obtained from microwave observations.
These latter quantities follow theoretically from the wave functions of the
electronic Schrödinger equation. The dipole moment, μ, of a species is of
crucial spectroscopic importance because the absorption coefficient, γ, that
determines laboratory microwave intensities, is proportional to its square (see
equation (3)). Furthermore, the treatment of the radiative transfer problem in
radioastronomy (see equations (9), (10), and (11)) is strongly dependent on μ.
The dipole moment of an ionic species, unlike that of a neutral one, depends on
the origin of the coordinate system. Spectroscopically the physically relevant
system is the center of mass coordinate system, in which the ion rotates and in
which its dipole moment appears to oscillate. Thus an ion's dipole moment
depends appreciably on the isotopic species considered, and in particular, ions
like HD^+, $O^{18}O^+$, or $HCCD^+$, whose symmetry is broken only by isotopic substitution,
have distinctly non-zero values of μ. The latter in debyes is simply 4.8 R,
where R is the distance (in Å) from the center of mass to the ion's geometrical
center [141], and this is typically .1-.3 D (even bigger in HD^+). Such
isotopically asymmetrical ions are potential, though difficult, candidates for
radioastronomy or laboratory rotational spectroscopy, but so far none of them
has been observed by these techniques. In the case of neutral species microwave
spectroscopy has traditionally provided a means for determination of dipole
moments to high precision through measurements of the Stark effect [142]. To
date this kind of measurements has not been reported for molecular ions because
of the difficulties of applying sufficiently large and sufficiently homogeneous
electric fields to the interior of the plasma regions in which ions are found.
The plasma is essentially a conductor, and high electric fields are usually
accompanied by excessively high current densities and power dissipation. Although
other methods independent of the Stark effect, e.g., measurements of absolute
intensities or the onset of power saturation effects, are available for obtaining
dipole moments from microwave spectroscopy, their expected accuracy for ions is
only about ±25%, even with the most careful work. A measurement of this precision
is probably not worth the great effort that would be required, because the dipole
moments of small molecules and (presumably) ions can be predicted much better
than this from good ab initio calculations (except when the overall magnitude of
μ is distinctly smaller than 1 D). While SCF level calculations on small
molecules produce values of μ which may deviate from measured ones by .5 D or
more, CI calculations of μ of a fairly routine nature are typically within .1-.2 D
of the true values. Haese and Woods [143] have obtained CI dipole moments (with
a basis set polarized by off-center bond functions) for HCN, HNC, HNN^+, HCO^+,
HBF^+, and $HCNH^+$, and their results for HCN and HNC differ by less than .1 D from
precision Stark effect values [144,145]. DeFrees et al. [130] have done a much
larger CI calculation with a larger basis set on the same species (excluding HBF^+
but including HOC^+) and obtain values differing by at most .11 D from those of
Haese and Woods [143]. The computed dipole moment for HOC^+ (2.80 D [130]) is

substantially less than that of HCO$^+$ (3.96 D [130] or 4.07 D [143]), which
according to equation (3) accounts for approximately a factor of two reduction of
the laboratory HOC$^+$ signal relative to that of HCO$^+$. The predicted low dipole
moment of HCNH$^+$ (0.23 D [143] or 0.31 D [130,146]) is probably the major
impediment to detection of the rotational spectrum of this astrophysically very
crucial ion (see equation (2)). Early dipole moment predictions for CO$^+$ [29],
HNN$^+$ [35], and NO$^+$ [58,59] have already been discussed. For HCS$^+$, CI level
calculations of μ have been reported by Gudeman et al. [41] (1.86 ± 0.3 D) and by
Chekir et al. [140] (2.2 D). Good theoretical calculations of electric dipole
moments have also been made for many other potentially observable small ions,
but in the interest of brevity a comprehensive review of them will not be given
here. Regretably, in view of the great physical significance of dipole moments,
many other excellent ab initio structural papers have neglected to report values
for them, even though the computational effort involved is minimal once the wave
function has been determined.

Because of its inherently high resolution microwave spectroscopy is particularly
well suited for the accurate determination of hyperfine coupling constants. Since
these constants may also be computed from the species' electronic wave function,
they provide one of the most concrete available tests of the quality of ab initio
calculations. Conversely the specific hyperfine patterns and values of coupling
constants from ab initio calculations can be of great help in assigning an unknown
microwave spectrum, as exemplified by the previously discussed identification of
the radioastronomical HNN$^+$ spectrum by means of its quadrupole hyperfine structure
[34-36]. The assignment of the NO$^+$ spectrum was also supported by observation of
^{14}N quadrupole splittings [60], and most other ions with a quadrupolar nucleus
(other than D) would be expected to show resolvable splittings in their lower J
transitions. In open-shell molecules or ions there are also magnetic hyperfine
terms in the Hamiltonian for any nucleus with a spin of 1/2 or more [147], and
these are theoretically especially sensitive to the distribution of the unpaired
electron(s). Piltch et al. [148] have reported a determination of the ^{13}C
magnetic hyperfine parameters in CO$^+$, the only open-shell ion yet studied in the
microwave region. They obtain b$_F$ = 1506 ± 15 MHz for the Fermi contact parameter
and t = 48.2 ± 0.7 MHz for the spin-dipolar hyperfine constant. These values
compare very well with those of Carrington et al. [149] (1511.5 ± 12.3 and
30.7 ± 13.8 MHz respectively), which were obtained from ion beam laser
spectroscopy of the electronic transitions of ^{13}CO$^+$ in the visible. Open-shell
ions have not only the possibility of magnetic hyperfine structure, but also
various spin fine structure terms in their model Hamiltonian operators. There
are spin-orbit interaction terms, spin-spin terms, spin-rotation terms, Λ-
doubling terms, etc., depending on the exact term symbol for the state observed.
In a number of cases values for these constants, of sufficient accuracy to make
microwave searching practical, are already known from high resolution optical
spectroscopy [150]. In cases where they are not available, especially in polyatomic open-shell ions, prediction of microwave transition frequencies and
planning of experiments become extremely difficult, even for ions which might be
chemically favorable. There is a high probability that quite a few of the
remaining interstellar U-lines belong to open-shell ions and radicals for which
no previous spectral data is available. Many such species would have spectra
with widely scattered lines and without any obvious pattern. At least in the
laboratory situation the Zeeman effect provides a powerful tool for recognizing
transitions of open shell species as such. Fine structure parameters are
potentially available from ab initio calculations and suitable for very
interesting comparisons of theory and experiment. The required calculations,
however, are generally much more difficult than ordinary structure or B$_0$
predictions (often requiring wave functions for several different electronic
states) and are seldom available for molecular ions. Hopefully future
theoretical work will attach more emphasis to computation of these parameters
(and magnetic hyperfine parameters) for open-shell species; such efforts could
have a major impact on radioastronomy and laboratory spectroscopy. For CO$^+$ ($X^2\Sigma$)

the spin-rotation constant, γ_0, which was known approximately from optical work
[28], has been precisely determined by the various microwave studies [26,44,148].
The isotope dependence of γ_0 (in CO^+, $^{13}CO^+$, and $^{18}CO^+$) was determined by Piltch
et al. [148], who showed that γ_0 was closely proportional to B_0 as predicted by
Mizushima [151]. Bogey et al. [74] obtained a good value for γ_1 (the spin
rotation constant in the excited vibrational state) and demonstrated that the
vibrational dependence of γ_v was also characterized by a proportionality to B_v.
This behavior had been observed earlier by Dixon and Woods [152] in the isoelectronic CN radical. Bogey et al. [74] also obtained B_1 and B_2, and thus B_e,
to microwave accuracy, and obtained an equilibrium bond distance, r_e = 1.11524 Å,
considerably more accurate than the optical one (1.11506 Å [27]). Piltch et al.
[148] observed in the isotope dependence of B_0 some preliminary evidence for a
breakdown of the Born-Oppenheimer approximation, in the sense that B_e is not
rigorously inversely proportional to reduced mass. These deviations have been
treated theoretically [153-155] and have been precisely measured in the neutral
CO molecule [156]. When accurate measurements of them become available for the
ion too, the comparison of the CO and CO^+ results should be very interesting.
Similar precise and detailed information on spin fine structure, hyperfine
constants, equilibrium bond distance, and Born-Oppenheimer breakdown effects
should eventually be obtainable for a large number of other diatomic ions by
microwave spectroscopy.

PRESSURE BROADENING

Except at the lowest pressures or highest frequencies the predominate contribution
to the linewidth of laboratory microwave transitions is collision broadening.
The halfwidth at half-maximum, $\Delta\nu$, is related to the mean lifetime between
collisions, τ, by the time-frequency uncertainty relation [157]

$$\Delta\nu = (2\pi\tau)^{-1} . \tag{23}$$

Furthermore, $\tau^{-1} = k_c N$, where N is the density of colliding molecules and k_c is
a rate constant for rotationally inelastic collisions. The experimental quantity
normally obtained is the pressure broadening parameter, $\Delta\nu/P$, which is the slope
of a plot of $\Delta\nu$ versus pressure. This parameter can be trivially converted to a
rate constant k_c using the preceding relations and the ideal gas law.
Determinations of $\Delta\nu/P$ for many neutral molecules, both self broadened and foreign
gas broadened, have been made and used to study the nature of the intermolecular
forces between colliding molecules [158]. For collisions of two strongly polar
neutrals the dipole-dipole force is normally most important, but in other cases
other multipole expansion forces, induction forces, or short range forces may
predominate. The first quantitative results for pressure broadening of an ion
transition were those of Anderson et al. [159] for the J = 0-1 transition of HCO^+
broadened by collisions with H_2 in a liquid nitrogen cooled discharge tube. They
obtained $\Delta\nu/P$ = 29.6 ± 3 MHz, which agreed with earlier qualitative conclusions
that ion transitions were three or four times broader than those of similarly
polar neutral species with the same perturber. Some predictions [12,34] before
ion microwave spectra were observed held that this factor might have been twenty
instead of three or four, and this would have had profoundly adverse consequences
in terms of the peak intensity (see equation (3)) and the possibility of
modulating ion transitions or distinguishing them from spurious standing wave
related signals. These early measurements [159] were done with the non phase-
locked version of the microwave spectrometer, and sensitivity considerations
prohibited testing other temperatures or foreign gases. A new and more extensive
collection of measurements on the J = 0-1 line of HCO^+ has been carried out by
Gudeman et al. [68] using the phase-locked, digitally programmed microwave source.
In that work the broadening ($\Delta\nu/P$) by H_2 at room temperature and by Ne, Ar, and
CO with liquid nitrogen cooling were also obtained, and the value for $\Delta\nu/P$ for H_2
at liquid nitrogen temperature was redetermined and found to be in good agreement
with the earlier result [159]. For comparison purposes $\Delta\nu/P$ for $HCN-H_2$ and $HCN-Ar$
at 77°K were also measured [68], and they were indeed three to four times less

than the corresponding HCO^+ values. There are several problems encountered in the ion measurements that restrict the precision obtainable in $\Delta\nu/P$. The low available signal to noise for weak ion lines, which are often not being observed in ideal conditions, causes some scatter in the plots of $\Delta\nu/P$. The range of pressures that can be studied is limited below by loss of the discharge and above by too severe overlapping of the various lobes of the modulated lineshape. The overall pressure can be measured well, but the actual steady state partial pressures of individual gases in the discharge are not measured. The experiment is always done with small and constant input of all gases except the desired collision partner, whose pressure is much larger and variable. The kinetic temperature is difficult to measure; in fact the temperatures of the ions and the perturbers are probably different. The linewidths for given input gas pressures and wall temperatures were observed to decrease slightly with doubling of the discharge current, as would be expected if the gas were being warmed to some extent by the discharge current. Broadening by collisions of the ions with free electrons is another process that must at least be considered. Such collisions have been shown theoretically [160] to have very much (10^4 times) greater cross sections than those between ions and neutrals, but the electron densities are many orders of magnitude less than gas densities in the discharge. Experimentally the decrease of the linewidth with increasing current constitutes strong evidence against any significant role for this process. Despite the several experimental problems the $\Delta\nu/P$ values obtained by Gudeman et al. [68] (with precisions of 10-20%) were found to be consistent, essentially within the experimental error, to an extremely simple model. The rate constants k_C deduced from $\Delta\nu/P$ as described earlier were found to be in very good agreement with the usual Langevin [161] rate constants

$$k_L = 2 \pi e (\alpha/\mu)^{1/2} , \tag{24}$$

where μ is the reduced mass of the colliding system and α is the polarizability of the neutral perturber. Thus the long range monopole-induced dipole force appears to completely dominate the broadening. In the simple picture suggested by the near equality of k_C and k_L collisions with impact parameters larger than the velocity dependent Langevin critical value [161]

$$b_0(v) = (4 e^2\alpha/\mu v^2)^{1/4} \tag{25}$$

have hyperbolic trajectories and experience no rotational quantum number changes, whereas those with smaller b_0's and spiraling trajectories experience very hard collisions and have near unit probability of a rotational quantum number change. This is very similar to the usual kinetic model for ion-molecule reactions, except that rotational excitation rather than chemical reaction is the end result of the strong collision. Pursuing the analogy to the kinetics of ion-molecule reactions further one might expect the so-called AADO theory of Su et al. [162] to be suitable for predicting the linewidth when the perturber is a very polar molecule. Another interesting case will be the broadening of CO^+ by CO, where there exists the possibility of resonant charge exchange. The simple predictability for $\Delta\nu/P$ for ions and most buffer gases has important consequences for laboratory spectroscopy. The $\Delta\nu$ in equation (3) can be reliably estimated in advance, which helps in predicting the feasibility of new detections. Furthermore, the ion linewidths are distinctly larger than those of almost any conceivable neutral in most buffers, so the linewidths become an excellent criterion for assignment of newly detected transitions to ions. This idea was relied on heavily in the assignment of HOC^+ [55] and excited vibrational states of HCO^+ [56].

The $\Delta\nu/P$ parameters are very closely related to the rotational excitation rates that play such a prominent role in the establishment of T_{ex} in the interstellar clouds, so the results of Gudeman et al. [68] also have important consequences for radioastronomy. The relevant quantities are $k(J \rightarrow J')$, the rate constants for excitation from rotational state J to state J'. The important collision gases are,

of course, H_2 and (to a lesser extent) He. Green [163] has done an important
quantum mechanical calculation of the $k(J \rightarrow J')$ for the case HNN^+-He. (Helium
is much simpler to treat theoretically than H_2.) First the intermolecular
potential was computed as a function of angle and distance with double zeta SCF
calculations, and then the k's were obtained in a close-coupling treatment of
the scattering. These theoretical values have been used by most of the astronomical analyses on ions discussed earlier [92,97-99,101,103,105,119,122].
Although not pointed out by Green [163], the sum, $k_{total} = \Sigma_{J'} k(J \rightarrow J')$, of his
theoretical state to state rate constants turns out to be very close to just k_L,
and since k_{total} represents essentially the same thing as k_c, the data of
Gudeman et al. [68] can be considered as the first experimental verification of
the results of Green [163] and a confirmation of the validity of the assumptions
that have been used to obtain column densities and abundances from astronomical
data. The individual $k(J \rightarrow J')$ values for a given J and different J''s are shown
by Green [163] to be more or less statistical, with small ΔJ's most probable and
the probability falling off monotonically with ΔJ. Green [163] points out that
these results are quite different from earlier ones of Green and Thaddeus [164]
on the analogous neutral system HCN-He. The value k_{total} is about four times
lower in the latter case (consistent with the laboratory $\Delta \nu/P$ measurements [68])
and the individual $k(J \rightarrow J')$'s show much different behavior, i.e., a strong
propensity for $\Delta J = 2$ or other even ΔJ's. This very non-statistical result was
attributed to sudden collisions with a very anisotropic potential, while the
statistical behavior in the ion case was attributed to the collision complex
associated with Langevin type mechanism [163].

DYNAMICAL INFORMATION AND DISCHARGE DIAGNOSTICS

A variety of different kinds of dynamical information are potentially available
from the study of lineshapes, intensities, widths, and frequency displacements of
microwave spectral lines, with the pressure broadening work providing just one
notable example. A great many others, collectively of immense significance, have
already been considered in our discussion of radioastronomical probing of ions in
the interstellar medium. In the present section the investigation of the
dynamical properties of molecular ions within laboratory discharge media will be
treated. The glow discharge itself is a very important and intriguing phenomenon
[165,166], and a thorough understanding of ion dynamics within it is of extreme
interest to the microwave spectroscopist, both from the point of view of predicting the feasibility of future microwave experiments on other ions and optimizing
their prospects for success and from the more general perspective of supporting
and enhancing the many other applications of laboratory discharges. The latter
include gas laser tubes, ion beam sources, plasma chemistry reactors, and sources
or cells for other kinds of spectroscopy. The relative intensities of transitions
originating in different vibrational and rotational states can be used to determine their relative populations. In the highly non-equilibrium environment of the
discharge plasma translational, rotational, vibrational, and electronic temperatures may all be different from each other and different for each different
species, especially for ions as opposed to neutrals. In polyatomics there may be
different temperatures for different normal modes of vibration, and worse yet, in
some degrees of freedom the distribution of population may not correspond to a
temperature at all. Rotational temperatures of molecular ions are of particular
interest in microwave spectroscopy, because the intensities of spectral lines
depend so strongly on them, but unfortunately they are extremely difficult to
determine using the technique. The explicit factor of T in equation (3) is
actually T_{rot} (analogous to T_{ex} in the astronomical context) and another factor of
T_{rot} is implicit in the rotational partition function dependence of the fraction
f. The relative intensities of different rotational lines of light ions, however,
are quite difficult to measure in the laboratory because they are so far apart —
in different spectral regions, requiring different sources, different harmonic
numbers, or different detectors. In radioastronomy calibration techniques are
available to obviate this difficulty, but in laboratory work no serious effort has
been made to determine T_{rot} for ions directly. The observed absorption

coefficients, γ, are most consistent with equation (3) if T_{rot} is not too much greater than the wall temperature, but this is a very crude comparison. Infrared relative intensity measurements are more straightforward, since different rotational lines are relatively close together, but those done so far [167-169] have been at considerably higher pressures than microwave work, so the results are difficult to transfer to the present context. Microwave spectroscopy is, on the other hand, well suited for the determination of vibrational temperatures by measurements of relative intensities of vibrational satellite lines, which are normally rather close to the ground vibrational state line and to each other. Vibration satellite relative intensities are to a good approximation just the same as relative state populations, since there is no variation of transition moment or Franck-Condon factors to worry about, and the other factors in equation (3) are sensibly independent of v. Gudeman and Woods [56] have obtained intensity ratios, γ_0/γ_v, of 123 ± 9, 30 ± 3, and 29 ± 4 for the (100), (02°0), and (001) satellites of HCO^+, respectively, in the DC discharge system with a gas mixture that was mostly argon. The quoted numbers are averages over the three different isotopic forms studied, and the uncertainties are just the isotopic variations. These correspond (using the <u>ab initio</u> vibrational frequencies [53]) to vibrational temperatures of 923 ± 12, 712 ± 22, and 995 ± 55 °K for ν_1, ν_2, and ν_3. As a comparison, Gudeman and Woods [56] also obtained vibrational temperatures of 1400, 460, and 960°K for the same modes of the isoelectronic neutral HCN in a predominantly argon discharge. The overall level of vibrational excitation is not grossly different for ion and neutral, but the different modes are much closer to being in equilibrium in HCO^+. The propensity for the bending vibration of HCN to be much less excited than the stretches is well known [170] and plays a major role in the mechanism of the HCN laser [171]. The more uniform excitation in HCO^+ may be attributable to the hard Langevin type collisions between HCO^+ and Ar suggested by the pressure broadening analysis. In any collision complex the perturbation of the argon might be expected to tend to randomize the vibrational energy within the ion. Shin [172] has predicted vibrational relaxation rates for ions (colliding with neutrals) much larger than for neutrals, and v-v intermolecular transfer in a Langevin strong collision is the implicated process. In the argon discharge this v-v channel is of course closed, and Gudeman and Woods [56] did find much lower (often unobservable) excited state populations in molecular discharges, where it is open. The observed state populations in a discharge depend both on excitation and relaxation processes, and these are unfortunately not separately determined, so unambiguous conclusions about either are elusive. Bogey et al. [74] have observed for CO^+ in their RF discharge in CO a relative intensity, γ_0/γ_1, of 5/1, which corresponds to a temperature of 2000°K. Thus vibrational temperatures in the RF discharge appear to be considerably higher than those in the DC discharge for ions, as they are already known to be for neutrals [73].

The determination of ion velocities from the Doppler effect is another important dynamical application of microwave spectroscopy. The center frequency of an ion line is shifted by the net drift motion of the ion in the axial electric field of a DC glow discharge. The effect was noticed in early studies of HCO^+, and the general magnitude of the shifts were determined in later work [70,81]. Gudeman et al. [67] have made an extensive set of very precise frequency measurements, aimed specifically at determining the pressure dependence of the Doppler shift in different gases and various conditions, for the J = 0-1 transition of HCO^+. Since the sought after shift is small compared to the linewidth, especially at higher pressures, every precaution must be taken to obtain sufficient precision in the frequency measurement. The signal averaged spectra, with points at very precisely known frequency increments, are subjected to baseline suppression and digital smoothing and then fit to a sum-of-Lorenztians model that incorporates the multiple-lobe pattern characteristic of the tone-burst modulation scheme [65]. Special care must also be taken to avoid possible measurement errors associated with either anomalous dispersion shifts or asymmetries in the tone-burst modulation side band structure. With a view to converting observed ion velocities into mobilities, K, the average electric fields along the axis of the positive column were also measured, by subtracting voltages observed on Langmuir probes inserted at the two ends of the absorbing region. Gudeman et al. [67] obtained frequency shift and electric field data in the 10-100 mTorr range for H_2 discharges with both fast and slow flow rates and for fast flowing Ar discharges. The effects of varying discharge current or the partial pressures of reactive gases were also tested. The precision obtained in the frequency measurements appeared to be adequate (±10 kHz or better) on the basis of the smoothness of the observed pressure variation, but the velocities were found to be more sensitive to parameters like flow rate, current, or CO partial pressure than expected. Shifts of ±50 to ±150 kHz (plus and minus corresponding to opposite discharge polarities) were seen in H_2 buffer gas discharges, while in Ar buffers shifts were in the range 0 to ±30 kHz and quite difficult to determine accurately. In fast flowing H_2 the shifts were almost independent of pressure, while in slow flowing H_2 they decreased by about a factor of 3 from the lowest to highest pressures studied. The reason for this differing behavior was not clear. In principle spectroscopic Doppler measurements of mobility are of special interest, since they constitute a quantum state specific determination of a transport property. The average mobility results of Gudeman et al. [67], however, are not very useful from that standpoint, because the values of K obtained in the experiment are from 3 to 7 times smaller than those predicted by the classical Langevin model [173]

$$K[cm^2/V\ sec] = \frac{38.6\ T}{(\alpha[\mathring{A}^3]\ \mu)^{1/2}\ P}\ ,\qquad\qquad (26)$$

which has been found to agree much more closely with other mobility measurements [173]. Gudeman et al. [67] suggest the gross discrepancy may be occasioned by the axial variation of the electric field in the striations that are known to be present in the positive column [174,175]. Since the current must be the same all along the tube, the electron (and ion) density must be proportionately larger where the electric field, and thus the velocity, is smaller. This effect biases the Doppler shift measurements in favor of the slowest ions, and tends to reduce the apparent mobility. If this is indeed the correct explanation, then the strong variation with parameters like flow rate may be due to changes in the striation profiles. The Doppler shift also provides another very useful tool for determining whether an unknown line is that of an ion, but the method may fail in a discharge gas like argon, because the shifts are too small to determine reliably for weak lines, as they were for HOC^+ [55]. Both the electric field and ion mobilities are low in argon. The extensive frequency data of Gudeman et al. [67] yield as a byproduct the most accurate measurement so far of the rest frequency of the X-ogen line. They obtain 89,188.542 ± 0.004 MHz by a combined fitting, extrapolation, and averaging of all the frequency versus pressure data for both discharge polarities and various conditions. Astronomical measurements of this frequency are not as precise as those for the rare isotopic species because of severe optical depth and self-absorption effects. Whenever Doppler broadening is larger than pressure broadening, the linewidth contains information about the distribution of velocities. In the 100 GHz range this situation only prevails at the lowest available pressures, but at higher frequencies the Doppler effect is increased while pressure broadening is not, so it should be possible to obtain both Doppler shifts and widths better at submillimeter frequencies. The Doppler width of a line in the positive column of a DC glow cannot, however, be directly converted into a translational temperature. Once again the spatial variation of the electric field within the striations must be considered. The observed width is due not only to the distribution of random velocities at a given point, but also to the distribution of different drift velocities at spatially different points. The ion random velocity may also vary from one part of a striation to another. The measurement of translational temperature from Doppler broadening should be more direct in other more homogeneous plasmas, e.g., in RF discharges or hollow cathode discharges. Further study will be required for a full understanding of the Doppler shifts and widths in the very complex milieu of the positive column of the DC glow. Hopefully such effort will lead to a viable means of obtaining reliable values of a fundamental transport property, the ion mobility, in selected vibrational and rotational states and also provide a useful probe of the complex wave phenomena in the glow discharge.

CONCLUSION

The potential of microwave spectroscopy for the study of molecular ions has now been clearly demonstrated, but its full realization is still an exciting prospect for the future. Radioastronomical observations of just a few ions have already greatly enriched our understanding of the interstellar medium, and each new detection will offer the promise of valuable new insights. Probably at least some of the many remaining U-lines will be of ionic origin, and the constantly improving receiver technology and expanding frequency range of both existing radiotelescopes and new ones now becoming operational will almost certainly lead to discovery of new ionic spectra. In the laboratory improved ion production methods and steady progress in the generation, control, and detection of microwave radiation should lead to observation of many new ions and new classes of ions: light hydride ions, non-linear ions, ions with internal tunneling motions, non-classical ions, open-shell polyatomic ions, negative ions, etc. Quantum chemical calculations have played a very important role in the development of this subject and have produced results that have compared in a most satisfying way to experiment. Future theoretical work, including work on more difficult open-shell systems and more detailed properties, can also have a dramatic impact and lead to very interesting comparisons to the many experimental results on structure, force field, and electronic properties of ions that microwave spectroscopy can be expected to produce.

ACKNOWLEDGMENTS

The author wishes to acknowledge that his research on microwave spectroscopy of molecular ions has been generously supported by the National Science Foundation Structural Chemistry Program, the Wisconsin Alumni Research Foundation, and the donors of the Petroleum Research Fund, administered by the American Chemical Society.

REFERENCES

[1]   R.J. Saykally and R.C. Woods, Ann. Rev. Phys. Chem. 32 (1981) 403-431.
[2]   D. Buhl and L.E. Snyder, Nature (London) 228 (1970) 267-269.
[3]   D.M. Rank, C.H. Townes, and W.J. Welch, Science 174 (1971) 1083-1101.
[4]   W. Klemperer, Nature (London) 227 (1970) 1230.
[5]   H. B. Jansen and P. Ros, Chem. Phys. Lett. 3 (1969) 140-143.
[6]   S. Forsén and B. Roos, Chem. Phys. Lett. 6 (1970) 128-132.
[7]   U. Wahlgren, B. Liu, P.K. Pearson, and H.F. Schaefer, Nat. Phys. Sci. 246 (1973) 4-6.
[8]   P.J. Bruna, S.D. Peyerimhoff, and R.J. Buenker, Chem. Phys. 10 (1975) 323-334.
[9]   W.P. Kraemer and G.H.F. Diercksen, Astrophys. J. 205 (1976) L97-L100.
[10] K.B. Jefferts, A.A. Penzias, R.W. Wilson, M. Kutner, and P. Thaddeus, Astrophys. Lett. 8 (1971) 43-44.
[11] D. Buhl and L.E. Snyder, Astrophys. J. 180 (1973) 791-800.
[12] E. Herbst and W. Klemperer, Astrophys. J. 188 (1974) 255-256.
[13] J.M. Hollis, L.E. Snyder, D. Buhl, and P.T. Giguere, Astrophys. J. 200 (1975) 584-593.
[14] L.E. Snyder and J.M. Hollis, Astrophys. J. 204 (1976) L139-L142.
[15] M. Morris, B. Zuckerman, B.E. Turner, and P. Palmer, Astrophys. J. 192 (1974) L27.
[16] E. Herbst and W. Klemperer, Astrophys. J. 185 (1973) 505-533.
[17] W.D. Watson, Astrophys. J. 183 (1973) L17-L20.

R. Claude Woods

[18] W.D. Watson, Astrophys. J. 188 (1974) 35-42.
[19] W.D. Watson, Rev. Mod. Phys. 48 (1976) 513-552.
[20] G.C. Dousmanis, T.M. Sanders, Jr., and C.H. Townes, Phys. Rev. 100 (1955) 1735-1755.
[21] R.C. Mockler and G.R. Bird, Phys. Rev. 98 (1955) 1837-1839.
[22] F.X. Powell and D.R. Lide Jr., J. Chem. Phys. 41 (1964) 1413-1419.
[23] M. Winnewisser, K.V.L.N. Sastry, R.L. Cook, and W. Gordy, J. Chem. Phys. 41 (1964) 1687-1691.
[24] R.C. Woods, Rev. Sci. Instrum. 44 (1973) 274-281.
[25] R.C. Woods, Rev. Sci. Instrum. 44 (1973) 282-288.
[26] T.A. Dixon and R.C. Woods, Phys. Rev. Lett. 34 (1975) 61-63.
[27] K.N. Rao, Astrophys. J. 111 (1950) 50-59.
[28] L.H. Woods, Phys. Rev. 63 (1943) 431-432.
[29] P.R. Certain and R.C. Woods, J. Chem. Phys. 58 (1973) 5837-5838.
[30] L.E. Snyder, J.M. Hollis, B.L. Ulich, F.J. Lovas, and D. Buhl, Bull. Am. Astron. Soc. 7 (1975) 497.
[31] L.E. Snyder, J.M. Hollis, F.J. Lovas, and B.L. Ulich, Astrophys. J. 209 (1976) 67-74.
[32] R.C. Woods, T.A. Dixon, R.J. Saykally, and P.G. Szanto, Phys. Rev. Lett. 35 (1975) 1269-1272.
[33] R.C. Woods, R.J. Saykally, T.A. Dixon, P.G. Szanto, and T.G. Anderson, 31st Annual Symposium on Molecular Spectroscopy, Columbus, Ohio, June 1976, paper TS2.
[34] B.E. Turner, Astrophys. J. 193 (1974) L83-L87.
[35] S. Green, J.A. Montgomery, Jr., and P. Thaddeus, Astrophys. J. 193 (1974) L89-L91.
[36] P. Thaddeus and B.E. Turner, Astrophys. J. 201 (1975) L25-L26.
[37] R.J. Saykally, T.A. Dixon, T.G. Anderson, P.G. Szanto, and R.C. Woods, Astrophys. J. 205 (1976) L101-L103.
[38] P. Thaddeus, M. Guélin, and R.A. Linke, Astrophys. J. 246 (1981) L41-L45.
[39] S. Wilson, Astrophys. J. 220 (1978) 739-741.
[40] J.W.C. Johns, J.M.R. Stone, and G. Winnewisser, J. Mol. Spectrosc. 38 (1971) 437.
[41] C.S. Gudeman, N.N. Haese, N.D. Piltch, and R.C. Woods, Astrophys. J. 246 (1981) L47-L49.
[42] D.J. DeFrees, G.H. Loew, and A.D. McLean, Astrophys. J. 254 (1982) 405-411.
[43] N.R. Erickson, R.L. Snell, R.B. Loren, L. Mundy, and R.L. Plambeck, Astrophys. J. 245 (1981) L83-L86.
[44] K.V.L.N. Sastry, P. Helminger, E. Herbst, and F.C. DeLucia, Astrophys. J. 250 (1981) L91-L92.
[45] G.F. Mitchell, J.L. Ginsberg, and P.J. Kuntz, Astrophys. J. Suppl. 38 (1978) 39.
[46] S.S. Prasad and W.T. Huntress, Astrophys. J. 239 (1980) 151.
[47] L.E. Snyder and D. Buhl, Bull. Am. Astron. Soc. 3 (1971) 388.
[48] R.J. Saykally, P.G. Szanto, T.G. Anderson, and R.C. Woods, Astrophys. J. 204 (1976) L143-L145.
[49] G.L. Blackman, R.D. Brown, P.D. Godfrey, and H.I. Gunn, Nature 261 (1976) 395.
[50] R.A. Greswell, E.F. Pearson, M. Winnewisser, and G. Winnewisser, Z. Naturforsch. 31A (1976) 221.
[51] P.F. Goldsmith, W.D. Langer, J. Ellder, W. Irvine, and E. Kollberg, Astrophys. J. 249 (1981) 524-531.
[52] E. Herbst, J.M. Norbeck, P.R. Certain, and W. Klemperer, Astrophys. J. 207 (1976) 110-112.
[53] P. Hennig, W.P. Kraemer, G.H.F. Dierckson, Max Planck Institüt für Physik und Astrophysik, Munich, 1978 (unpublished report).
[54] R.H. Nobes and L. Radom, Chem. Phys. 60 (1981) 1-10.
[55] C.S. Gudeman and R.C. Woods, Phys. Rev. Lett. 48 (1982) 1344-1348.
[56] C.S. Gudeman and R.C. Woods, to be published.

[57] R.C. Woods, C.S. Gudeman, R.L. Dickman, P.F. Goldsmith, G.R. Hugenin, W.M. Irvine, Å. Hjalmarson, L.-Å Nyman, and H. Olofsson, Astrophys. J. (in press, 1983).
[58] Ch. Jungen and H. Lefebvre-Brion, J. Mol. Spectrosc. 33 (1970) 520.
[59] F.P. Billingsley, Chem. Phys. Lett. 23 (1973) 160.
[60] W.C. Bowman, E. Herbst, and F.C. DeLucia, J. Chem. Phys. 77 (1982) 4261-4262.
[61] J.F. Verdiek and C.D. Cornwell, Rev. Sci. Instrum. 32 (1961) 1383.
[62] C.H. Townes and A.L. Schawlow, Microwave Spectroscopy (McGraw-Hill, New York, 1955) p. 19.
[63] R.C. Woods, Disc. Faraday Soc. 71 (1981) 57-62.
[64] M.A. Heald and C.B. Wharton, Plasma Diagnostics with Microwaves (Wiley, New York, 1965) p. 10 and p. 121.
[65] H.M. Pickett, Ap. Optics 19 (1980) 2745.
[66] H.M. Pickett, Rev. Sci. Instrum. 48 (1977) 706.
[67] C.S. Gudeman, N.D. Piltch, and R.C. Woods, to be published.
[68] C.S. Gudeman, N.D. Piltch, and R.C. Woods, to be published.
[69] M. Bogey, C. Demuynck, and J.L. Destombes, Mol. Phys. 43 (1981) 1043-1050.
[70] K.V.L.N. Sastry, E. Herbst, and F.C. De Lucia, J. Chem. Phys. 75 (1981) 4169-4170.
[71] K.V.L.N. Sastry, P. Helminger, E. Herbst, and F.C. De Lucia, Chem. Phys. Lett. 84 (1981) 286-287.
[72] G.A. Blake, P. Helminger, E. Herbst, and F.C. De Lucia, Astrophys. J. (in press, 1983).
[73] R. Bustreel, C. Demuynck, J.L. Destombes, and G. Journel, Chem. Phys. Lett. 67 (1979) 178-182.
[74] M. Bogey, C. Demuynck, and J.L. Destombes, Mol. Phys. 46 (1982) 679-681.
[75] W.C. King and W. Gordy, Phys. Rev. 90 (1953) 319.
[76] F.C. van den Heuvel and A. Dymanus, Chem. Phys. Lett. 92 (1982) 219-222.
[77] R.D. Brown, P.D. Godfrey, D.C. McGilvery, and J.G. Crofts, Chem. Phys. Lett. 84 (1981) 437-439.
[78] N. Erickson, J.H. Davis, N.J. Evans II, R.B. Loren, L. Mundy, W.L. Peters, III, M. Scholtes, and P.A. VandenBout, in B.H. Andrew (Ed.), Interstellar Molecules (Reidel, Dordrecht, 1980) pp. 25-30.
[79] M.L. Meeks (Ed.), Methods of Experimental Physics, Vol. 12, part B, Radiotelescopes (Academic, New York, 1976).
[80] G. Winnewisser, E. Churchwell, and C.M. Walmsley, in G.W. Chantry (Ed.), Modern Aspects of Microwave Spectroscopy (Academic, London, 1979) pp. 312-503.
[81] R.C. Woods, Richard J. Saykally, T.G. Anderson, T.A. Dixon, and P.G. Szanto, J. Chem. Phys. 75 (1981) 4256-4260.
[82] J.M. Hollis, L.E. Snyder, F.J. Lovas, and D. Buhl, Astrophys. J. 209 (1976) L83-L85.
[83] M. Guélin and P. Thaddeus, Astrophys. J. 227 (1979) L139-L141.
[84] W.D. Watson, in J. Andouze (Ed.), CNO Isotopes in Astrophysics (Reidel, Dordrecht, 1976).
[85] M. Guélin, W.D. Langer, R.L. Snell, and H.A. Wootten, Astrophys. J. 217 (1977) L165-L168.
[86] D.G. York and J.B. Rogerson, Jr., Astrophys. J. 203 (1976) 378.
[87] T.G. Anderson, T.A. Dixon, N.D. Piltch, R.J. Saykally, P.G. Szanto, and R.C. Woods, Astrophys. J. 216 (1977) L85-L86.
[88] L.E. Snyder, J.M. Hollis, D. Buhl, and W.D. Watson, Astrophys. J. 218 (1977) L61-L64.
[89] W.D. Watson, L.E. Snyder, and J.M. Hollis, Astrophys. J. 222 (1978) L145-L147.
[90] W.D. Langer, R.W. Wilson, P.S. Henry, and M. Guélin, Astrophys. J. 225 (1978) L139-L142.
[91] B.E. Turner and B. Zuckerman, Astrophys. J. 225 (1978) L75-L79.
[92] B.E. Turner and P. Thaddeus, Astrophys. J. 211 (1977) 755-771.
[93] L.E. Snyder, W.D. Watson, and J.M. Hollis, Astrophys. J. 212 (1977) 79-83.

[94]   O.E.H. Rydbeck, Å, Hjalmarson, G. Rydbeck, J. Elldér, H. Olofsson, and
        A. Sume, Astrophys. J. 243 (1981) L41-L45.
[95]   T.B.H. Kuiper, E.N. Rodriguez Kuiper, and B. Zuckerman, in B.H. Andrew
        (Ed.), Interstellar Molecules (Reidel, Dordrecht, 1980) pp. 31-32.
[96]   N.Z. Scoville, in B.H. Andrew (Ed.), Interstellar Molecules (Reidel,
        Dordrecht, 1980) pp. 33-38.
[97]   P.J. Huggins, T.G. Phillips, G. Neugebauer, M.W. Werner, P.G. Wanier, and
        D. Ennis, Astrophys. J. 227 (1979) 441-445.
[98]   A. Wootten, N.J. Evans II, R. Snell, and P. Vanden Bout, Astrophys. J. 225
        (1978) L143-L148.
[99]   A. Wootten, R. Snell, and N.J. Evans II, Astrophys. J. 240 (1980) 532-546.
[100]   R.B. Loren and A. Wootten, Astrophys. J. 242 (1980) 568-575.
[101]   R.L. Snell, Astrophys. J. Supp. 45 (1981) 121-175.
[102]   A. Baudry, F. Combes, M. Perault, and R. Dickman, Astron. Astrophys. 85
        (1980) 244-248.
[103]   A. Baudry, J. Cernicharo, M. Perault, J. de la Noë, and D. Despois,
        Astron. Astrophys. 104 (1981) 101-115.
[104]   D.F. Dickinson, E.N. Rodriguez Kuiper, A. St. Clair Dinger, and T.B.H.
        Kuiper, Astrophys. J. 237 (1980) L43-L45.
[105]   Y. Fukui, N. Kaifu, M. Morimoto, and T. Miyaji, Astrophys. J. 241 (1980)
        147-154.
[106]   R.A. Linke, A.A. Stark, and M.A. Frerking, Astrophys. J. 243 (1981) 147-
        154.
[107]   A.A. Stark and R.S. Wolff, Astrophys. J. 229 (1979) 118-120.
[108]   L.J. Rickard and P. Palmer, Astrophys. J. 243 (1981) 765-768.
[109]   A.A. Penzias, Astrophys. J. 228 (1979) 430-434.
[110]   A.A. Penzias, Science, 208 (1980) 663-669.
[111]   R.W. Wagoner, Astrophys. J. 179 (1973) 343.
[112]   A.A. Stark, Astrophys. J. 245 (1981) 99-104.
[113]   P.G. Wannier, A.A. Penzias, R.A. Linke, and R.W. Wilson, Astrophys. J.
        204 (1976) 26.
[114]   M. Guélin, J. Cernicharo, and R.A. Linke, Astrophys. J. 263 (1982) L89-L93.
[115]   A. Wootten, R. Snell, and A.E. Glassgold, Astrophys. J. 234 (1979) 876-880.
[116]   N.G. Adams and D. Smith, Astrophys. J. 248 (1981) 373-379.
[117]   W.T. Huntress, Jr., and V.G. Anicich, Astrophys. J. 208 (1976) 237.
[118]   E. Herbst, Astron. Astrophys. 111 (1982) 76-80.
[119]   A. Wootten, R.B. Loren, and R.L. Snell, Astrophys. J. 255 (1982) 160-175.
[120]   R.L. Snell and A. Wootten, Astrophys. J. 228 (1979) 748-754.
[121]   E.J. O'Donnell and W.D. Watson, Astrophys. J. 191 (1974) 89.
[122]   M. Guélin, W.D. Langer, and R.W. Wilson, Astron. Astrophys. 107 (1982)
        107-127.
[123]   W.T. Huntress, Jr., S.S. Prasad, P.R. Kemper, R.D. Cates, and M.T. Bowers,
        Astron. Astrophys. 114 (1982) 275-277.
[124]   T. McAllister, Astrophys. J. 225 (1978) 857-859.
[125]   A.J. Illies, M.F. Jarrold, and M.T. Bowers, J. Chem. Phys. 77 (1982)
        5847-5848.
[126]   J. Kraitchman, Am. J. Phys. 21 (1953) 17.
[127]   C.C. Costain, J. Chem. Phys. 29 (1958) 864.
[128]   P.G. Szanto, T.G. Anderson, R.J. Saykally, N.D. Piltch, T.A. Dixon, and
        R.C. Woods, J. Chem. Phys. 75 (1981) 4261-4263.
[129]   C.S. Gudeman and R.C. Woods, to be published.
[130]   D.J. DeFrees, G.H. Loew, and A.D. McLean, Astrophys. J. 257 (1982) 376-382.
[131]   I.H. Hillier and J. Kendrick, J. Chem. Soc. Chem. Commun. (1975) 526.
[132]   J.A. Montgomery, Jr. and C.E. Dykstra, J. Chem. Phys. 71 (1979) 1380.
[133]   G. Winnewisser, A.G. Maki, and D.R. Johnson, J. Mol. Spectrosc. 39 (1971)
        149.
[134]   E.F. Pearson, R.A. Greswell, M. Winnewisser, and G. Winnewisser,
        Z Naturforsch. A 31 (1970) 1394.
[135]   P.K. Pearson, H.F. Schaefer, and U. Wahlgren, J. Chem. Phys. 62 (1975) 350.

[136] P. K. Pearson, G.L. Blackman, H.F. Schaefer, B. Roos, and U. Wahlgren, Astrophys. J. 184 (1973) L19-L22.
[137] C.S. Gudeman and R.C. Woods, to be published.
[138] D.R. Lide and C. Matsumura, J. Chem. Phys. 50 (1969) 3080.
[139] P.J. Bruna, S.D. Peyerimhoff, and R.J. Buenker, Chem. Phys. 27 (1978) 33.
[140] S. Chekir, F. Pauzat, and G. Berthier, Astron. Astrophys. 100 (1981) L14-L15.
[141] N.N. Haese and R.C. Woods, J. Chem. Phys. 73 (1980) 4521-4527.
[142] J.E. Wollrab, Rotational Spectra and Molecular Structure (Academic, New York, 1967) Ch. 8.
[143] N.N. Haese and R.C. Woods, Chem. Phys. Lett. 61 (1979) 396-398.
[144] B.N. Bhattacharya and W. Gordy, Phys. Rev. 119 (1960) 144.
[145] G.L. Blackman, R.D. Brown, P.D. Godfrey and H.I. Gunn, Nature 261 (1976) 395.
[146] P.S. Dardi and C.E. Dykstra, Astrophys. J. 240 (1980) L171.
[147] R.A. Frosch and H.M. Foley, Phys. Rev. 88 (1952) 1337.
[148] N.D. Piltch, P.G. Szanto, T.G. Anderson, C.S. Gudeman, T.A. Dixon, and R.C. Woods, J. Chem. Phys. 76 (1982) 3385-3388.
[149] A. Carrington, D.R.J. Milverton, and P.J. Sarre, Mol. Phys. 35 (1978) 1505.
[150] K.P. Huber and G. Herzberg, Molecular Spectra and Molecular Structure IV. Constants of Diatomic Molecules (Van Nostrand, New York, 1979).
[151] M. Mizushima, Theory of Rotating Diatomic Molecules (Wiley, New York, 1975), pp. 239-245.
[152] T.A. Dixon and R.C. Woods, J. Chem. Phys. 67 (1977) 3956-3964.
[153] R.M. Herman and A. Asgharian, J. Mol. Spectrosc. 19 (1966) 305.
[154] P.R. Bunker, J. Mol. Spectrosc. 42 (1972) 478.
[155] J.K.G. Watson, J. Mol. Spectrosc. 45 (1973) 99.
[156] A.H.M. Ross, R.S. Eng, and H. Kildal, Optics Commun. 12 (1974) 433.
[157] C.H. Townes and A.L. Schawlow, Microwave Spectroscopy (McGraw-Hill, New York, 1955) p. 343.
[158] J.E. Wollrab, Rotational Spectra and Molecular Strcture (Academic, New York, 1967) pp. 327-330.
[159] T.G. Anderson, C.S. Gudeman, T.A. Dixon, and R.C. Woods, J. Chem. Phys. 72 (1980) 1332-1336.
[160] S.S. Bhattacharyya, B. Bhattacharyya, and M.V. Narayan, Astrophys. J. 247 (1981) 936-940.
[161] G. Gioumousis and D.P. Stevenson, J. Chem. Phys. 29 (1958) 294.
[162] T. Su, E.C.F. Su, and M.T. Bowers, J. Chem. Phys. 69 (1978) 2243-2250.
[163] S. Green, Astrophys. J. 201 (1975) 366-372.
[164] S. Green and P. Thaddeus, Astrophys. J. 191 (1974) 653-657.
[165] F. Llewellyn-Jones, The Glow Discharge (Methuen, London, 1966).
[166] A. von Engel, Ionized Gases (Oxford, London, 1965) Ch. 8.
[167] T. Oka, Phys. Rev. Letters 45 (1980) 531-534.
[168] P. Bernath and T. Amano, Phys. Rev. Letters 48 (1981) 20-22.
[169] M. Wong, P. Bernath, and T. Amano, J. Chem. Phys. 77 (1982) 693-696.
[170] F.C. De Lucia and P.A. Helminger, J. Chem. Phys. 67 (1977) 4262-4267.
[171] D.R. Lide and A.G. Maki, App. Phys. Lett. 11 (1967) 62.
[172] H. Shin, in P.J. Ausloos (Ed.), Ion-Molecule Reactions in the Gas Phase (ACS, Washington, 1966).
[173] E.W. McDaniel and E.A. Mason, Mobility and Diffusion of Ions in Gases (Wiley, New York, 1973).
[174] G. Francis, in S. Fluggel (Ed.), Handbuch der Physik, Vol. 22 (Springer-Verlag, Berlin, 1956).
[175] A. Garscadden, in M.N. Hirsch and H.J. Oskam (Ed.), Gaseous Electronics (Academic, New York, 1978).

Molecular Ions: Spectroscopy, Structure and Chemistry
Terry A. Miller and V.E. Bondybey (editors)
© North-Holland Publishing Company, 1983

HIGH-RESOLUTION INFRARED SPECTROSCOPY OF MOLECULAR IONS

Alan Carrington and T. P. Softley
Department of Chemistry
University of Southampton
Hampshire
SO9 5NH

Techniques and results for the study of the high-
resolution infrared spectra of molecular ions are
reviewed. The methods described are far-infrared
laser magnetic resonance, difference-frequency
infrared spectroscopy, Fourier-transform spectroscopy, and ion beam/laser beam spectroscopy.

INTRODUCTION

This review is concerned with developments in infrared spectroscopy which have
occurred during the past six years. Molecular ions are important because of their
roles in both laboratory and interstellar chemistry, and because they pose some
particularly interesting and unique problems in molecular structure theory.
Because of the high chemical reactivity of ions the traditional methods of spectroscopy have been applied with only limited success, the main exception to this
being the study of electronic emission spectra of ions (mainly diatomic) in
electrical discharges [1]. Infrared spectroscopy has previously been possible
only by trapping the ionic species in solid matrices, and although the results
have been valuable, they are limited because of the absence of free molecular
rotation and the information that provides. The position is now changing rapidly
because of the use of lasers, and also because of combinations of the techniques
of molecular spectroscopy and mass spectrometry.

This review is restricted to high-resolution infrared studies, by which we mean
the study of individual vibration-rotation levels. The number of ions studied
at this level of resolution is still very small, but the techniques which have
been developed show promise of being quite general and, hopefully, this review will
stimulate further development. It will also become clear that the methods which
are being applied are capable of illuminating certain aspects of molecular physics
which are not always so accessible in neutral molecules, particularly phenomena
which involve molecular dynamics (for example, photodissociation and predissociation phenomena). Much of the spectroscopy described in this review overlaps the
fields of intermolecular forces, reactive scattering, multiphoton dissociation and
astrophysics, and also gives promise of increasing our understanding of molecular
processes previously studied only by mass spectrometry.

Three of the techniques described in this review are directed towards the study
of vibration-rotation or electronic spectra of molecular ions. The fourth technique described, far-infrared laser magnetic resonance, is essentially a method
for studying rotational transitions in molecules which are paramagnetic. As such
it is closely related to microwave spectroscopy and electron paramagnetic resonance,
as well as to other branches of infrared spectroscopy. We have chosen to include
laser magnetic resonance in this review, partly because it does involve the use
of infrared radiation, but mainly because the technique may well be successfully
extended to the vibrational spectra of molecular ions, as has occurred with
neutral free radicals [2].

LASER MAGNETIC RESONANCE

The laser magnetic resonance technique makes use of the Zeeman effect to tune
pure rotational transitions (far-infrared LMR) or vibration-rotation transitions
(mid-infrared LMR) into resonance with a suitable fixed-frequency laser line.
An external magnetic field perturbs the energies of the magnetic sublevels,
lifting the 2J + 1 degeneracy. For certain values of the field, transitions
between particular sublevels of the upper and lower J states are brought into
coincidence with the laser frequency.

The general principles and applications of far-infrared LMR have recently been
reviewed by Evenson, Saykally, Jennings, Curl and Brown [3] while mid-infrared
LMR has been reviewed by McKellar [4]. No mid-infrared LMR transitions have
yet been reported for molecular ions, but the far-infrared LMR spectra of HBr^+,
DBr^+ and HCl^+ have been observed [5,6]. The spectrometer used by Ray, Lubic
and Saykally [6] for their recent measurements on HCl^+ is illustrated in Figure 1.

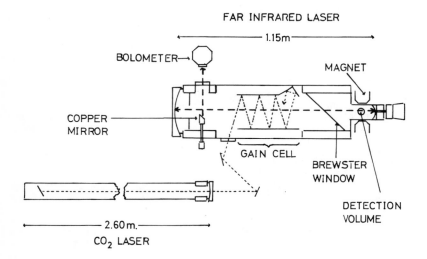

FAR INFRARED LASER

Figure 1
Schematic diagram of an optically pumped far-infrared LMR spectrometer

It is an intracavity spectrometer in which the sample region is located within
the laser cavity, enabling multiple passage of the radiation through the sample.
The 1.15 m Fabry-Perot cavity is divided into two sections by a 12.5 μm poly-
propylene beam splitter. One section contains the lasing gas which is transvers-
ely pumped in a 0.58 m gain cell region by up to 70 watts of single line infrared
radiation from a 3 m carbon dioxide laser. Variations of the laser gas and pump
laser frequency have been used to obtain nearly 1000 fixed-frequency laser lines
in far-infrared LMR spectrometers. The ions are generated in the second vacuum-

sealed region of the laser cavity by a positive column discharge maintained
between two brass electrodes. A flowing mixture of 1% HBr in He at 1 torr is
used to generate HBr$^+$, whilst the optimum conditions for generation of HCl$^+$ are
2 m torr HCl in 1.5 torr He. A current density of 20 mA is used for both species.

A magnetic field of up to 2 tesla is provided by a 15 inch electromagnet and is
measured with an NMR gaussmeter. The magnetic field is sinusoidally modulated
at an amplitude of 5 mT and frequency 2.5 kHz. The polarisation of the laser
radiation may be selected by rotating the Brewster window. A small proportion of
the laser power is coupled out using a 45^0 copper mirror and detected by means
of a liquid helium-cooled gallium-doped germanium bolometer; the absorption signal
is demodulated by means of a phase-sensitive detector.

HBr$^+$ was the first molecular ion to be detected by far-infrared LMR. Both HBr$^+$
and HCl$^+$ have $^2\pi$ ground states and an energy level diagram for HBr$^+$, obtained
from optical data [7] is shown in Figure 2.

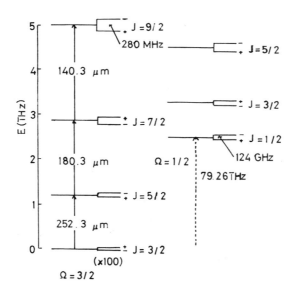

Figure 2
Energy level diagram for the $^2\pi$ ground state of HBr$^+$

In the initial experiment [5] a solenoid was used to provide a longitudinal field
of maximum 0.5 tesla. With this arrangement only $\Delta M_J = \pm 1$ transitions could be
observed because the vector oscillating electric field vector and static magnetic field
were always perpendicular. In the ground vibrational state two components of
the $J = \frac{3}{2} \rightarrow \frac{5}{2}$ ($\Omega = \frac{3}{2}$) transition and one component of the $J = \frac{5}{2} \rightarrow \frac{7}{2}$ ($\Omega = \frac{3}{2}$)
transition were observed. For example, the $M_J = -\frac{3}{2} \rightarrow -\frac{1}{2}$ component of the
$J = \frac{3}{2} \rightarrow \frac{5}{2}$ transition was tuned into resonance with the 251.1 μm laser line
of CH_3OH. In addition three components of the $J = \frac{5}{2} \rightarrow \frac{7}{2}$ transition in the
$v = 1$ state were identified, whilst in DBr^+ one component of the $J = \frac{3}{2} \rightarrow \frac{5}{2}$
($\Omega = \frac{3}{2}$) transition was observed.

Each Zeeman component should be split into 16 lines, the splitting arising from
magnetic and electric quadrupole hyperfine interaction for the ^{79}Br and ^{81}Br
isotopes, combined with the Λ-doubling of the $^2\pi_{\frac{3}{2}}$ state. All sixteen components
of the $J = \frac{3}{2} \rightarrow \frac{5}{2}$, $M_J = -\frac{3}{2} \rightarrow -\frac{1}{2}$ transition were observed.

Using a transverse magnetic field, so that both $\Delta M = \pm 1$ and $\Delta M = 0$ transitions
are possible, 230 lines have now been observed and are currently being analysed.
The spectroscopic line width of 10 MHz is determined by pressure broadening and
since the proton hyperfine splitting is estimated to be ~ 1 MHz it is not
observed.

The energy level diagram for HCl^+ is similar to that of HBr^+ and three Zeeman
components of the $J = \frac{5}{2} \rightarrow \frac{7}{2}$ ($\Omega = \frac{3}{2}$) transition have been observed, each split
into 8 lines. The $H^{37}Cl^+$ component of this transition has not been observed,
but it has been detected in the $J = \frac{7}{2} \rightarrow \frac{9}{2}$ transition. The data has been used
to obtain Λ-doubling, magnetic hyperfine, electric quadrupole and rotational
constants by least squares regression techniques involving the $^2\pi$ Hamiltonian
given by Brown, Kaise, Kerr and Milton [8].

The minimum detectable absorption coefficient is estimated [3] to be $\sim 3 \times 10^{-10}$
for far-infrared LMR, compared with $\sim 10^{-7}$ for the difference-frequency laser
method described later. The high sensitivity is obviously advantageous, but
the method also suffers from a few disadvantages. Only open-shell paramagnetic
molecules can be studied and the transitions must lie close to fixed-frequency
laser lines, the magnetic tuning range being typically 10^4MHz/tesla. Moreover
magnetic tuning is only possible for molecules in which the electron spin is
appreciably coupled to the electric dipole moment. The analysis of the spectra
is fairly complicated, the use of high magnetic fields requiring that a large
number of terms be included in the effective Hamiltonian if accurate zero-field
constants are to be obtained. Nevertheless it is to be expected that LMR spectra
of other ions will be detected in the near future, and the detection of vibration-
rotation spectra by mid-infrared LMR is also possible.

INFRARED DIFFERENCE-FREQUENCY LASER SPECTROSCOPY

The difference-frequency laser method has been used to measure the infrared
spectra of H_3^+ [9 - 11], HeH^+ [12] and NeH^+ [13]. The principles of the method
are simple; tunable coherent infrared radiation is passed through a discharge cell
in which the ions are produced, and the absorption of radiation is measured
directly.

The apparatus used by Oka [11] is shown in Figure 3. The infrared source is
essentially the difference-frequency system developed by Pine [14,15] and later
used by Oka to measure the weak overtone bands of PH_3 [16] and the electric
quadrupole fundamental band of D_2 [17]. The output from a single mode argon ion
laser, typically 80 mW at either 5145 Å or 4880 Å, is combined on a dichroic
mirror with 20 - 60 mW of power from an argon ion laser-pumped C.W. dye laser,
and is passed into a temperature-controlled lithium niobate crystal. The non-
linear response of the crystal generates several microwatts of radiation at an
infrared difference frequency $\omega_1 = \omega_3 - \omega_2$, where ω_3 and ω_2 are the frequencies

of the argon ion and dye lasers respectively. The resultant radiation is tunable over the range 2400 - 4400 cm^{-1} by tuning the dye laser from yellow to red. Frequency modulation of amplitude 400 MHz and frequency 2.5 kHz is achieved by modulating the argon ion laser frequency. The band width is determined by the dye laser bandwidth and is about 10 MHz.

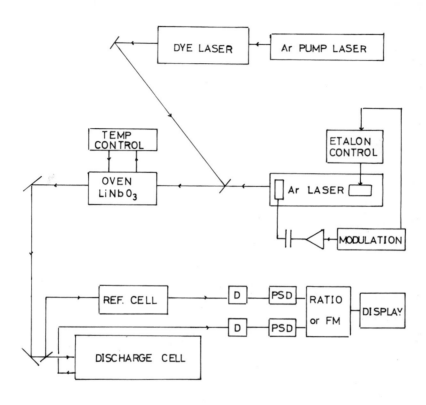

Figure 3
Difference-frequency infrared laser spectrometer

The ions are formed in a multiple reflection discharge cell of length 2 m, cooled
by either liquid nitrogen or ice-water. A summary of the experimental conditions
used for the three ions is presented in Table 1.

Ion	Gas mixture	Current density	Ions/cm^3	Cooling
H_3^+	H_2, 1 torr	60 mA/cm^2	3×10^{10}	N_2 or ice
HeH$^+$	He/H$_2$(100:1), 0.45 torr	30 mA/cm^2	10^9	N_2
NeH$^+$	Ne, 1.5 torr	30 mA/cm^2	2×10^{10}	ice

Table 1

For liquid nitrogen cooling the rotational temperatures are expected to be
$\sim$ 200 K, while for ice-water cooling rotational temperatures of $\sim$ 500 K are
estimated. The infrared radiation traverses the discharge cell 16 times, giving
a total absorption path length of 32 m, and the signal is processed using a
phase-sensitive detector.

Frequency calibration is achieved using C_2H_4, H_2CO, N_2O or H_2S standards and the
absolute accuracy of the measurements is $\pm$ 0.002 cm^{-1}. The spectra are Doppler-
limited, however, resulting in linewidths of 100 - 300 MHz.

Using the difference-frequency laser method Oka made the first direct spectroscopic
observation of H_3^+. This triangular-shaped molecule is particularly important in
being the simplest well-bound polyatomic system, and is probably also of importance
in interstellar ion-molecule reactions. The 30 lines observed in the range
2450 - 3050 cm^{-1} have been assigned by Watson to the ν_2 fundamental, and have
been fitted to 13 vibration-rotation constants with a least-squares method.
Ab-initio calculations by Carney and Porter [18] were used to assist in the
identification of the lines, and the lower-order constants show good agreement
between experiment and theory.

Wong, Bernath and Amano [13] have observed 11 transitions in ^{20}NeH$^+$ and 8
transitions in ^{22}NeH$^+$ which were assigned to the fundamental vibration-rotation
band of the $^1\Sigma^+$ ground state on the basis of calculations by Rosmus and Reinsch
[19]. Nine vibration-rotation transitions of HeH$^+$ in its $^1\Sigma^+$ ground state were
observed by Bernath and Amano [12] and provide complementary information to the
data of Tolliver, Kyrala and Wing [20] and Carrington, Buttenshaw, Kennedy and
Softley [21] obtained through ion-beam methods and described later. Bernath
and Amano find a systematic deviation of $\sim$ 0.3 cm^{-1} between their measurements
and the calculations of Bishop and Cheung [22].

The observation of these species indicates that the difference-frequency method
has considerable potential for systematic infrared spectroscopy of molecular ions
in discharges. Estimates of 10^{-2} to 10^{-3} have been given for the sensitivity of
the apparatus. The use of a diode laser might improve the sensitivity by two
orders-of-magnitude [23] but the difference-frequency laser has the considerable
advantage of a very wide tuning range. The scanning speeds used in the above
experiments ranged from 8 to 30 cm^{-1}/hr, very much faster than the ion beam

methods described later. A present disadvantage, however, is the Doppler-limited linewidth; insufficient infrared power is available for saturation-type experiments which could overcome this problem. Nevertheless the technique may prove to be very useful for in-situ monitoring of molecular ions in ion-molecule reactions, and will also enhance the chances of detecting ionic species in interstellar gas clouds.

FOURIER TRANSFORM SPECTROSCOPY

High resolution infrared spectra of a large number of neutral molecules have been recorded in recent years by the method of Fourier transform infrared spectroscopy (FTIR). The spectra are obtained relatively quickly, trace quantities down to 1 part in 10^9 have been observed, and the instrumental resolution has been improved such that the observed linewidths are Doppler limited. Despite this, the only gas phase spectrum of an ion reported using this method is that of ArH^+ recorded by Brault and Davis [24].

The FTIR technique may be used to record either emission or absorption spectra of molecules. In the case of absorption spectra, infrared radiation from a conventional broad band source passes through an absorption cell containing the molecule and the transmitted radiation passes to an interferometer. For emission studies the emitted infrared radiation from an appropriate sample cell is analysed by the interferometer.

A Michelson interferometer is most commonly used and a schematic diagram [25] is shown in Figure 4.

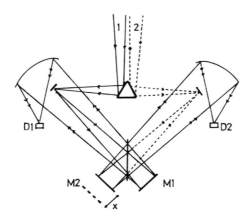

Figure 4
Schematic diagram of a Michelson interferometer
used for Fourier transform spectroscopy

The radiation transmitted or emitted by the sample enters at input 1, is split by amplitude division, and falls on two mirrors M_1 and M_2. The position of M_2 may be altered so as to introduce a variable difference in optical path length between the two beams. The beams are subsequently recombined and the intensity of radiation is measured at D_1 and D_2 as a function of the path length difference (x). The phase delay resulting from the path length difference gives rise to interference between the two beams and the intensities at D_1 and D_2 are given by

$$I_{D1}(x) = I(0) + \int_0^\infty T(\nu)S(\nu)\cos(2\pi\nu x)d\nu$$

$$I_{D2}(x) = I(0) - \int_0^\infty T(\nu)S(\nu)\cos(2\pi\nu x)d\nu$$

where $T(\nu)$ is the transmission efficiency of the spectrometer, $S(\nu)$ is the spectral intensity recoverable by the instrument and $I(0)$ is the mean intensity. The intensity of the input radiation as a function of frequency may be obtained by applying an inverse Fourier transformation,

$$I(\nu) = S(\nu)T(\nu) = \int_0^\infty I(x)\cos(2\pi\nu x)dx$$

where $I(x) = \frac{1}{2}(I_{D1}(x) - I_{D2}(x))$. In theory in order to obtain the spectrum $I(\nu)$, it is required to vary the optical path length difference x from 0 to ∞ and to measure the intensity at infinitesimally small intervals in this range. In practice a finite range of x and step size Δx must be chosen, at the expense of lower resolution and a smaller frequency range. The maximum spectral frequency is limited to $\nu_{max} = 1/2\Delta x$, while the best resolution possible is limited by the maximum path length difference available, D_{max}, to $\delta\nu_{min} = 1/D_{max}$. Background thermal radiation is generally superimposed on the signal, but may be cancelled out by using a dual input system. Input 2 in Figure 4 receives background radiation only, from near the sample source. The 180° phase difference between the two inputs results in a cancellation of background signals at D_1 and D_2.

There are three principal advantages of using Fourier transform techniques [26], as follows:

(a) Multiplex advantage

The entire spectrum can be recorded in the time taken for a grating spectrometer to measure a single spectral element. If the limiting noise of the system is independent of the signal level, the signal-to-noise ratio is improved by a factor of $M^{\frac{1}{2}}$ over a system where the M spectral elements are recorded sequentially in the same total elapsed time.

(b) Throughput advantage

In a dispersion system there is a loss of radiation due to passage through gratings and slits. The Fourier transform spectrometer does not possess these components and the throughput is limited only by the size and quality of the mirrors.

(c) Frequency advantage

The accuracy of the line frequencies is determined by the accuracy of the measurement of x. Calibration is normally achieved using a laser and the frequencies are found to be accurately reproducible over a long period of time. The reproducibility does depend upon the maintenance of constant pressure and temperature conditions to avoid fluctuations in the refractive index of the radiation path.

The multiplex and throughput advantages should be of considerable benefit when attempting to measure the spectra of species in very low concentration, such as ions in discharges.

The emission spectrum of ArH$^+$ was recorded using the McMath Solar Fourier transform spectrometer at the Kitt Peak National Observatory [27]. The spectrometer employs a Michelson interferometer of the Connes type with a maximum path difference of 1 m. A resolving limit of 0.013 cm^{-1} was obtained, over the range 1840 - 8500 cm^{-1}, which is of the same order as the Doppler width. A hollow cathode discharge was used as the source and was operated in flowing argon at a pressure of 3 torr. The cathode was a 30 mm long hollow cylinder made of carbon, with an inside diameter of 8 mm; the discharge current was 0.3 amps. The pressure of hydrogen as an impurity proved sufficient to produce ArH$^+$, and the addition of hydrogen did not increase the signal strength.

Frequency calibration was made on the basis of the positions of Ar(I) lines also present in the spectrum, and was accurate to 0.001 cm^{-1}. The relative ArH$^+$ line positions were measured to an accuracy within the range 0.006 to 0.0003 cm^{-1} depending on the signal-to-noise ratio. Five vibration-rotation bands of the $^1\Sigma^+$ ground state were observed, with extensive P and R branches. For the 1-0 band the highest rotational components observed were R(35) and P(27) but the number of lines observed in each band decreased with increasing vibrational quantum number. The lines were fitted to a set of rotational constants (B, D, H, L) for each vibrational band, and the band constants then used to derive a set of Dunham coefficients. The vibrational constants ω_e = 2710.916 and $\omega_e x_e$ = 61.63 cm^{-1} obtained from the analysis agree well with theoretical values of 2723 and 56 cm^{-1} [28]. An approximate excitation temperature of 3800 K was estimated from the line positions and intensities.

Attempts to measure the spectrum of ArD$^+$ by adding D$_2$ or D$_2$O to the flowing argon produced a few lines but an unequivocal assignment was not possible.

ION BEAM METHODS

I Introduction

The first infrared spectrum of a gaseous molecular ion was reported in 1976 by Wing, Ruff, Lamb and Spezeski [29] who used the interaction of an infrared CO laser with a beam of HD$^+$ ions. Subsequently the spectra of D$_3^+$ [30], H$_2$D$^+$ [31], HeH$^+$ [20] and further lines of HD$^+$ [32] have been observed by the same group. In a related series of experiments, Carrington and his colleagues have observed the infrared spectra of HD$^+$ [33,34], CH$^+$ [35], H$_3^+$ [36] and four isotopes of HeH$^+$ [21,37]. The transitions observed by the former group involve the lowest vibrational levels, while the latter group have observed levels close to the dissociation limits; the experimental methods are therefore complementary. The general principle of both methods is that a high velocity, well collimated beam of ions interacts colinearly with a single frequency infrared laser beam. Changes in the accelerating potential of the ion beam enable vibration-rotation transitions of the ion to be tuned into resonance with a suitable laser line, by means of the Doppler shift. Indirect methods are used to detect population transfer at resonance.

The ions are generally produced by electron impact ionisation of a neutral molecule, although plasma discharge sources may also be used. The ions are extracted by applying a 1-10 kv positive potential to the source and are collimated into a narrow beam by a series of electrostatic lenses. The resultant beam possesses a number of properties offering considerable advantages for spectroscopic study, which we now discuss [38].

(a) Ions present in the beam enjoy a collision-free environment; thus the internal energy distribution of the ions in the beam will reflect directly the energetics of the formation processes in the source. As a result, ions with considerable electronic, vibrational or rotational excitation are often present in the beam.

(b) A high degree of spatial control of the ion beam is possible using collimating and deflecting lenses, enabling very efficient interaction with the laser beam to occur.

(c) Unambiguous identification of the ions in the beam can be made through use of the now standard techniques of mass spectrometry. In addition ions of different charge-to-mass ratio can be separated from each other, ion beam intensity measurements are straightforward, and even very weak beams of only a few ions per second can be detected by means of high-gain electron multipliers.

(d) The individual ions in the beam travel at velocities which are typically in the range 10^5 to 10^6 ms^{-1}, depending upon the beam potential and mass of the ion. The effective laser frequency observed by an ion moving at velocity v parallel or antiparallel to the laser beam is given by the relativistic Doppler-shift formula,

$$\nu_{effective} = \nu_{laser} \times \left\{ \frac{1 \pm (v/c)}{1 \mp (v/c)} \right\}^{\frac{1}{2}}$$

where the upper and lower signs refer to antiparallel and parallel orientations respectively, and c is the velocity of light. By sweeping the source potential, frequency scanning may be achieved with a fixed-frequency laser line.

(e) Sub-Doppler resolution is obtainable as a result of an effect known as kinematic compression or velocity bunching. This effect may be understood [39] by considering two different ions, a and b, of the same mass and which have velocity components in the z direction:

$$v_a = 0$$
$$v_b = (2kT/m)^{\frac{1}{2}}$$

After acceleration in the z direction through a potential difference V the ions have final velocity z-components:

$$v'_a = (2eV/m)^{\frac{1}{2}}$$
$$v'_b = (v_b^2 + (2eV/m))^{\frac{1}{2}} = (v_b^2 + v'^2_a)^{\frac{1}{2}} \approx v'_a + \frac{v_b^2}{2v'_a}$$

Hence

$$\frac{v'_b - v'_a}{v_b - v_a} \approx \frac{1}{2}(kT/eV)^{\frac{1}{2}} = R$$

For T = 2000K and V = 10 kV the value of R is 2.1 x 10^{-3}; consequently the velocity difference and therefore Doppler shift between the different ions is greatly reduced. In practice the resolution is improved by factors of 10 to 100 over normal Doppler-limited spectra as a result of this velocity bunching effect.

Instantaneous ion densities in typical molecular ion beams are estimated to be in
the range 10^4 to 10^6 ions/cm^3, which is too low for spectroscopy to be carried
out with conventional light sources. Even with the use of high-powered CW
lasers it is necessary to use indirect methods to detect the absorption of
radiation and three methods have been used to record infrared spectra:

(a) *Charge transfer* After interaction with the laser beam the ions are partly
    neutralised by charge exchange with an appropriate target gas. The cross-
    section for charge exchange is dependent upon the vibrational state of the
    ion and therefore the population transfer between vibrational states at
    resonance with the laser can be detected by a corresponding change in the
    degree of charge neutralisation. This method of detection has been used
    by Wing and his colleagues in all of their studies.

(b) *Photodissociation* The molecular ions in the beam interact with two photons.
    The first photon is resonant with a vibration-rotation transition, whilst the
    second has sufficient energy to photodissociate the upper level, but insuff-
    icient to photodissociate directly the lower level. Resonant population
    transfer is thus detected as an increase in the number of photofragment ions.
    Carrington and his colleagues have used this method to study the spectrum of
    HD$^+$, employing an infrared CW carbon dioxide laser to drive the resonant
    vibration-rotation transitions and also to photodissociate the upper states.
    The upper state of a transition must therefore lie within one infrared photon
    of the dissociation continuum.

(c) *Predissociation* In this method the upper state of the transition prediss-
    ociates during the passage time of the ion through the apparatus. The
    transition is therefore detected by the appearance of an increase in fragment
    ions at resonance. The infrared predissociation spectra of CH$^+$, HeH$^+$ and
    H$_3$$^+$ have been detected in this manner by Carrington and his coworkers.

It is noteworthy that the resonant signal will be comparable with or larger than
the background signal in cases (b) and (c), whereas in method (a) the change in
parent ion beam current after partial neutralisation is relatively small at
resonance. Techniques (b) and (c) are therefore far more sensitive than (a),
but can only be used to study levels near to the dissociation limit.

II Charge exchange methods

The apparatus used by Wing and his group which involves charge exchange detection
is shown in Figure 5. The electron bombardment type source is found to be sup-
erior to the plasma discharge source in that it forms ions with less vibrational
excitation. Consequently the off-resonance population differences between the
initial and final states are greater, enhancing the signal-to-noise ratio. The
vibrational populations for HD$^+$ from the electron bombardment source are determined by the corresponding Franck-Condon factors for ionisation of the neutral
molecule, while the rotational distribution is essentially Boltzmann at 320 K.
Ion beam currents vary from 3×10^{-7} A for HD$^+$ (produced by admitting a mixture
of H$_2$ and D$_2$ through a hot palladium leak into the source) to 3×10^{-13} A for
D$_3$$^+$ (from D$_2$). After focussing, the ion beam enters a constant potential interaction region, maintained at a pressure of 10^{-6} torr, in which it is crossed at
an angle of 11 mrad by an infrared laser beam. The continuous wave CO laser
employed is capable of producing 150 lines in the 5.2 to 6.2 μm region. Single
line selection is achieved using a diffraction grating and powers of a few mW
to 1 W are obtained per line. The source potential is swept by a microcomputer,
usually in 0.1 V steps with an integration time of 8 sec per step, in order to
Doppler-tune transitions into coincidence with a single selected laser line. A
fast scan mode in which 15 laser lines run simultaneously is also used to increase
the search rate.

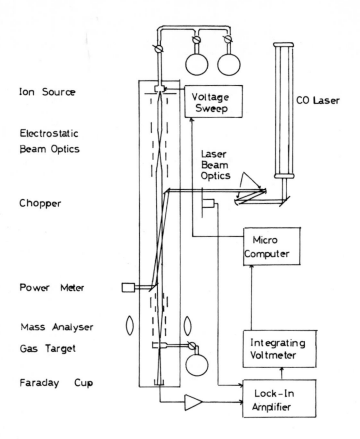

Figure 5
Ion beam/laser beam apparatus used for charge-exchange
detection by Wing and coworkers

After the interaction region, crossed electric and magnetic fields are used to mass-select the ion of interest. Subsequently the ion current is attenuated by passing it through a gas target such as N_2, Ar, H_2 or He, and the remaining ion current is measured at a Faraday cup. The current alters by a few p.p.m. at resonance; hence the laser beam is chopped at 1 kHz and the a.c. Faraday cup signal synchronously detected.

Twenty-five vibration-rotation transitions of HD^+ in its $^2\Sigma_g^+$ ground state, involving v = 0 to 5 and N = 0 to 5 have now been observed [29, 32]. All the transitions obey the harmonic oscillator selection rule $\Delta v = \pm 1$, and the measured frequencies are accurate to ± 0.0007 cm^{-1}, the main limitations being voltage calibration and uncertainty in the laser frequencies. The observed linewidths are in the region 7 to 25 MHz, and become smaller at higher source potentials because of increased kinematic compression effects. The experimental results differ by only 0.001 to 0.002 cm^{-1} from the most recent non-adiabatic calculations by Bishop and Cheung [40] and Wolniewicz and Poll [41]. The agreement between theory and experiment is sufficiently good to corroborate the 1973 adjusted value of the proton/electron mass ratio used in the theory. Hyperfine and spin-rotation structure is observed for each line with two or three strong components and one to four weaker components; the splittings range from 12 to 45 MHz.

Five P branch transitions involving v = 0 to 2 and N = 8 to 13 have been observed for $^4HeH^+$, with linewidths of 5 to 8 MHz and a quoted accuracy of 0.002 cm^{-1} in line positions. Since HeH^+ has a closed shell ground state, electron spin hyperfine interactions do not occur and the resonance lines do not exhibit splittings. Bishop and Cheung have calculated adiabatic energy levels for HeH^+ in its 3 lowest vibrational states [22] and the theoretical transition frequencies agree with the measured values to within ~ 0.2 cm^{-1}.

Spectra of D_3^+ [31] and H_2D^+ [30] have also been recorded using a carbon monoxide infrared laser and charge exchange detection. After scanning 10 cm^{-1} of the D_3^+ spectrum 8 lines have been observed, whilst a scan of 75 cm^{-1} of the H_2D^+ spectrum has revealed 9 lines. Assignment of these lines is difficult because of the incompleteness of the data, and the complexity of the H_3^+ vibration-rotation spectrum revealed by Oka's studies [9 - 11]. Four transitions of D_3^+ have been assigned to specific components of the ν_2 fundamental, and it is anticipated that further recording of the spectra will facilitate a complete assignment. The wide range capability of the difference-frequency infrared laser is, however, a great advantage in problems of this type.

III Photodissociation and predissociation methods

The apparatus employed by Carrington and his colleagues for photodissociation and predissociation detection of infrared transitions in ion beams is shown in Figure 6 [33]. In essence a tandem mass spectrometer (Vacuum Generators ZAB 1F) has been adapted for spectroscopic purposes. Ions are formed by electron impact or plasma discharge and the extracted ion beam is mass-analysed by a 55° electromagnetic sector. The beam is focussed at the intermediate slit and the ion current may be measured immediately after the slit using an off-axis electron multiplier. The ion beam then passes into an electrostatic analyser (ESA) which selects a particular fragment ion formed from the parent ion between the two analysers. The fragment ion current may be measured either by means of a Faraday cup, or with a second off-axis electron multiplier. Prior to the intermediate slit the ion beam passes through a tube of 40 cm length, known as the drift tube. Potentials in the range -500 to +500 volts may be applied to this tube, and the ESA is able to separate fragment ions formed in the drift tube from those formed elsewhere at earth potential due to their different kinetic energies. The path of the ion beam is evacuated to $\sim 7 \times 10^{-8}$ torr.

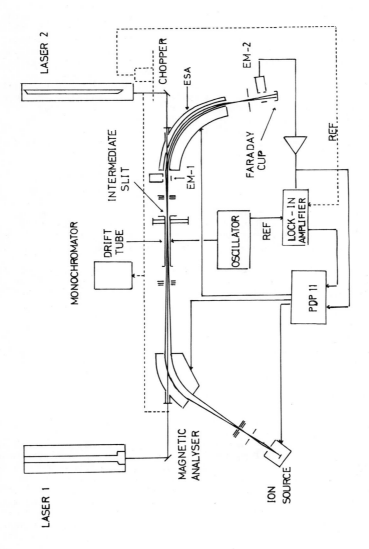

Figure 6

Tandem mass spectrometer system used by Carrington and coworkers to study
photodissociation and predissociation infrared spectra

Two lasers are available for spectroscopic studies. Laser 1 is of sealed tube design and may be operated with CO as the lasing gas to give maximum powers of 3W CW in a single line, or with CO_2 giving powers of up to 25 W per line. The laser beam is reflected into parallel alignment with the ion beam and is focussed at the intermediate slit. Laser 2 may be operated in a sealed tube or flowing gas mode with CO_2 as the lasing gas, and has a maximum output power of 60 W CW in a single line. The laser beam is aligned antiparallel with the ion beam and focussed into the drift tube, although when both lasers are operated simultaneously a small angle is created between the two laser beams to avoid interference.

In experiments using the photodissociation method both lasers are used simultaneously. The vibration-rotation transitions are Doppler tuned into resonance with laser 1, operated at low power to minimise power broadening, and the transitions are detected by dissociation of the upper state using laser 2 at high power to maximise the sensitivity. For predissociation detection either laser 1 or laser 2 is used depending on whether parallel or antiparallel alignment of the laser and ion beams are required. Doppler tuning and signal detection is achieved using one of four different scanning modes, the choice depending on the widths of the lines and the spectral range to be searched.

(a) *Laser line scan* The source potential is fixed at a convenient value, the magnetic field set to transmit the desired parent ion, and the ESA set to select the appropriate photofragment ion. The laser beam is chopped at 700 Hz and the fragment ion synchronously detected. The photofragment signal is measured for irradiation with a number of consecutive laser lines, thus giving data points at 1 to 2 cm^{-1} intervals. Interpolation between these points may be achieved by repeating the process for a few widely spaced source potentials. The method is only suitable for measuring lines with widths greater than 1 cm^{-1}, such as occur for HeH^+ and its isotopes [21,37].

(b) *Source potential scan with laser beam chopping* A PDP-11 minicomputer is used to scan the source potential over the range 2 to 10 kV, typically in 0.1 to 1 V steps. The magnetic field is simultaneously scanned by the computer to transmit the desired parent ion, and the ESA is also scanned to transmit the fragment ion; the computer repeatedly reads the fragment ion current and adjusts the sectors for maximum intensity. The laser beam is chopped and the signal demodulated as before. In cases where the fragment ion is particularly light (e.g. H^+ ions) a positive voltage is applied to the drift tube and reacceleration to earth potential increases the fragment ion kinetic energy and thus improves the detection sensitivity. This mode of operation can be used for measuring absorption lines with widths in the range 0.0003 to 0.5 cm^{-1}.

(c) *Source potential scan with drift tube modulation* This method is similar to (b) but instead of chopping the laser beam, a square-wave modulation of frequency up to 10 kHz and amplitude up to 10 volts is applied to the drift tube. Voltage modulation is, through the Doppler effect, equivalent to frequency modulation; hence the ESA voltage is synchronously modulated and a phase-sensitive detector used to demodulate the fragment ion signal. The method is more sensitive than laser beam chopping for the narrowest lines but is not generally applicable if the linewidths are greater than 100 MHz.

(d) *Drift tube scan with drift tube modulation* In cases where the lines are narrow and the line positions are accurately predicted or known, such as in HD^+, spectra can be recorded by fixing the source potential and magnetic field and scanning the drift tube potential over a few hundred volts. Modulation of the drift tube potential is accompanied by simultaneous modulation and scanning of the ESA potential, and fragment ions formed at the drift tube potential are detected.

In addition to the above modes of operation it is possible to measure the
kinetic energy spectrum of the fragment ions, by fixing the source potential,
magnetic field and drift tube voltage, and scanning the ESA voltage.

The experimental conditions and modes of operation for the ions studied to date
are summarized in Table 2. We now discuss the spectra of each of the four ions
studied so far.

HD^+

Figure 7 illustrates the principles underlying the measurements of vibration-
rotation transitions in HD^+ using photodissociation detection. Transitions
between v = 16 and v = 18 are Doppler-tuned into resonance with an appropriate
CO_2 line of laser 1. The v = 18 levels are photodissociated by infrared photons
from laser 2, operated at high power, and the resonant increase in the population
of v = 18 is observed by an increase in the D^+ fragment ion current. The observation of these transitions shows that the higher vibrational levels of HD^+ are
significantly populated, as was suggested by several theoretical calculations
(for example, those of Tajeddine and Parlant [42]). The spectra also demonstrate
the high sensitivity of the technique, signal-to-noise ratios of 100:1 for a
1 second integration time being observed. Spectra can be observed using one laser
only, a second photon from laser 1 being absorbed in the photodissociating step.
Transitions from v = 14 to 17 have been observed recently by Carrington and
Kennedy [43] using laser lines of CO. A total of 9 rotational components of the
v = 18 - 16 band [33], 7 components of the v = 14 - 17 band [43] and 1 component
of the v = 15 - 17 band [33] have now been observed. Linewidths down to 7 MHz
are obtained, permitting the resolution of doublet splittings arising from
proton hyperfine interaction. Transition frequencies are determined to
± 0.0005 cm^{-1} and the results provide a severe test of the recent non-adiabatic
calculations of Wolniewicz and Poll [41]. The agreement between theory and
experiment is excellent for transitions involving low rotational quantum number N,
but becomes less good as N increases. It would be of particular interest to detect
transitions involving the highest vibrational levels, v = 20 and 21; attempts
made so far have been unsuccessful.

HeH^+

The addition of the centrifugal energy term to the rotationless potential energy
curve of the $^1\Sigma^+$ ground state of HeH^+ gives rise to an effective potential curve
with a maximum above the dissociation limit at long internuclear distances. As
a result, a number of discrete "quasibound" levels exist above the dissociation
limit but below the maximum in the potential energy curve. Carrington and his
coworkers have observed a number of transitions in HeH^+ and its isotopic
variations from bound levels below the dissociation limit to quasibound levels
above, using the CO_2 laser. The transitions are detectable because the quasibound levels predissociate by barrier penetration and an increase in H^+ fragment
ion current is detected at resonance. The observed linewidths are inversely
proportional to the quasibound lifetimes. Predictions of two bound to quasibound
transitions in $^4HeH^+$, lying in the CO_2 laser region, were made from calculations
by Dabrowski and Herzberg [44] and Kolos and Peak [45], and were observed by
Carrington, Buttenshaw, Kennedy and Softley [21]. More recently 6 transitions
of $^4HeD^+$, 1 transition of $^3HeH^+$ and 2 transitions of $^3HeD^+$ have been observed
[37] following theoretical predictions of Fournier [46]. Agreement between
experiment and theory ranges from 0.05 to 1.05 cm^{-1} whilst the theoretical and
observed linewidths also agree fairly well. Table 3 collects together all of the
available data for vibration-rotation transitions in $^4HeH^+$, and nicely illustrates
the complementary nature of the different techniques which have been used.

Table 2

Ion	Source	Gas(torr)	Beam(Amps)	Fragment	Tuning mode	Detection
HD^+	E.I.	$HD \sim 10^{-3}$	10^{-6}	D^+	(c), (d)	Photodissociation
HeH^+	E.I./plasma	He/H_2 10:1 $\sim 10^{-3}$	10^{-8} to 10^{-9}	H^+	(a), (b)	Predissociation
CH^+	E.I.	C_2H_4 $\sim 10^{-4}$	5×10^{-9}	C^+	(b)	Predissociation
H_3^+	E.I.	H_2 $\sim 10^{-3}$	10^{-7}	H^+	(c)	Predissociation

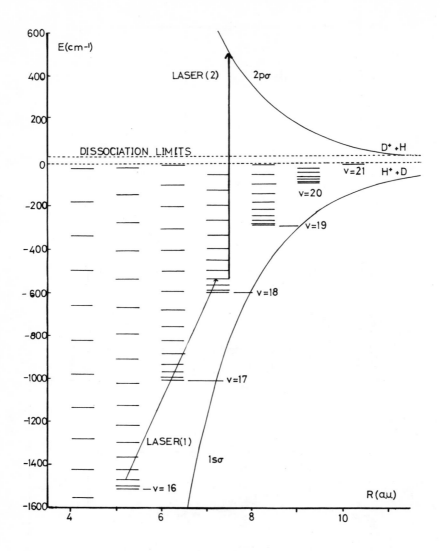

Figure 7
Principles of the two-photon infrared dissociation technique for HD$^+$

High-Resolution Infrared Spectroscopy of Molecular Ions67

v',N'	v",N"	Obs.freq.(cm^{-1})	Obs.width(cm^{-1})	Calc.freq.(cm^{-1})	Calc.width(cm^{-1})	Ref. (exp)	Ref. (theory)
1,5	0,4	3157.2967	-	3157.632	-	23	22
1,4	0,3	3121.0765	-	3121.417	-	23	22
1,3	0,2	3077.9919	-	3078.332	-	23	22
1,2	0,1	3028.3750	-	3028.715	-	23	22
1,1	0,0	2972.5732	-	2972.913	-	23	22
1,0	0,1	2843.9035	-	2844.242	-	23	22
1,1	0,2	2771.8059	-	2772.140	-	23	22
1,2	0,3	2695.0500	-	2695.381	-	23	22
1,3	0,4	2614.0295	-	2614.198	-	23	22
1,11	0,12	1855.905	-	1856.152	-	20	22
1,12	0,13	1751.971	-	1752.198	-	20	22
2,8	1,9	1896.992	-	1897.139	-	20	22
2,9	1,10	1802.349	-	1802.492	-	20	22
2,10	1,11	1705.543	-	1705.684	-	20	22
7,11	5,12	938.2	4.1	939.46	5.17	21	44,45
6,13	5,12	979.90	0.16	981.76	0.13	21	44,45

CH$^+$

Calculated potential energy curves for the electronic states correlating with the
two lowest dissociation limits of CH$^+$ are shown in Figure 8 [47,48].

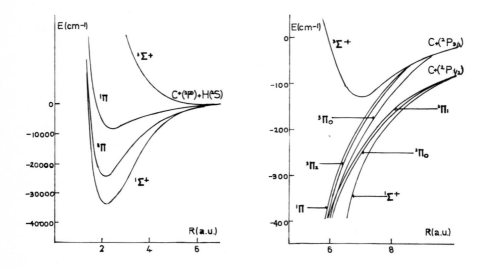

Figure 8
Calculated potential energy curves for the lowest electronic states of CH$^+$

Two types of predissociation could be exhibited by this system:

(a)   Rotationally quasibound levels are expected to exist for some or all of the
      electronic states involved and transitions to quasibound levels of the $^1\pi$
      state have indeed been observed in the visible region by Helm, Cosby, Graff
      and Moseley [49].

(b)   Electronic states correlating with the upper fine-structure limit may have
      rovibrational levels lying above the lower fine-structure limit. Prediss-
      ociation of these levels may be induced by spin-orbit or Coriolis coupling
      to electronic states correlating with the lower dissociation limit.

Carrington, Buttenshaw, Kennedy and Softley [35] have observed infrared transitions in CH$^+$ beams, the upper states predissociating to yield C$^+$ ions. Carrington
and Softley [50] are now in the process of carrying out a systematic search over
the region spanned by the CO$_2$ laser, 875 to 1095 cm^{-1}. Each laser line is selected
in turn and the source potential scanned from 2 to 10 kV with parallel and antiparallel alignment of the ion and laser beams. Using ^{12}CO$_2$ and ^{13}CO$_2$ about 60%
of the total range can be covered by Doppler tuning. 78 lines have been observed
in the 90 cm^{-1} scanned so far, with linewidths varying from 12 to 4000 MHz; many

of the lines exhibit doublet splittings ranging from 20 to 670 MHz. No unequiv-ocal assignment of these lines has yet been made, but the predissociation mechanism is believed to be electronic rather than rotational. Because of heavy mixing of the electronic states in the vicinity of the dissociation limits, it is likely that the only firm selection rules are $\Delta\Omega = 0, \pm 1$, $\Delta J = 0, \pm 1$, and + parity $\leftrightarrow$ - parity. The number of possible transitions is therefore large, particularly if the $^1\pi$, $^3\pi$ and $^1\Sigma$ states are all appreciably populated close to the dissociation limits. One particular group of about 20 transitions are characteristically weak and sharp, and show doublet splittings in the range 100 to 670 MHz. We believe that for this group the upper state might well be the $^3\Sigma^+$ state which, as Figure 8 shows, is predicted to have a shallow minimum. The proton Fermi contact interaction is expected to be large for levels belonging to this state, and could be responsible for the observed doublet splittings.

In the present absence of sufficiently accurate calculations, the assignment of the spectrum is a difficult task. Nevertheless the spectrum contains unique information about the structure of this important molecular ion close to the dissociation limits, and both the measurements and attempts at analysis are continuing.

H_3^+

A remarkable infrared spectrum of the H_3^+ ion has been reported by Carrington, Buttenshaw and Kennedy [36]. A beam of H_3^+ ions formed by flowing H_2 through an electron impact ionisation source, is Doppler-tuned into resonance with CO_2 laser lines, and H^+ ions formed by predissociation are detected. The apparatus is operated in mode (c). More than 300 lines were reported in the initial communication and Carrington and Kennedy [51] are now in the process of system-atically recording the complete spectrum from 880 cm^{-1} to 1095 cm^{-1}. At the time of writing a range of 40 cm^{-1} has been recorded; the data is far from complete but the salient features at present are as follows:

(a)   After scanning a total of 40 cm^{-1} (which are not consecutive), over 6000 lines have been recorded, with their positions and relative intensities measured.

(b)   The observed linewidths range from 2 to 100 MHz; broader lines are also present but the drift tube modulation technique discriminates in favour of the sharper lines.

(c)   1 cm^{-1} segments of the spectrum have been recorded at different points throughout the range 880 to 1095 cm^{-1}. The numbers of lines in these 1 cm^{-1} segments range from 100 to 180, and the signal-to-noise ratios range from 30:1 to 1.5:1 for a 1 second integration time.

(d)   Adjustment of the ratio of the ESA voltage to the source potential yields a scan of the kinetic energy spectrum of the H^+ ions formed in the resonant predissociation transitions. Preliminary studies indicate that most of the predissociating levels observed lie within 20 cm^{-1} of a dissociation limit.

(e)   The dimensions of the apparatus and calculated ion velocities lead to the conclusions that the initial states involved in the resonance transitions must have lifetimes of 5 x 10^{-6} secs or greater, and the final (predissocia-ting) states must have lifetimes in the range 10^{-6} to 10^{-9} secs.

(f)   A plot of the total line intensity per cm^{-1} versus the frequency (in cm^{-1}) is highly structured. This plot, which approximates to a low resolution spectrum, shows five peaks, the frequencies of which correspond closely to $\Delta v = 0$, $\Delta N = +2$ transitions of the H_2 molecule. The accumulation of more data may reveal further peaks in this constructed spectrum.

(g) Limited searches for resonance lines of H_3^+ predissociating to give H_2^+
     have, so far, proved to be negative.

Various conclusions about the nature and origin of this spectrum may be drawn,
but they must be regarded as tentative. Less than 20% of the complete spectrum
(which probably contains more than 30,000 lines in the CO_2 laser region) has
been recorded, and the acquisition of more data may change some of the conclusions
and reveal other features. The present position, however, seems to be as
follows:

  (i)   The H_3^+ beam contains ions with up to 2 or 3 eV of internal energy
        greater than that corresponding to the lowest dissociation limit,
        H_2 (v = 0, N = 0) + H^+.

  (ii) Metastable levels of H_3^+ exist which are close in energy to the
        H_2 (v, N) + H^+ dissociation limits. The metastability may arise from
        centrifugal barriers associated with the end-over-end rotation of the
        H_2 (v, N) .... H^+ complex.

  (iii) Transitions between long-lived (> 5 µs) and short-lived (< 1 µs) levels
        (both upwards and downwards in energy) give rise to the peaks in the
        pseudo-low resolution spectrum described above. The ΔN = ± 2,0
        selection rule arises not because the spectrum is necessarily quadrupolar
        in character, but because of the nuclear spin symmetry restrictions for
        the H_2 molecule, which still hold for the H_3^+ ion.

  (iv) In addition to the peaks in the pseudo-low resolution spectrum, there is
        a background spectrum of lines arising from transitions between the
        highest bound levels of H_3^+ and the four lowest dissociation limits
        (N = 0 to 3).

When the complete spectrum has been recorded it will be possible to use
convolution techniques to construct spectra at any desired resolution level.
Such spectra should be extremely informative and if feature (f) above is
reproduced satisfactorily, a more detailed analysis will depend upon accurate
calculations of the vibration-rotation levels of the H_2 .... H^+ complex for
given v, N states of the hydrogen molecule. Finally we note that Carrington
and Kennedy [51] have also observed similar spectra of the D_3^+, D_2H^+ and H_2D^+
ions.

ACKNOWLEDGEMENTS

A.C. thanks the Royal Society for a Research Professorship and T.P.S. thanks
the University of Southampton for a Research Studentship. We are indebted to
the Science and Engineering Research Council for their support of the work on
ion beam spectroscopy described in this review.

REFERENCES

[1]   G. Herzberg, Quart. Rev. Chem. Soc. 25 (1971) 201

[2]   R. J. Saykally and R. C. Woods, Ann. Rev. Phys. Chem. 32 (1981) 403

[3]   K. M. Evenson, R. J. Saykally, D. A. Jennings, R. F. Curl and J. M. Brown, in: Chemical and Biochemical Applications of Lasers (Academic Press, New York, 1980) p. 95.

[4]   A. R. W. McKellar, Discussions Faraday Soc. 71 (1981) 63

[5]   R. J. Saykally and K. M. Evenson, Phys. Rev. Letters 43 (1979) 515

[6]   D. Ray, K. G. Lubic and R. J. Saykally, Mol. Phys. 46 (1982) 217

[7]   R. F. Barrow and A. D. Gaunt, Proc. Phys. Soc. 66 (1953) 617

[8]   J. M. Brown, M. Kaise, C. M. L. Kerr and D. J. Milton, Mol. Phys. 36 (1978) 553

[9]   T. Oka, Phys. Rev. Letters 45 (1980) 531

[10] T. Oka, Phil. Trans. Roy. Soc. London A303 (1981) 543

[11] T. Oka, in: Laser Spectroscopy (Springer, 1981) p. 320

[12] P. Bernath and T. Amano, Phys. Rev. Letters 48 (1982) 20

[13] M. Wong, P. Bernath and T. Amano, J. Chem. Phys. 77 (1982) 693

[14] A. S. Pine, J. Opt. Soc. Am. 64 (1974) 1683

[15] A. S. Pine, J. Opt. Soc. Am. 66 (1976) 97

[16] P. Bernard and T. Oka, J. Mol. Spectry. 75 (1979) 181

[17] A. R. W. McKellar and T. Oka, Can. J. Phys. 56 (1978) 1315

[18] G. D. Carney and R. N. Porter, J. Chem. Phys. 65 (1976) 3547

[19] P. Rosmus and E. A. Reinsch Z. Naturforsch 35a (1980) 1066

[20] D. E. Tolliver, G. A. Kyrala and W. H. Wing, Phys. Rev. Letters 43 (1979) 1719

[21] A. Carrington, J. A. Buttenshaw, R. A. Kennedy and T. P. Softley, Mol. Phys. 44 (1981) 1233

[22] D. M. Bishop and L. M. Cheung, J. Mol. Spectry. 75 (1979) 462

[23] J. Reid, J. Shewchun, B. K. Garside and E. A. Ballik, Appl. Opt. 17 (1978) 300

[24] J. W. Brault and S. P. Davies, Physica Scripta 25 (1982) 268

[25] A. F. M. Moorwood, Infrared Phys. 17 (1977) 441

[26] J. L. Koenig, Acc. Chem. Res. 14 (1981) 171

Alan Carrington & T.P. Softley

[27]   J. W. Brault, in Proc. of the Workshop on future Solar observations -
       Needs and Constraints (Florence, 1978)

[28]   P. Rosmus, Theor. Chim. Acta. 51 (1979) 359

[29]   W. H. Wing, G. A. Ruff, W. E. Lamb and J. J. Spezeski, Phys. Rev.
       Letters 36 (1976) 1488

[30]   J. W. Farley, W. E. Lamb and W. H. Wing, Phys. Rev. Letters 45 (1980) 535

[31]   J. T. Shy, J. W. Farley and W. H. Wing, Phys. Rev. A24 (1981) 1146

[32]   W. H. Wing, private communication (1981)

[33]   A. Carrington and J. A. Buttenshaw, Mol. Phys. 44 (1981) 267

[34]   A. Carrington, J. A. Buttenshaw and R. A. Kennedy, J. Mol. Structure 80
       (1982) 47

[35]   A. Carrington, J. A. Buttenshaw, R. A. Kennedy and T. P. Softley, Mol. Phys.
       45 (1982) 747

[36]   A. Carrington, J. A. Buttenshaw and R. A. Kennedy, Mol. Phys. 45 (1982) 753

[37]   A. Carrington, R. A. Kennedy, T. P. Softley and P. A. Fournier, to be
       published.

[38]   A. Carrington, Proc. Roy. Soc. London A367 (1979) 433

[39]   S. L. Kaufman, Opt. Comm. 17 (1976) 309

[40]   D. M. Bishop and L. M. Cheung, Phys. Rev. A16 (1977) 640

[41]   L. Wolniewicz and J. D. Poll, J. Mol. Spectry. 72 (1978) 264

[42]   M. Tadjeddine and G. Parlant, Mol. Phys. 33 (1977) 1797

[43]   A. Carrington and R. A. Kennedy, to be published

[44]   I. Dabrowski and G. Herzberg, Trans. N.Y. Acad. Sci. 38 (1977) 14

[45]   W. Kolos and J. M. Peek, Chem. Phys. 12 (1976) 381

[46]   P. A. Fournier, unpublished work

[47]   S. Green, P. S. Bagus, B. Liu, D. McLean and M. Yoshimine, Phys. Rev. A5
       (1972) 1614

[48]   J. F. Bazet, C. Havel, R. McCarroll and A. Riera, Astr. Astrophys. 43
       (1975) 223

[49]   H. Helm, P. C. Cosby, M. M. Graff and J. T. Moseley, Phys. Rev. A25 (1982)
       304

[50]   A. Carrington and T. P. Softley, unpublished work

[51]   A. Carrington and R. A. Kennedy, unpublished work

Molecular Ions: Spectroscopy, Structure and Chemistry
Terry A. Miller and V.E. Bondybey (editors)
© North-Holland Publishing Company, 1983

THE H_3^+ ION

Takeshi Oka

Department of Chemistry
Department of Astronomy and Astrophysics
The University of Chicago
Chicago, Illinois 60637

The great many experimental and theoretical studies on
the fundamental H_3^+ molecular ion, the simplest stable
polyatomic system, are reviewed. Emphasis is placed
upon the studies which are related to the infrared spectrum of this molecular ion recently observed by the
author. The subjects reviewed are: Discovery of H_3^+;
Ion-Molecule Reactions; Ab-initio Calculations; H_3^+ in
D.C. Glow Discharges; H_3^+ in Interstellar Space; Other
Experiments; and Spectroscopy of H_3^+. This review is
not meant to be exhaustive, but rather heavily biased
by the author's interest.

I. Discovery of H_3^+

The existence of the molecular ion H_3^+ was first reported by J.J. Thomson in 1912
[1] in a paper titled "Further Experiments on Positive Rays". It is a long paper
(45 pages) in which he describes details of the improvement of his apparatus
which increased the brightness of the positive rays and his new findings using
the apparatus. On the 33rd page of this paper he states "Existence of H_3. — On
several plates taken when the discharge-tube contains hydrogen, the existence of
a primary line for which m/e = 3 has been detected. There can, I think, be little
doubt that this line is due to H_3, ····· the existence of this substance is interesting from a chemical point of view, as it is not possible to reconcile its
existence with the ordinary conceptions about valency, if hydrogen is regarded as
always monovalent". One of the plates obtained by him is shown in Fig. 1. The
H_3^+ line is seen much weaker than the H^+ or H_2^+ lines. He carefully eliminates
other possibilities for this line such as the carbon atom with four charges etc.
This reasoning is clearly stated in his monograph "Rays of Positive Electricity" [2].

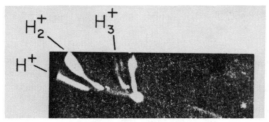

Fig. 1 "A photograph showing the m/e = 3 line
together with the H^+ and the H_2^+ line. Other
lines are due to the oxygen atom and molecular
and to the mercury atom" J.J. Thomson [1].

Thomson's paper was followed in 1916 by
the paper of A.J. Dempster in Chicago
[3] in which he replaced the low pressure (3 mTorr) and high-voltage (20 KV)
discharge of Thomson's experiment with
an electron bombardment (800 V) of hydrogen and showed that much higher yield
of H_3^+ could be obtained at higher pressure of H_2. One example of his results
is shown in Fig. 2. Thus he has shown
that H_3^+ is more abundant than H^+ or H_2^+
under certain conditions.

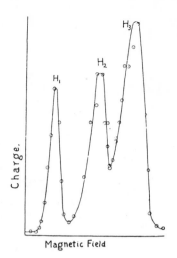

Fig. 2.   Result of A.J. Dempster [3].

Twenty years later, with the discovery
of deuterium some confusion was introduced for the interpretation of the
m/e = 3 line. Thus Harold Urey wrote in
his review article on deuterium [4] "As
early as 1911 Sir J.J. Thomson observed
particles of molecular weight three in
his apparatus for the detection of isotopes. These molecules were of two
varieties, one transient and probably
H_3^+ and the other permanent and probably
HD^{+}". J.J. Thomson himself wrote in
1934 [5], "the evidence seems to me to
leave little doubt that the gas I called
H_3^+ more than twenty years ago is the same as that which is now called heavy hydrogen". He went on to study very carefully the two kinds of molecular ions which
can produce the m/e = 3 line [6][7].

Probably because of this confusion, sometimes Dempster is quoted as the discoverer
of H_3^+ [8].

II.   Ion-Molecule Reactions

At the end of his paper [3], Dempster stated that his electrons of 800 Volts produce "positive molecules" without decomposition and they "are able to dissociate
the gas" to form H_3^+. Smyth [9] studied the primary and secondary products of
ions in hydrogen using electrons with much less energy and determined the relative concentrations of H^+, H_2^+, and H_3^+ for many experimental conditions. Hogness
and Lunn [10] did similar studies, and it is in their paper that the celebrated
reaction,

$$H_2 + H_2^+ \rightarrow H_3^+ + H, \qquad\qquad (1)$$

appears as the first example of ion-molecule reactions. Since then a great number
of people worked on this reaction using a variety of experimental techniques. A
summary of these results are found in reviews [11,12,13]. The recent detailed
study on the effect of vibration and translational energy on the dynamics of this
reaction by Lee and coworkers [14] is particularly noteworthy. This paper also
contains extensive references for this reaction. Many theoretical papers have also been published on this reaction since the early work of Eyring, Hirschfelder,
and Taylor [15] who gave the cross section with remarkable accuracy.

It is now well established, both experimentally [16,17] and theoretically [18],
that H_2 and H_2^+ in Eq. (1) approach each other without having to surmount

potential barrier (other than the barrier of centrifugal energy), and that the large cross section of $\sim$100 $\AA^2$ [19] is determined essentially by the long range inductive force. Thus if we consider the charge-(induced dipole) interaction with the potential of $V = -\alpha e^2/2r^4$ we obtain the Langevin cross section [20]

$$\sigma = 2\pi e(\alpha/\mu)^{1/2} / v = \pi e(2\alpha/E)^{1/2} \qquad (2)$$

where μ is the reduced mass. Using the polarizability of H_2, $\alpha = 0.79 \times 10^{-24}$ cm^3 [21] and the kinetic energy $E = 3kT/2$ at room temperature we obtain $\sigma \sim 76 \AA^2$ in agreement with the observed value. The orientation dependence of the potential is relatively small because of the small value of $\alpha' = 2(\alpha_{11} - \alpha_{\perp})/3 = 0.14 \times 10^{-24}$ cm^3. There also exists charge-quadrupole interaction with the potential of $V = Qe^2 P_2(\cos \theta)/r^3$ where Q is the quadrupole moment of H_2 and $P_2(\cos \theta)$ is the Legendre polynomical. Although this potential is comparable in magnitude with the charge-(induced dipole) interaction discussed earlier, an average over θ makes this effect less important [22].

In general for an interaction with the potential $V = -\beta/r^n$ (n > 2), the Langevin cross section is given as

$$\sigma = \pi n(n-2)^{-\frac{n-2}{n}} (\beta/\mu v^2)^{\frac{2}{n}} \qquad (3)$$

(see § 18 of Landau and Lifshitz [23]).

The exothermicity of the reaction (1) is calculated to be 1.703 eV (see next section). This exothermicity and the large cross section make the reaction very efficient both in the laboratory discharge and in space and produces abundant H_3^+ in such an environment.

The other important ion-molecule reaction involving H_3^+ is the reaction

$$H_3^+ + X \rightarrow HX^+ + H_2 \qquad (4)$$

in which a proton jumps from H_2 to molecule X. This reaction also has a large Langevin cross section [24,25] and is exothermic because the proton affinity of H_2 4.4 eV is smaller than that of other neutral molecules, for example CO (6.1 eV), HCN (7.5 eV), N_2 (4.9 eV), C_2H_2 (6.5 eV), H_2O (7.3 eV), NH_3 (8.5 eV), CH_4 (5.4 eV) [24]. This reaction thus produces HCO^+, H_2CN^+, HN_2^+, $C_2H_3^+$, H_3O^+, NH_4^+, and CH_5^+ efficiently.

III.   Ab Initio Calculations

By the 1930s there was sufficient evidence of the abundance of H_3^+ [25] but "experiments have not been able to tell us more about H_3^+ than that it exists" [26]. J.J. Thomson's query about the valency of hydrogen had not been answered. Henry Eyring was quoted [27] to have said that the H_3^+ problem was "the scandal of modern chemistry". From 1936-1938 Eyring, Hirschfelder, and others have published a series of five papers on the theory of H_3 and H_3^+. In the part V of this series, Hirschfelder [28] showed that (a) triangular structure is more stable than the linear structure, (b) the formation energy of H_3^+ from two hydrogen atoms and a proton is 184 kcal, and therefore (c) the reaction $H_2 + H_2^+ \rightarrow H_3^+ + H$ is very exothermic. After this pioneering work, theory of H_3^+ seems to have laid dormant for more than two decades (except for a few papers quoted by Joshi [29]) until 1963-64 when the burst of activity started using modern computers. It was shown that the equilateral triangle configuration has the energy minimum [30,31] and much more accurate values of the formation energy and bond distance were obtained.

The other important physical quantities to be predicted were the vibrational frequencies of H_3^+. With the equilateral triangle configuration with D_{3h} symmetry, H_3^+ has the totally symmetric A_1' mode ν_1 and the doubly degenerate E' mode ν_2 (Fig. 3); the former is Raman active and the latter is infrared active. The detection of latter was being attempted intensively by several spectroscopists. Theoretical predictions for the vibrational frequencies are listed in Table I together with those for the formation energy and the bond length.

Table I. Theoretical Calculations of H_3^+

	ΔH (hartree)	r (bohr)	ν_1 (cm^{-1})	ν_2 (cm^{-1})	Ref.
Hirschfelder 1938	-1.293	1.79	1550	1100	[28]
Huff and Ellison 1963,65	-1.415	1.76	3454	2326	[32]
Christoffersen 1964	-1.3326	1.6575	3354	2790	[26]
Pearson et al. 1966	-1.3185	1.66	3610	–	[33]
Schwartz and Schaad 1967	-1.3376	1.6504	3301	–	[34]
Borkman 1970	-1.3392	1.640	3450	2850	[35]
Salmon and Poshusta 1973	-1.3412	1.650	3272	2735	[36]
Carney and Porter 1974	-1.3441	1.6500	3471	2814	[37]
Carney and Porter 1976	-1.33519	1.6585	3185	2516	[38]
Dykstra et al. 1978	-1.34278	1.6504	3347	–	[39]
Carney 1980			3221	2546	[40]

Experiment

Petty and Moran 1970			3350	2800	[41]
Oka 1980				2521	[42]

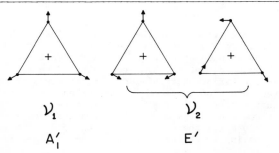

ν_1

A_1'

ν_2

E'

Fig. 3. Normal vibration of H_3^+.

It is seen in Table I that the theoretical prediction progressed in three steps; first the pioneering work of Hirschfelder [28] in 1938, second the use of computers in 1964-74 starting from Christofferson [26] which predicted the structure and formation energy accurately and predicted ν_2 at around 2800 cm^{-1}, and third the introduction by Carney and Porter [38] of the effect of anharmonicity by using Simons-Parr-Finlan expansion [43]. The agreement between their predicted value of ν_2 and the observed value [42] is remarkable. Carney and Porter's theory not only predicted the vibrational frequency but also the rotational constants and the vibration-rotation constants of H_3^+ and D_3^+ [42,44,45]. Carney and Porter also gave predicted frequencies of deuterated species H_2D^+, HD_2^+ [46].

If we use the most recent theoretical formation energy of -1.342784 hartree calculated by Dykstra et. al [39], we obtain the following enthalpy relation where ΔH_e is the equilibrium value and ΔH_o is the zero point value.

		ΔH_e (eV)	ΔH_o (eV)
H_3^+ = 3P + 2e		-36.539	-36.030
= 2H + P		- 9.342	- 8.833
= H_2 + P		- 4.593	- 4.354
= H_2^+ + H		- 6.548	- 6.182
H_3^+ + H = H_2 + H_2^+		- 1.799	- 1.703

Theoretical studies have been done [28,26,47-50] for electronic excited states but none are predicted to be stable with respect to vibration, except for a linear symmetric $^3\Sigma_u^+$ state with the bond length of 2.457 a.u. = 1.300 Å [51,52]. The formation energy of this state was calculated to be -1.11568 a.u. which means that the H_3^+ ($^3\Sigma_u^+$) state is more stable than $H_2^+(^2\Sigma_g^+)$ + H(^{2}S) by only 0.365 eV. The three harmonic vibrational frequencies in this state were predicted to be ν_1= 1233 cm^{-1}, ν_2 = 715 cm^{-1} and ν_3 = 826 cm^{-1} [52].

Theoretical calculations [34,47,53,54] show that the double charged ion H_3^{++} is unstable and explain the negative search for this ion [55]. On the other hand theory shows that the positive hydrogen clusters $H^+(H_2)_n$ [56] are quite stable. A recent paper by Shaefer and others [57] has many references on this problem.

IV. H_3^+ in Discharges

Because of the high efficiency of the ion molecule reaction H_2 + H_2^+ → H_3^+ + H, the H_3^+ ions exist abundantly in discharged hydrogen as suggested from the earlier experiment of Dempster [3]. The direct mass spectrometric monitoring of hydrogen glow discharge confirms the abundance of H_3^+ in glow discharge both in the positive column [58-61] and in the negative glow [56]. The volume density n_+ of positive molecular ions in a D.C. glow discharge can be estimated from that of electrons n_e which is given in terms of the current density I/S and the electron velocity v_e as

$$n_e = I/Sev_e \qquad (5)$$

The electron drift velocity v_e in discharge is well documented. Thus for example using Fig. 61 and 126 of [62] we obtain the electron drift velocity of $v_e \sim 7 \times 10^6$

78 *Takeshi Oka*

cm/sec and the ion volume density of $n_+ = n_e \sim 3\times10^{10}/cm^3$ for the discharge current of $I = 300$ mA in a tube with the diameter of $R = 1.8$ cm and the pressure of $P = 1$ Torr. Since the electron drift velocity is dependent only on the electric field E and independent of the current I, the number of ions can be increased simply by increasing the current [61]. The fraction of H_3^+ among the positive ions increases with the pressure. Yamane [60] reported it to increase from 38% at 0.2 Torr to 62% at 1.6 Torr while others gave much higher values [56,61].

The H_3^+ ion produced in the discharge is accelerated towards the cathode by the D.C. discharge field and to the wall by the ambipolar diffusion through H_2. The mobility of H_3^+ in H_2 has been measured by many people [62-64]. In the early papers [65,66], the results were interpreted as due to the migration of the H_2^+ ion until in 1960 Varney [67] pointed out that it must be due to H_3^+ and this was confirmed by the experiment of Barnes et. al [68]. The zero field reduced mobility K_o which is expressed in terms of the H_3^+ drift velocity v_d, the electric field E (in V/cm), temperature T (in K), and pressure p (in Torr) as

$$K_o = \lim_{E \to o} \frac{v_d}{E} \frac{P}{760} \frac{273}{T} \qquad (6)$$

was obtained to be $K_o = 11.1 \pm 0.6$ cm^2/Vsec by Albritton et. al. [69] which agrees with earlier values [70-73]. It is interesting that this value is smaller than the value for the H^+ drift velocity (which was measured to be 16.0 ± 0.8 cm^2/Vsec [69]) contrary to theoretical prediction [74]. This discrepancy is explained [67] as due to the proton transfer process $H_3^+ + H_2 \to H_2 + H_3^+$ in which a proton jumps from the accelerated H_3^+ to the unaccelerated neutral H_2 to form a new H_3^+ ion and thus lower the drift velocity. A measurement of the temperature dependence of the mobility [64] supports this explanation. The H_3^+ drift velocity is calculated to be $v_d = 1.6\times10^4$ E cm/sec for p = 1 Torr and T = 500K. Thus for an electric field of $E = 10$V/cm, the Doppler shift in the 4µ infrared region is calculated to be $\sim$400 MHz comparable to the thermal Doppler width for H_3^+ of $\sim$300 MHz.

The diffusion coefficient D is related to the mobility $K = v_d/E$ through the Einstein relation [75].

$$D = \frac{kT}{e} K \qquad (7)$$

The Chapman Enskog theory gives [76]

$$D = \frac{3}{16N} (\frac{2\pi kT}{\mu})^{1/2} \frac{1}{\Omega} \qquad (8)$$

where the collision integral Ω is a weighted average of the cross section and N is the total number density of the gas in discharge. If we approximate Ω by the Langevin cross section σ given in Eq. (2) and use $E = 3kT/2$, we obtain from Eqs. (7) and (8)

$$K = \frac{3}{16N} (\frac{3}{2\pi\mu\alpha})^{1/2} \qquad (9)$$

which gives numerically $K_o = 12.48/(\mu\alpha)^{1/2}$ cm^2/Vsec if the polarizability α is in $\mathring{A}^3$ and the reduced mass μ is in atomic mass unit. More detailed study [77] gives 13.87 as the coefficient. Eq. (9) gives $K_o = 12.8$ cm^2/Vsec for H_3^+ in H_2 which agrees with the experimental value; this agreement may be fortuitous because the more sophisticated calculation [74] gives much higher value of 22.0 cm^2/Vsec.

In the discharge the positive ions and electrons diffuse together towards the wall with the same diffusion coefficient D which is twice of D given in Eq. (7) (ambipolar diffusion) [64,78,79]. The velocity of the ambipolar diffusion v$_a$ is given by [64]

$$v_a = -D_a \frac{1}{n(r)} \frac{dn(r)}{dr} \qquad (10)$$

where n(r) = n$_e$(r) = n$_+$(r) is the common number density of electron and positive ion as a function of radial coordinate r and is approximately given [64] by n(r) = AJ$_o$ (2.405 r/R) where R is the radius of the tube. If we approximate Eq. 10 by v$_a^o$= -2D/R we obtain by using the reduced mobility of 11.1 cm^2/Vsec, v$_a$ = 5.33x 10^{-3} T^2/PR. For T = 500K, p = 1 torr and R = 1 we obtain v$_a$ = 1.3x10^3 cm/sec which is comparable to the ion drift velocity for an electric field of $\sim$0.1 Volt/ cm.

The pressure broadening parameter $(\Delta\nu)_p$ can be calculated from $\Delta\nu = n\bar{\sigma}v/2\pi P$ and the Langevin cross section given in Eq. (2) to be

$$\Delta\nu = \frac{e(\alpha/\mu)^{1/2}}{kT} \qquad (11)$$

V. H$_3^+$ in Interstellar Space

When ion-molecule reactions were recognized by Herbst and Klemperer [80] and by Watson [81] as an important mechanism to produce the many interstellar polyatomic molecules observed by radioastronomers in dense molecular clouds [82], the H$_3^+$ ion emerged as one of the most important molecular ions in this scheme. It acts as the protonator of neutral molecules and atoms by the universal reaction H$_3^+$ + X $\rightarrow$ HX$^+$ + H$_2$ given earlier in Eq. (4). Thus the protonated ions HCO$^+$ [83,84] and HN$_2^+$ [85,86] observed by radio-astronomers are produced by this reaction from CO and N$_2$; the direct attachment of protons to these molecules by radiative associa- tion is much less efficient.

The interstellar H$_3^+$ also plays the crucial role in deuterium fractionation through the proton transfer reaction [87],

$$H_3^+ + HD \rightleftharpoons H_2 + H_2D^+ \qquad (12)$$

in which a proton jumps from H$_3^+$ to HD to form H$_2$D$^+$. The effect of such a process in reducing the H$_3^+$ mobility was discussed earlier. If we use the calculated vi- brational frequencies of H$_2$D$^+$ [46] and H$_3^+$ [38] and the observed vibrational fre- quencies of HD and H$_2$ [88] and use the Born-Oppenheimer approximation in the sense that electronic equilibrium energies are the same for H$_3^+$ and H$_2$D$^+$ and for H$_2$ and HD, we obtain ΔH = 56 cm^{-1} = 81 K as the exothermicity of Eq. (12) due to the lowering of the zero-point vibration. This is a small energy but because of the low kinetic energy of interstellar molecules ($\sim$30 K) in the dense cloud, the reac- tion favors the products on the right hand side of Eq. (12). Subsequent reactions starting from H$_2$D$^+$ thus formed produces DCO$^+$ [89] and DN$_2^+$ [90] with anomalous abundances. In the extreme case of the very cool (T = 5$\sim$ 10K) dark dust clouds L63 and L134N the apparent ratio DCO$^+$/HCO$^+$ exceeds unity [91].

The formation of H$_3^+$ in dense clouds starts from the ionization of H$_2$ into H$_2^+$ by cosmic rays. The efficiency of the ionization is given by the cosmic ray ionization flux ζ, the probability per second that an H_2 molecule will be ionized. Both calculation and analysis of observation shows that $\zeta_{H_2} \simeq 10^{-17}$ sec^{-1} [92]. The crucial numbers in this calculation are the total energy density 1.3×10^{-12} erg cm^{-3} and the energy spectrum of cosmic ray [93], the cross section for the H_2 ionization by a proton [94] and the H_2 ionization potential of 15.4 eV [88]. It is seen that a 100 MeV proton can penetrate a cloud with the H_2 column density of 10^{24}/cm^2 and produce $\gtrsim 10^6$ H_2^+ in its path. These ionized H_2^+ is immediately ($\ll$ 1 year) converted to H_3^+ by the ion-molecule reaction of Eq. (1). Thus the ionization of H_2 is the rate determining process for the formation of H_3^+.

The destruction of H_3^+ occurs mainly by the reaction $H_3^+ + CO \rightarrow HCO^+ + H_2$ because of the large abundance of CO in dense clouds. The electronic recombination reaction has much larger cross section but is less important because of the high $CO2/n_e$ ratio of $\sim 10^4$ [80]. Equating the rates of production and destruction we have

$$\frac{[H_3^+]}{[H_2]} = \frac{\zeta}{k[CO]} \qquad (13)$$

where the rate constant k for the destruction reaction is $\sim 10^{-9}$ cm^2 sec^{-1} [95]. Thus we see H_3^+ is more abundant in the carbon depleted clouds in which carbon atoms are frozen on dust grains. These are the clouds for which deuterium fractionation is very effective. For a typical CO abundance of 10 cm^{-3} [80] we obtain the ratio $[H_3^+]/[H_2] \sim 10^{-9}$. More detailed calculations by de Jong, Dalgarno, and Boland [96] gives similar values for dark clouds with a large magnitude of extinction.

Detection of H_3^+ in interstellar space by its spectrum is certainly a very exciting and challenging possibility. Since the electronic excited states are unstable (sec. III), the use of the infrared ν_2 fundamental band recently observed in the laboratory [42] seems to be the most promising method of detection. The laboratory spectrum shows that a H_3^+ column density of 10^{14}/cm^2 is sufficient to see the H_3^+ spectrum in absorption using the state-of-art high resolution infrared telescope. If we use the ratio $[H_3^+]/[H_2] \sim 10^{-9}$ this requires the H_2 column density of $\sim 10^{23}$/cm^2. A more difficult condition to be met is the existence of sufficiently bright infrared sources behind the molecular cloud which is satisfied by very few clouds. More detailed discussions of the infrared detection of H_3^+ can be found in Ref. [97]. One interesting fact is that the small ratio of $[H_3^+]/[H_2] \sim 10^{-9}$ is approximately compensated by the large intensity ratio of the dipole transition to quadrupole transition. Thus observation of H_3^+ in absorption should be as difficult as that of H_2. Observation of infrared emission of H_2 [98] on the other hand is much easier than that of H_3^+ because of the long life time of H_2 in upper state. The detection of interstellar H_3^+ may also be done by using its electron recombination spectrum [99] either in the microwave region or in the optical region. The latter, the beautiful Rydberg spectrum of H_3, has recently been studied in the laboratory by Herzberg and his coworkers both in the visible region [100-102] and in the infrared region [103].

Although H_3^+ is of equilateral triangle configuration without a permanent dipole moment and thus no rotational spectrum is allowed in the normal sense, a microwave spectrum is still possible. Firstly, the deutereated species H_2D^+ has an effective dipole moment of 0.6 Debye due to the mismatch of its centers of mass and charge, and thus leads to a microwave spectrum [104]. Secondly, the centrifugal distortion of H_3^+ causes a small dipole moment and hence pure rotational transition [105]. Such rotational transitions cool the rotational temperature of H_3^+ in relatively short time of $10^4 \sim 10^6$ sec [106]. Such transitions may be observable in the far infrared region.

VI. Other Experiments

Since H$_3^+$ is easy to produce a great many experiments have been done using this molecular ion. Many of them are on chemical reactions using mass spectrometers, ion cyclotron resonance, crossed beams, photoionization etc.   A few of them were discussed in Section II. Many results have been published on the mobility of this ion in various gases as discussed in Section IV.

VIa.   Collision Experiments

Many collision experiments have been reported. Thus Petty and Moran [41] obtained approximate vibrational frequencies of H$_3^+$ experimentally for the first time from inelastic energy losses in the collision of H$_3^+$ with Ne. Their values were in agreement with the then available theoretical values of Christoffersen [26] and of Schwartz and Schaad [34] (see Table I) but later shown to have been too high and was reinterpreted by Carney [107].   Goh and Swan [108] studied collisional dissociation of 700 eV $\sim$ 2KeV H$_3^+$ ion with He and determined the formation energy of H$_3^+$ to be -1.33 ± 0.01 a.u. in agreement with the theoretical value (see Table I). Fragmentation of 10KeV H$_3^+$ colliding with H$_2$ has been studied by Vogler and Meierjohann [109]. Many other papers published on this subject can be traced from references quoted in the papers mentioned above. H$_3^+$ has been used for bombardment of solid surfaces. In particular bombardment of alkali metal surface and back scattering of H$^-$ has been studied in detail [110]. H$_3^+$ bombardment is used for sputtering of solid surfaces such as graphite [111] and silicon carbide [112].

VIb.   Recombination with Electron

The dissociative recombination of H$_3^+$ with electron has been the subject of many studies. As in the case of ion mobility experiments the abundance of H$_3^+$ in hydrogen plasma confused interpretation of the earlier results of dissociative recombination of H$_2^+$ [112]. The more recent results of microwave afterglow [114] and the trapped-ion method [115] gave results which allow clear-cut interpretation. The method of merged beam recently applied to this study by McGowan and others [116] has provided information on the recombination of molecular ions with electrons of very low center of mass energy $\lesssim$ 0.08 eV ($\sim$1000 K) which are important for application to astrophysical problems; it is also capable of providing detailed information on the recombination process because of its high resolution of the electron energy. It has been shown [116] that the electron-ion dissociative recombination cross section is inversely proportional to the center of mass electron energy as predicted by theory also for such a low energy region [117]. The dissociation cross section was reported to be 1.0×10^{-13} cm^2 and its rate constant at 120K as 6.6×10^{-7} cm^3/sec. Recently, the branching ratio of two reactions

$$\text{H}_3^+ + e \rightarrow \text{H} + \text{H} + \text{H} \qquad\qquad (14a)$$

and

$$\rightarrow \text{H}_2 + \text{H} \qquad\qquad (14b)$$

was measured [117a] and the former was shown to dominate in the energy range 0.01-0.50 eV. Herzberg's extensive spectra of H$_3$ and D$_3$ Rydberg states [100-102] provide detailed information on the radiation emitted during the recombination process.

VIc.   Structure of H$_3^+$

Although the equilateral triangle configuration of H$_3^+$ had been well established theoretically for a number of years as we saw in Section IV, the experimental evidence did not arrive until 1978 when Gaillard et al. demonstrated this by a

beautiful and conceptually very simple experiment [118]. In their experiments
carried out in three laboratories, Argonne in U.S.A., Lyon in France, and Rehovot
in Israel, accelerated H_3^+ ions (2 ∿ 4 MeV) were passed through thin carbon foils
(100 ∿ 200 Å). The carbon foil strips the two electrons of H_3^+ and the stripped
three protons emerge. Since the ion is highly accelerated and the carbon foil is
thin, the stripping process occurs adiabatically, that is, the moments and the
relative positions of the three protons remain the same. After passing through
the foil the three now free protons fly apart due to Coulomb repulsion. Observing these three protons at appropriate distances they could literally "see"
individual H_3^+. The photographs of the "exploded" H_3^+ obtained by the Rehovot
group are shown in Fig. 4. Their results confirmed the theoretical prediction
that H_3^+ is indeed equilateral triangle and gave the most probable value of the
internuclear distance to be 0.97 ± 0.03 Å (Argonne), 0.95 ± 0.06 Å (Lyon), and
1.2 ± 0.2 Å (Rehovot). These are thermally averaged values and considerably
larger than the theoretical equilibrium value of 0.873 Å (see Table I). Detailed
discussion of these results can be found in Gemmell's review papers [119,120].
Much work has been done on the behavior of H_3^+ in the thin foil, such as the study
of dissociation observed by Lyman α emission [121], the transmission of H_2^+ and
H_3^+ [122], relative population of H levels as a function of the principal quantum
number n [123], and so on.

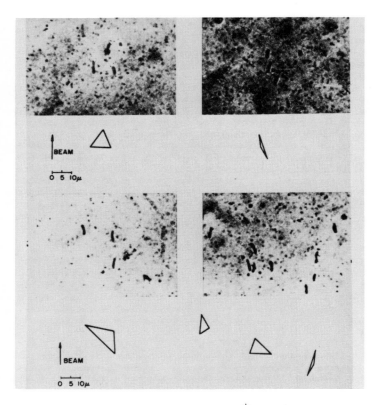

Fig. 4. Photographs of exploded H_3^+. Underneath each
photograph is a reconstructed normal projection   of the
exploded molecular ion.

VII. Spectroscopy of H$_3^+$

In spite of the great many experimental papers published over nearly seven decades since the discovery of H$_3^+$, the definitive spectrum of H$_3^+$ was not discovered until 1980 [42]. In the early years when the great many emission lines of H$_2$ in the optical region were not well understood, some of them were thought to be due to H$_3^+$. Thus as early as in 1927, observation of emission spectrum of triatomic hydrogen was claimed [124,125]. After the successful interpretation of most of the lines in terms of electronic transitions of H$_2$, however, the problem of the H$_3^+$ spectrum was not considered for four decades. The absorption spectrum of H$_3^+$ was also discussed in relation with the diffuse interstellar lines [126], the opacity of the sun [127] and the atmospheric absorption [128].

By 1967 it was clear that since H$_3^+$ does not have stable electronic states (see Section III), the only observable high resolution spectrum of H$_3^+$ should be its vibrational spectrum. Thus in 1967 Herzberg describes his attempt at observing infrared emission lines of the ν_2 degenerate fundamental band of H$_3^+$ from a hydrogen discharge together with J.W.C. Johns [129]. He persisted and in 1979 discovered a group of emission lines near 3600 cm^{-1} which he believed to be from a triatomic deuterium species [130]. In this experiment he used, together with Lew and Sloan, a high resolution infrared Fourier transform spectrometer to collect and analyze infrared emission from a liquid N$_2$ cooled hollow cathode deuterium discharge tube. This spectrum and similar but weaker band from hydrogen discharge were later assigned by Watson to be due to electronic $^2A_1' - ^2E'$ perpendicular band of neutral D$_3$ and H$_3$ [102]. These and other emission spectra of H$_3$ and D$_3$ observed in the optical region [100,101,102] are emitted when H$_3^+$ recombines with an electron in a hollow cathode.

The project of observing the first spectrum of H$_3^+$ was initiated at the end of 1975. The project was based on an estimate that the strongest Doppler limited _absorption_ line of the ν_2-fundamental band of H$_3^+$ would absorb several percent of infrared radiation after traversing 50m through the positive column of a hydrogen discharge. Crucial in this estimate were the electron drift velocity in hydrogen discharge [62], Saporoschenko's observation that H$_3^+$ is the dominating ion in H$_2$ discharge [72], Albritton et al.'s observation that H$_3^+$ thermalizes with H$_2$ [69] (see Section IV), the transition dipole moment of 0.156 Debye calculated by Carney and Porter [37], and the fact well known among laser spectroscopists that the rotational and translational temperatures of molecules in CO$_2$ lasers are not very much higher than that of the cooling water.

Technically the project was based on two recent developments. Firstly, the use of a long positive column of a glow discharge had been introduced to microwave spectroscopy by Woods and his collaborators [131,132]. Their successful observation of protonated ions such as HCO$^+$ [133] and HN$_2^+$ [134] suggested strongly that H$_3^+$ was also observable in glow discharge. Secondly, a widely tunable laser infrared source in the region of 2.2 $\sim$ 4.2μ had been developed by Pine [135] using the difference frequency radiation between Ar laser and dye laser radiation. The wide frequency coverage, and high spectral purity of this source which allowed the search for spectrum with Doppler limited resolution were essential in the successful observation of the H$_3^+$ spectrum. More detail of the apparatus can be found in Ref. [136].

The first spectral line of H$_3^+$ was found on April 25, 1980. By May 9, 10 lines were observed and the approximate spectral pattern and their response to varying discharge condition suggested strongly that indeed the ν_2-band of H$_3^+$ had been obtained but individual lines were not assigned with certainty. The 10 lines were analyzed by Watson on the night of May 9 using his program developed for Herzberg's H$_3$ and D$_3$ spectra. The good fit of the observed spectrum to the theoretical calculation without any extraneous lines except for the hydrogen atom Brackett α line confirmed the observation conclusively. By the beginning of June, 15 lines had been observed and assigned and a paper was sent on June 6. A few weeks later there was a telephone call from Wing saying that he had obtained eight lines of D_3^+ using the entirely different method of ion beam spectroscopy. The two experimental papers [42,45] and the theoretical paper by Carney and Porter [44] were published side by side.

Thirty lines of H_3^+ have been obtained so far. The spectral pattern from a liquid nitrogen cooled plasma and an ice-water cooled plasma are shown in Fig. 5.

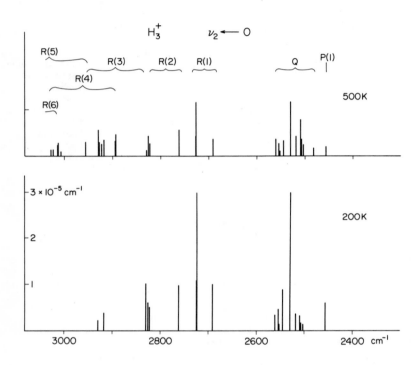

Fig. 5. Observed spectral pattern of the H_3^+ ν_2 fundamental band. The top pattern corresponds to the water cooling of discharge and the bottom to liquid nitrogen cooling.

The observed relative intensities of the absorption lines indicate that the rotational temperatures of H_3^+ in the discharges are about 200K and 500K, respectively. This agrees also with the translational temperature estimated from the width of the Doppler broadened absorption lines. The lack of any obvious regularity or symmetry in the spectral pattern is due to an extraordinary large ℓ-resonance in the ν_2 state with the ℓ-doubling constant as large as $q_\ell = -5.383$ cm^{-1}. The large spacing between the Q branch lines and the R(1) lines is due to the missing of the J=0, K=0 level because of the Pauli principle, and provides the best experimental evidence that the equilibrium configuration of H_3^+ is indeed equilateral triangle and the three protons are equivalent.

There have been several important developments in the last one year. Wing and his collaborators observed more lines of D_3^+ although conclusive assignments are yet to be made. They also reported several lines of H_2D^+ [137] which shows interesting K-doubling structure. Observation of <u>emission</u> infrared spectrum of H$_3^+$ has been claimed by Ding and his collaborators [138] although the assignments are not conclusive. The relation of such spectra and the many emission lines of H$_2$ and H$_3$ in the infrared region [139] remains to be seen. Carrington, Buttenshaw, and Kennedy [140] observed extremely rich spectral lines of H$_3^+$ near the dissociation limit using the ion-beam spectroscopy. This work is described by Carrington in this book.

The successful observation of the H$_3^+$ spectrum has opened various exciting possibilities. Firstly, we can apply the same method to observation of spectra of other molecular ions. In particular many protonated ions, for which very little spectroscopic information is available, can be studied. Spectra of HeH$^+$ [141] and NeH$^+$ [142] have been reported using the same spectrometer. Applications of this method to other protonated ions such as H$_3$O$^+$, NH$_4^+$, HCO$^+$, etc. will be done although it is much more difficult because of the strong absorption lines of neutrals in the discharge. Secondly, the observed infrared spectrum of H$_3^+$ gives us a means to monitor this ion in discharge and in chemical reaction. Thus we should be able to measure the mobility of this ion <u>in situ</u> in a discharge from the Doppler shift of the spectral line, and its concentration and translational, rotational and vibrational temperature from relative and absolute intensities of the absorption lines. Thirdly, the observed line positions allow us to look for this fundamental ion in space [97]. All these will be attempted actively in the next several years.

I would like to thank Nathan N. Haese for his critical reading of this paper.

References

1. J.J. Thomson, Phil. Mag. <u>24</u>, (1912) 209

2. J.J. Thomson, "Rays of Positive Electricity", Longmans, Green, New York (1916)

3. A.J. Dempster, Phil. Mag. <u>31</u>, (1916) 438

4. H.C. Urey and G.K. Teal, Revs. Modern Phys. <u>7</u>, (1935) 34

5. J.J. Thomson, Nature, <u>133</u>, (1934) 280

6. J.J. Thomson, Phil. Mag. <u>17</u>, (1934) 1025

7. J.J. Thomson, "Pieter Zeeman Verhandelingen", pp. 355-363, Martinus Nijhoff (1935)

8. J.L. Franklin, "Ion-Molecule Reactions", Plenum Press, New York (1972)

9. H.D. Smyth, Phys. Rev. <u>25</u>, (1925) 452

10. T.R. Hogness and E.G. Lunn, Phys. Rev. <u>26</u>, (1925) 44

11. E.W. McDaniel, V. Cerniak, A. Dalgarno, E.E. Ferguson, L. Friedman, "Ion-Molecule Reactions", Wiley-Interscience, New York (1970)

12. "Ion-Molecule Reactions", edit. J.L. Franklin, Plenum Press, New York (1972)

13. E.E. Ferguson, Ann. Rev. Phys. Chem. <u>26</u>, (1975) 17

14. S.L. Anderson, F.A. Houle, D. Gerlich and Y.T. Lee, J. Chem. Phys. <u>75</u>, (1981) 2154

86 Takeshi Oka

15. H. Eyring, J.O. Hirschfelder, and H.S. Taylor, J. Chem. Phys. 4, (1936) 479

16. A.B. Lees and P.K. Rol, J. Chem. Phys. 61, (1974) 4444

17. C.H. Douglas, D.J. McClure, and W.R. Gentry, J. Chem. Phys. 67, (1977) 4931

18. R.D. Poshusta and D.F. Zelik, J. Chem. Phys. 58, (1973) 118

19. D.P. Stevenson and D.O. Schissler, J. Chem. Phys. 29, (1958) 282

20. P. Langevin, Ann. Chim. Phys. 5, (1905) 245

21. J.O. Hirschfelder, C.F. Curtiss and R.B. Bird, "Molecular Theory of Gases and Liquids", John Wiley and Sons, New York (1967)

22. T. Su and M.T. Bowers, Int. J. Mass Spectrom. Ion Phys. 17, (1975) 309

23. L.D. Landau and E.M. Lifshitz, "Mechanics" Nauka, Moscow (1965)

24. R. Walder and J.L. Franklin, Int. J. Mass Spectrom. Ion Phys. 36, (1980) 85

25. H.D. Smyth, Rev. Mod. Phys. 3, (1931) 347

26. R.E. Christoffersen, J. Chem. Phys. 41, (1964) 960

27. G. Handler and J.R. Arnold, J. Chem. Phys. 27, (1957) 144

28. J.O. Hirschfelder, J. Chem. Phys. 6, (1938) 795

29. B.D. Joshi, J. Chem. Phys. 44, (1966) 3627

30. R.E. Christoffersen, S. Hagstrom, and F. Prosser, J. Chem. Phys. 40, (1964) 236

31. H. Conroy, J. Chem. Phys. 40, (1964) 603; 41, (1964) 1341

32. F.O. Ellison, N.T. Huff and J.C. Patel, J. Am. Chem. Soc. 85, (1963); N.T. Huff and F.O. Ellison, J. Chem. Phys. 42, (1965) 364

33. A.G. Pearson, R.D. Poshusta, and J.C. Browne, J. Chem. Phys. 44 (1966) 1815

34. M.E. Schwartz and L.J. Schaad, J. Chem. Phys. 47, (1967) 5325

35. R.F. Borkman, J. Chem. Phys. 53, (1970) 3153

36. L. Salmon and D. Poshusta, J. Chem. Phys. 59, (1973) 3497

37. G.D. Carney and R.N. Porter, J. Chem. Phys. 60, (1974) 4251

38. G.D. Carney and R.N. Porter, J. Chem. Phys. 65, (1976) 3547

39. C.E. Dykstra, A.S. Gaylord, W.D. Gwinn, W.C. Swope, and M.F. Schaeffer III, J. Chem. Phys. 68, (1978) 3951; C.E. Dykstra and W.C. Swope, J. Chem. Phys. 70, (1979) 1

40. G.D. Carney, Mol. Phys. 39, (1980) 923

41. F. Petty and T.F. Moran, Chem. Phys. Lett. 5, (1970) 64

42. T. Oka, Phys. Rev. Lett. 45, (1980) 531

43. G. Simons, R.G. Parr, and J.M. Finlan, J. Chem. Phys. $\underline{59}$, (1973) 3229

44. G.D. Carney and R.N. Porter, Phys. Rev. Lett. $\underline{49}$, (1980) 537

45. J.T. Shy, J.W. Farley, W.E. Lamb, Jr., and W.H. Wing, Phys. Rev. Lett. $\underline{45}$, (1980) 535

46. G.D. Carney and R.N. Porter, Chem. Phys. Lett. $\underline{50}$, (1977) 327

47. H. Conroy, J. Chem. Phys. $\underline{51}$, (1969) 3979

48. K. Kawaoka and R.F. Borkman, J. Chem. Phys. $\underline{54}$, (1971) 4234

49. C.W. Bauschlicher, S.V. O'Neal, R.K. Preston, H.F. Shaefer, and C.F. Bender, J. Chem. Phys. $\underline{59}$, (1973) 1286

50. A.A. Wu and F.O. Ellison, J. Chem. Phys. $\underline{48}$, (1968) 1491, 5032

51. L.J. Schaad and W.V. Hicks, J. Chem. Phys. $\underline{61}$, (1974) 1934

52. R. Ahlrich, C. Votava, and C. Zirz, J. Chem. Phys. $\underline{66}$, (1977) 2771

53. C.M. Rosenthal and E.B. Wilson, Jr., J. Chem. Phys. $\underline{53}$, (1970) 388

54. R.L. Somorjai and C.P. Yue, J. Chem. Phys. $\underline{53}$, (1970) 1657

55. E.H. Berkowitz and H. Stocker, J. Chem. Phys. $\underline{55}$, (1971) 4606

56. P.F. Dawson and A.W. Tickner, J. Chem. Phys. $\underline{37}$, (1962) 672

57. Y. Yamaguchi, J.F. Gaw, H.F. Shaefer III, J. Chem. Phys. in press

58. H.D. Beckey and H. Dreeskamp. Z. Naturforsch. $\underline{99}$, (1954) 735

59. H. Dreeskamp, Z. Naturforsch. $\underline{129}$, (1957) 876

60. M. Yamane, J. Chem. Phys. $\underline{49}$, (1968) 4624

61. N.N. Haese, Thesis, University of Wisconsin (1981)

62. A. von Engel, "Ionized Gases" Oxford Press (1965)

63. E.W. McDaniel, "Collision Phenomena in Ionized Gases" John Wiley and Sons, New York (1964)

64. E.W. McDaniel and E.A. Mason, "The Mobility and Diffusion of Ions in Gases", John Wiley and Sons, New York (1973)

65. N.E. Bradbury, Phys. Rev. $\underline{40}$, (1932) 508

66. A.M. Tyndall, "The Mobility of Positive Ions in Gases" Cambridge University Press, New York (1938)

67. R.N. Varney, Phys. Rev. Lett. $\underline{5}$, (1960) 559

68. W.S. Barnes, D.W. Martin and E.W. McDaniel, Phys. Rev. Lett. $\underline{6}$, (1961) 110

69. D.L. Albritton, T.M. Miller, D.W. Martin and D.W. McDaniel, Phys. Rev. $\underline{171}$, (1968) 94

70. L.M. Chanin, Phys. Rev. $\underline{123}$, (1961) 526

71. G. Sinnott, Phys. Rev. 136, (1964) A370

72. M. Saporoschenko, Phys. Rev. 139, (1965) A349

73. J. Dutton, F.L. Jones, W.D. Rees and E.M. Williams, Phil. Trans. Roy. Soc. London A259, (1966) 299

74. E.A. Mason and J.T. Vanderslice, Phys. Rev. 114, (1959) 497

75. A. Einstein, Ann. Phys. 17, (1905) 549

76. S. Chapman and T.G. Cowling, "The Mathematical Theory of Non-Uniform Gases" Cambridge University Press, New York (1970)

77. G. Heiche and E.A. Mason, J. Chem. Phys. 53 (1970) 4687

78. W. Schottky, Phys. Z. 25, (1924) 635

79. E.M. Lifshitz and L.P. Pitaevsky, "Physical Kinetics" Pergamon Press, (1981)

80. E. Herbst and W. Klemperer, Ap. J. 185, (1973) 505

81. W.D. Watson, Ap. J. 183, (1973) L17

82. P. Thaddeus, Phil. Trans. R. Soc. Lond. A303, (1981) 469

83. D. Buhl and L.E. Snyder, Nature 227, (1970) 862

84. W. Klemperer, Nature 227, (1970) 1230

85. B.E. Turner, Ap. J. 193 (1974) L83

86. S. Green, J.A. Montgomery, Jr., and P. Thaddeus, Ap. J. 193, (1974) L89

87. W.D. Watson, Rev. Modern Phys. 48, (1976) 513

88. K.P. Huber and G. Herzberg, "Constants of Diatomic Molecules", Van Nostrand Reinhold Co., New York (1979)

89. J.M. Hollis, L.E. Snyder, F.J. Lovas and D. Buhl, Ap. J. 209, (1976) L83

90. L.E. Snyder, J.M. Hollis, D. Buhl, and W.D. Watson, Ap. J. 218, (1977) L61

91. B.E. Turner and B. Zuckerman, Apl. J. 225, (1978) L79

92. L. Spitzer, Jr. "Physical Processes in the Interstellar Medium", John Wiley and Sons, New York (1978)

93. J.A. Simpson, private communication

94. D.R. Bates and G. Griffing, Proc. Phys. Soc. London A66, (1953) 961

95. J.A. Burt, J.L. Dunn, M.J. McEwan, M.M. Sutton, A.E. Roche, and M.I. Schiff, J. Chem. Phys. 52, (1970) 6062

96. de Jong, A. Dalgarno, and W. Boland, Astron. Astrophys. 91, (1980) 68

97. T. Oka, Phil. Trans. R. Soc. Lond. A303, (1981) 543

98. T.N. Gautier, III, U. Fink, R.P. Treffers, and H.P. Larson, Apl. J. 207, (1976) L129

99. E.E. Salpeter and R.C. Malone, Ap. J. <u>167</u>, (1971) 27

100. G. Herzberg, J. Chem. Phys. <u>70</u>, (1979) 4806

101. I. Dabrowski and C. Herzberg, Can. J. Phys. <u>58</u>, (1980) 1238

102. G. Herzberg and J.K.G. Watson, Can. J. Phys. <u>58</u>, (1980) 1250

103. G. Herzberg, H. Lew, J.J. Sloan, and J.K.G. Watson, Can. J. Phys. <u>59</u>, (1980) 428

104. A. Dalgarno, E. Herbst, S. Novick and W. Klemperer, Ap. J. <u>183</u>, (1973) L131

105. T. Oka, T. Shimizu, F.O. Shimizu, and J.K.G. Watson, Ap. J. <u>165</u>, (1971) L15

106. F.S. Pan and T. Oka, unpublished

107. G.D. Carney, J. Chem. Phys. <u>71</u>, (1979) 1036

108. S.C. Goh and J.B. Swan, Phys. Rev. <u>A24</u>, (1981) 1624

109. M. Vogler and B. Meierjohann, J. Chem. Phys. <u>69</u>, (1978) 2450

110. P.J. Schneider, K.H. Berkner, W.G. Graham, R.V. Pyle and J.W. Stearns, Phys. Rev. <u>23B</u>, (1981) 941

111. R. Yamada, K. Nakamura and M. Saidoh, J. Nucl. Mater. <u>98</u>, (1981) 167

112. K. Sone, M. Saidoh, K. Nakamura, R. Yamada, Y. Murakami, T. Shikama, M. Fukutomi, M. Kitagawa, M. Okuda, J. Nucl. Mater. <u>98</u>, (1981) 270

113. J.N. Bardsley and M.A. Biondi, Adv. At. Mol. Phys. <u>6</u>, (1970) 1

114. M.T. Leu, M.A. Biondi, and R. Johnson, Phys. Rev. <u>A8</u>, (1973) 413

115. D. Mathur, S.U. Khan, and J.B. Hasted, J. Phys. B, <u>11</u>, (1978) 3615

116. J. Wm. McGowan, R. Caudano, and J. Keyser, Phys. Rev. Lett. <u>36</u>, (1976) 1447

117. J. Wm. McGowan, P.M. Mul, V.S. D'Angelo, J.B.A. Mitchell, P. Defrance, and H.R. Froelich, Phys. Rev. Lett. <u>42</u>, (1979) 373

117a. J.B.A. Mitchell, J.L. Forand, C.T. Ng, D.P. Levac, R.E. Mitchel, P.M. Mul, W. Clacys, A. Sen and J. Wm. McGowan, to be published.

118. M.J. Gaillard, D.S. Gemmell, G. Goldring, I. Levine, W.J. Pietsch, J.C. Poizat, A.J. Ratkowski, J. Remillieux, Z. Vager, and B.J. Zabransky, Phys. Rev. <u>A17</u>, (1978) 1797

119. D.S. Gemmell, Chemical Reviews, <u>80</u>, (1980) 301

120. D.S. Gemmell, "Physics of Electronic and Atomic Collisions", S. Datz editor, North-Holland Publishing Co., p. 841 (1982)

121. H.G. Berry, T.J. Gay and R.L. Brooks, Ann. Isr. Phys. Soc. <u>4</u>, (1981) 217; IEEE Trans. Nucl. Sci. <u>NS28</u>, (1981) 1174

122. N. Cue, N.V. DeCastro-Faria, M.J. Gaillard, J.C. Poizat, and J. Remillieux, Phys. Lett. <u>72A</u>, (1979) 104

123. B. Andresen, S. Hultberg, B. Jelenkovic, L. Liljeby, S. Mannervik, and E. Veje, Phys. Scr. 19, (1979) 335

124. H.S. Allen and I. Sandeman, Proc. Roy. Soc. A114, (1927) 293

125. C.J. Brasefield, Phys. Rev. 31, (1928) 52

126. G. Herzberg, Comment, Pontif, Acad. Sci. 2, (1968) 8

127. J.L. Linsky, Solar Phys. 11, (1970) 198

128. R.W. Patch, J. Chem. Phys. 57, (1972) 2594

129. G. Herzberg, Trans. Roy. Soc. Can. 5, (1967) 3

130. G. Herzberg, "Interstellar Molecules" IAU Symposium; B.H. Andrew (ed.), D. Reidel Publishing Co., Dordrecht, Holland, p. 231 (1980)

131. R.C. Woods, Rev. Sci. Instrum. 44, (1973) 282

132. T.A. Dixon and R.C. Woods, Phys. Rev. Lett. 34, (1975) 61

133. R.C. Woods, T.A. Dixon, R.J. Saykally and P.G. Szanto, Phys. Rev. Lett. 35, (1975) 1269

134. R.J. Saykally, T.A. Dixon, T.G. Anderson, P.G. Szanto, and R.C. Woods, Ap. J. Lett. 205, (1976) 101

135. A.S. Pine, J. Opt. Soc. Amer. 64, (1974) 1683; 66, (1976) 97

136. T. Oka, "Laser Spectroscopy V", p. 320, Edit. A.R.W. McKella, T. Oka, and B.P. Stoicheff, Springer-Verlag, Berlin, Heidelberg, New York (1981)

137. J.T. Shy, J.W. Farley, and W.H. Wing, Phys. Rev. A24, (1981) 1146

138. U. Steinmetzger, A. Redpath and A. Ding, Seventh Colloquium on High Resolution Molecular Spectroscopy, 14-18 September, Reading, 1981

139. G. Herzberg and Ch. Jungen, J. Mol. Spectrosc. in press

140. A. Carrington, J. Buttenshaw, and R. Kennedy, Mol. Phys. 45, (1982) 753

141. P. Bernath and T. Amano, Phys. Rev. Lett. 48, (1982) 20

142. M. Wong, P. Bernath, and T. Amano, J. Chem. Phys. 77, (1982) 693

Molecular Ions: Spectroscopy, Structure and Chemistry
Terry A. Miller and V.E. Bondybey (editors)
© North-Holland Publishing Company, 1983

SPECTROSCOPY OF MOLECULAR IONS
IN NOBLE GAS MATRICES

Lester Andrews

Department of Chemistry
University of Virginia
Charlottesville, Virginia 22901
U.S.A.

Matrix-isolated molecular ions have been prepared for spectroscopic study by chemical and physical methods. Chemical reactions produce bound ion pairs like $Li^+O_2^-$, and photophysical processes give isolated cations and anions trapped in the matrix. Mercury arc photolysis of matrix samples containing isolated cations and anions rearranges, dissociates or neutralizes isolated cations; the latter requires photodetachment of electrons from anions in the matrix.

INTRODUCTION

Molecular ions are of considerable chemical and physical interest for comparing the spectroscopic and bonding properties of neutral molecules with their positive and negative molecular ions. The study of molecular ions in the gas phase by photoelectron, photo-ionization mass and ion cyclotron resonance spectroscopies can be complemented by infrared and optical absorption spectra of the molecular ion trapped in a solid inert gas host. Further ion studies with tunable infrared lasers will be greatly aided by the vibrational data obtained for molecular ions in noble gas solids. The ion-matrix interaction is of fundamental and practical interest as matrix spectra of ions are related to the gas phase.

Charged species in matrices form two general classes described in the literature as "isolated" and "chemically bound" with respect to the counterion. The first ionic species characterized in matrices, $Li^+O_2^-$, is of the later type where the lithium cation and the superoxide anion are Coulombically bound together [1,2], and charge-transfer occurs because this electrostatic attraction more than makes up for the difference between the ionization energy of lithium and the electron affinity of oxygen. The next molecular ions identified in matrices, $B_2H_6^-$ and C_2^-, are of the "isolated" type where the cation is separated by an undetermined number of matrix atoms from the anion [3,4]. These ionic systems have been characterized as "Coulomb ion pairs" which exist because of essentially zero overlap between the wavefunctions for the electron on the recipient molecule and the cation that provided the electron [5]. Clearly, the formation of ions of the "isolated" type requires that the ionization energy of a precursor atom or molecule in the matrix host be supplied by an external source, which produces a cation and an electron. The electron may be trapped by another molecule or fragment elsewhere in the matrix, forming a negative ion.

These early studies of $Li^+O_2^-$, $B_2H_6^-$ and C_2^- employed different methods of production, chemical reaction, photoionization of sodium in the sample or photoionization of the C_2H_2 precursor with and without added cesium, respectively. Different matrix spectroscopic techniques, infrared absorption, electron spin resonance, and optical absorption, were used for detection. Experimental methods for the production and investigation of molecular ions in noble gas matrices will be described in the next section.

EXPERIMENTAL METHODS

The most straightforward method for producing ions in matrices is to react alkali metal atoms with an electron acceptor molecule during condensation with excess argon at 15-20 K. This method, developed by Andrews and Pimentel in a study of the lithium atom-nitric oxide reaction [6], necessarily gives a chemically bound ion pair due to the Coulombic attraction between anion and cation required to sustain charge transfer. In the case of $M^+O_2^-$ species, vibrational spectra clearly demonstrate an M^+ effect on the O_2^- stretching mode and the bound nature of the ion pair [7].

The first technique used to prepare isolated ions involved mercury arc photoionization of sodium atoms in the sample to provide electrons for capture by molecules elsewhere in the matrix [3]. This method, which has been discussed by Kasai [8], produces "Coulomb ion pairs" with Na^+ separated from the anion by at least several layers of argon atoms such that the anion is not affected by the alkali cation.

Another important technique for preparing isolated ions employed vacuum-ultraviolet photoionization of a precursor molecule with a LiF-filtered hydrogen or argon resonance lamp. In the C_2H_2 experiments of Milligan and Jacox, hydrogen-resonance photolysis produced C_2 absorption at 238.2 nm and new bands at 520.6 and 472.5 nm. The latter absorptions were enhanced in subsequent studies with cesium atoms added to the sample, which supported their identifi- cation as C_2^- [4]. Brus and Bondybey later explained that the C_2^- anion was produced as a "Coulomb ion pair" with C_2H_2 by direct photoionization of C_2H_2, with a red shift in the ionization energy owing to solvation of the charged products in the matrix [9] followed by electron capture by the C_2 photolysis product [5].

Radiolysis, a well-known method for producing free radicals, has been used to generate ions for infrared matrix-isolation study. Proton currents of 20-40 μA were extracted from a radiofrequency discharge through hydrogen, accelerated to 2keV, and directed at a condensing matrix sample containing a reagent molecule [10]. The first studies on CCl_4 produced CCl_3^+ at 1037 cm^{-1}, which was stable to photolysis, and photosensitive absorptions at 927, 502 and

374 cm^{-1}, which were attributed to CCl_4^+ and Cl_3^+, in addition to a very large yield of CCl_3 radical at 898 cm^{-1} [10,11].

Recently, argon discharge tubes have become a major method for producing charged species in matrices. Argon is excited by a microwave discharge while passing through a quartz tube with a 1-mm orifice directed at the sample. Jacox has proposed that the charged products are formed upon collision with excited metastable argon atoms from the discharge tube [12] whereas Wight, et al. have attributed the formation of charged species to photoionization by argon resonance radiation eminating from the open discharge tube owing to the absence of charged products when the 1-mm orifice was placed in the side of the discharge tube [13]. Support for this proposal is found in studies by Smardzewski using a capillary array in the discharge tube to deactivate metastable argon atoms which reduced the charged product yield by the transmission of the array [14].

Further studies in this laboratory employed 3-mm and 10-mm orifice discharge tubes; the yield of charged species increased on going from the 1-mm to the 3-mm orifice tube, and the 10-mm open tube produced the highest yield of photolytically stable charged species and a lower yield of photosensitive charged species [15]. The 10-mm open discharge tube operates at lower argon pressure which provides enhanced emission from Ar^+ and higher argon excited states; this tube functions as a windowless resonance lamp with the major output at 11.6-11.8 eV and substantial radiation between 13 and 15 eV [16]. The windowless discharge lamp and sample configuration used in this laboratory is shown in Figure 1. Reagent molecules are subjected to the intense vacuum ultraviolet radiation from the open discharge tube during condensation with excess argon at 15 K, which traps molecular ions for spectroscopic study.

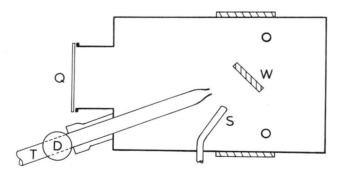

Figure 1. Vacuum vessel base cross section used for absorption spectroscopy of molecular ions produced in matrix photoionization experiments. The diagram shows the position of the rotatable 15 K cold window--W, optical windows--O, gas sample deposition line--S, open discharge tube vacuum ultraviolet source--T, microwave discharge cavity--D, and quartz photolysis window--Q.

Another chemical method for producing new charged species, developed by Ault and Andrews, involves the reaction of salt vapor with a suitable precursor molecule during condensation with excess argon at 15 K. Matrix reactions with NaCl and Cl_2 produced $Na^+Cl_3^-$ and with CsCl and HCl gave $Cs^+HCl_2^-$ [17-19]. This technique is particularly useful for the synthesis of less stable polyhalides like $Cs^+F_3^-$, $Cs^+BrF_2^-$ and $Cs^+SiF_5^-$ for spectroscopic study [20-22].

An electrical discharge technique for producing and trapping charged species in solid argon has been very recently developed by Kelsall and Andrews [22]. Potentials in the +30 to +2000 volts d.c. range were applied to a stainless steel electrode positioned in the center of the cold window and a discharge was maintained in the condensing matrix gas. This method was particularly effective for the production and trapping of CF_3^+.

In recent years, resonance two-photon ionization has been used to produce cations for gas phase spectroscopic study [24]. This technique is particularly efficient owing to the unusually high cross-section for reabsorption into the ionization continuum for favorable cases, such as the naphthalene and biphenyl aromatic molecules. Although the gas phase technique employs pulsed laser excitation tuned into a particular excited vibronic line, continuous focussed high-pressure mercury arc irradiation of certain matrix isolated molecules in a broad absorption band for a longer period performs two-photon ionization of matrix-isolated molecules; added halocarbons trap the photoelectrons, which prevents neutralization and allows the cations to remain isolated in the matrix [25].

Spectroscopic measurements on matrix-isolated species have typically involved conventional spectroscopic methods with special sampling techniques developed to obtain the measurement with a particular spectroscopy. Thus, infrared and optical absorption experiments require salt and sapphire or quartz optics for transmission of the examining radiation. Laser-Raman or laser-induced fluorescence experiments use a tilted metal wedge for sample collection; the laser beam is directed vertically at the sample surface and scattered or emitted light is focussed into a monochromator and detection system for analysis [26,27].

We now turn to the spectroscopy and characterization of a number of interesting molecular ions in noble gas matrices using these techniques.

ALKALI METAL SUPEROXIDES - $M^+O_2^-$

The five alkali superoxide molecules were formed by cocondensation reaction of the elements and characterized by sharp, weak intraionic $(O-O)^-$ stretching modes and strong symmetric and antisymmetric interionic $M^+-O_2^-$ stretching modes in their infrared spectra; the vibrational assignments were verified by isotopic substitution [2,7,28]. The isosceles triangular structure for $Li^+O_2^-$ was dictated by the scrambled oxygen isotopic spectrum, which

demonstrated equivalent oxygen atoms, and the $Li^+O_2^-$ ion-pair
characterization was first suggested from agreement between its O-O
stretching mode and the Raman value for O_2^- from crystals [1,2].
The ionic model for $Li^+O_2^-$ was confirmed by the matrix-Raman
spectrum of $Li^+O_2^-$, which gave a very strong intraionic $(O-O)^-$
stretching mode signal within 0.5 cm^{-1} of the weak infrared absorption with identical isotopic data [26,29]. Infrared and Raman
spectra of the mixed oxygen isotopic $Li^+O_2^-$ species are compared in
Figure 2. Further support for an $M^+O_2^-$ ionic model of polarizable
ion pairs was derived from the small M^+ effect on the O_2^-
vibrational frequency, ν_1, which ranged from 1094 cm^{-1} for $Na^+O_2^-$ to
1114 cm^{-1} for $Cs^+O_2^-$ as given in Table I. Note the larger
dependence of the interionic ν_2 and ν_3 modes on M^+.

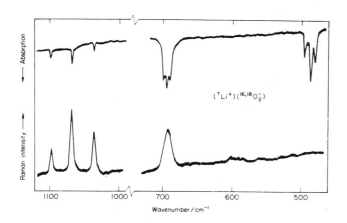

Figure 2. Infrared and Raman spectra of lithium superoxide, $Li^+O_2^-$
using lithium-7 and 30% $^{18}O_2$, 50% $^{16}O^{18}O$, 20% $^{16}O_2$ with Ar/O_2=100.
Raman spectrum recorded using 200 mW of 488.0 nm excitation and
long-wavelength pass dielectric filter in 1000 cm^{-1} region.

Complementary ESR studies have verified the ionic nature of these
$M^+O_2^-$ molecules [30,31]. The magnitude of the sodium hyperfine
splitting in $Na^+O_2^-$ demonstrated the bound ion-pair nature of the
species and that the Na^+ ion is located equidistant from the two
oxygen atoms. Optical absorption studies on $M^+O_2^-$ matrix systems
exhibited an absorption near 250 nm, in excellent agreement with the
spectrum of O_2^- doped into alkali halide crystals [32]. Finally,
the LiO_2 molecule has been the subject of a number of theoretical
calculations [33] which have verified the charge-transfer model and
isosceles triangular geometry for LiO_2 deduced from the original
infrared matrix spectrum.

Table I.   Fundamental frequencies (cm$^{-1)}$) assigned to the intraionic and interionic modes of the alkali metal superoxide molecules
in solid argon at 15 K.

Molecule	v_1	v_2	v_3
$^6LiO_2$	1097.4	743.8	507.3
$^7LiO_2$	1096.9	698.8	492.4
NaO_2	1094	390.7	332.8
KO_2	1108	307.5	---
RbO_2	1111.3	255.0	282.5
CsO_2	1115.6	236.5	268.6

OZONIDES - $M^+O_3^-$ and O_3^-

Following the alkali metal-oxygen matrix reactions, an extensive
study of alkali metal-ozone reactions was performed in this
laboratory [34,35].   The infrared spectra were characterized by very
intense bands near 800 cm^{-1} depending upon the alkali atom, and
weaker bands near 600 cm^{-1} for the heavier alkali metal reagents.
The very strong 800 cm^{-1} bands were assigned to v_3 and the weak 600
cm^{-1} absorptions were attributed to v_2 of O_3^- in the $M^+O_3^-$ species,
produced by the charge-transfer alkali metal-ozone reaction.   The
small variation in v_3 ozonide modes as a function of M^+ demonstrates
that the cation is adjacent to the ozonide ion as required by the
energetics of the charge-transfer reaction.   Similar ESR studies
done on the sodium-ozone reaction product observed a quartet
subsplitting due to sodium hyperfine structure which verified the
ionic nature of the $Na^+O_3^-$ species [36].   Furthermore, a comparison
of the calculated and experimental anisotropic part of the sodium
hyperfine splitting tensor determined the location of the Na^+ ion to
be above the O_3^- plane and equidistant from the two terminal oxygen
atoms; the latter conclusion was also reached from the infrared
spectrum of scrambled oxygen isotopic $Na^{+\ 16,18}O_3^-$ species [34].

The infrared v_3 absorption and a visible electronic band system for
the ozonide ion were reported by Jacox and Milligan using the
photoreaction of N_2O, O_2 and alkali atoms [37].   These workers
proposed a mechanism involving the photoproduction and reaction of
O^- with O_2 to give O_3^-.   Andrews and Tevault have argued against
this mechansim in favor of O atom reaction with $M^+O_2^-$ to give $M^+O_3^-$
since the bound ion-pair $M^+O_3^-$ species and not "isolated" O_3^- was in
fact formed in the photolysis experiments [38].   The intimate
involvement of alkali metal atoms must be considered in interpreting
alkali metal matrix-reaction systems.

The "isolated" O_3^- ion has been produced by proton radiolysis and
argon resonance photoionization of argon-oxygen samples [39,13].   It
is interesting to note that the v_3 fundamental of O_3^- without an
adjacent cation, 804.3 cm^{-1}, is in agreement with values for the

$Na^+O_3^-$, $Ca^+O_3^-$, $Sr^+O_3^-$ and $Ba^+O_3^-$ species but this mode for ion pairs with Li, K, Rb, Cs and Mg varies from 787 to 844 cm^{-1} [34,40,41]. It has been suggested that the Na^+ cation, in the out-of-plane position determined from the ESR study, may not perturb the O_3^- vibration whereas coplanar $M^+O_3^-$ arrangements may lead to slight metal ion perturbations of the O_3^- vibrations [39].

Matrix Raman studies on $M^+O_3^-$ are of interest in part because of resonance enhancement of the scattering intensity. Absorption spectra of $M^+O_3^-$ species in noble gas matrices revealed strong vibronic absorptions in the visible region. The 12-component series observed for the sodium ozonide ion-pair in solid argon yielded a band origin at 18 182 cm^{-1}, and harmonic vibrational frequencies of 878 $\pm$ 8 and 834 $\pm$ 8 cm^{-1} for the oxygen-16 and oxygen-18 species, respectively [42]. Accordingly, argon and krypton ion laser lines are ideal for a resonance Raman study of the $M^+O_3^-$ species in solid argon where the low temperature matrix retards decomposition of photosensitive molecules and quenches fluorescence so that the resonance Raman spectrum can be observed. Blue excitation of $Cs^+O_3^-$ at 488.0 nm yielded a regular progression of fundamental and overtone bands at 1018, 2028, 3024 and 4014 cm^{-1} with decreasing intensities. The $Cs^{+18}O_3^-$ species produced a progression out to $5\nu_1$; bands were observed at 962, 1915, 2859, 3795 and 4724 cm^{-1} [35]. The regularly decreasing intensity pattern for an overtone progression is characteristic of the resonance Raman effect.

The use of different exciting lines for a $Na^+O_3^-$ sample produced a profile similar to the absorption spectrum. The 647.1 nm line gave a weak 1011 cm^{-1} fundamental and excitation at 568.2 nm produced a strong fundamental at 1011 cm^{-1} and a weak overtone at 2013 cm^{-1}. The 530.9 nm line produced an intense fundamental, first overtone and a weak second overtone at 3001 cm^{-1}. Excitation at 514.5 nm yielded intense fundamental, intense first and second overtones and a weak third overtone at 3977 cm^{-1}; the 488.0 nm line produced a similar spectrum. The 457.9 nm line gave the intense fundamental and two moderately intense overtones. This increase in overtone intensity relative to fundamental intensity as the exciting wavelength enters the electronic absorption is characteristic of resonance Raman spectra.

DIHALIDES - $M^+X_2^-$

The alkali dihalide species are of interest as transient inter-mediates in the alkali metal atom-halogen molecule reaction and for comparison to V centers in irradiated alkali halide crystals. Alkali metal reactions with F_2 produced a strong Raman band between 452 and 475 cm^{-1} depending on the alkali metal counterion and a strong ultraviolet absorption at 310 nm. The Raman bands were assigned to the intraionic $(F-F)^-$ mode and the alkali metal dependence was attributed to interaction with the interionic $M^+-F_2^-$ mode; the 310 nm absorption was assigned to the $\sigma \rightarrow \sigma^*$ transition for F_2^- in the $M^+F_2^-$ species [43,44].

Matrix reactions of chlorine and alkali metal atoms gave extraordinarily intense Raman signals in the 225-264 cm^{-1} shifted region, again depending on the counterion, which are assigned to the intraionic (Cl-Cl)$^-$ stretching mode. The alkali metal dependence is due to interaction with the interionic M^+-Cl_2^- stretching mode which is predicted to be higher in the $Na^+Cl_2^-$ case, forcing the (Cl-Cl)$^-$ mode down to 225 cm^{-1}, and for the heavier $K^+Cl_2^-$ species, the interionic mode is predicted near 200 cm^{-1}, forcing the (Cl-Cl)$^-$ vibration up to 264 cm^{-1}.

Owing to the orange color of these matrix samples, resonance enhancement of the intensity was suspected and intense overtone series were observed with argon laser excitation [45]. In the $Cs^+Cl_2^-$ case, 457.9 nm excitation produced an extremely strong 259 cm^{-1} fundamental and seven overtones with regularly decreasing intensities and increasing resolution of chlorine isotopic splittings out the progression. A very strong optical absorption at 350 nm is responsible for the resonance Raman intensity enhancement for the $M^+Cl_2^-$ species [44]. This near-ultraviolet absorption for the dichloride ion in the $M^+Cl_2^-$ species is in excellent agreement with the 350 nm maximum in the photodissociation cross-section of the gaseous Cl_2^- ion [46]. The matrix-isolated $M^+Cl_2^-$ species is stabilized sufficiently by the ion-pair arrangement in the matrix to enable resonance Raman spectra of Cl_2^- to be obtained without shifting the electronic spectrum of the Cl_2^- anion.

Resonance Raman and optical spectra have been observed for the $M^+Br_2^-$, $M^+I_2^-$ and $M^+I_3^-$ species [47-50]. Red laser excitation produced resonance Raman spectra for I_2^- in the $M^+I_2^-$ ion pair. Spectra for I_3^- in the $M^+I_3^-$ species compare very favorably with solution spectra [51].

BIHALIDE IONS - HX_2^- and $M^+HX_2^-$

Perhaps the most interesting controversy in matrix-isolation spectroscopy over the last 15 years has involved the bichloride species. Noble and Pimentel first reported strong new absorptions at 696 and 956 cm^{-1} after passing Ar/HCl/Cl_2 mixtures through a coaxial microwave discharge tube and onto a 14 K surface. Chlorine isotopic splittings on the 956 cm^{-1} band required two equivalent chlorine atoms and the H/D frequency ratio for the strong absorption, 696/464=1.50, indicated a centrosymmetric species which was identified as ClHCl radical [52]. Milligan and Jacox observed the same two absorptions following hydrogen resonance photolysis of Ar/HCl samples and mercury arc photolysis of Ar/HCl/Cs mixtures and suggested that these two absorptions might instead be contributed by the (ClHCl)$^-$ anion [53]. Similar studies on HBr systems were interpreted as BrHBr radical or anion [54,55]. The formation of bihalide anions was suggested to involve dissociative electron capture of HX into H and X$^-$ with bihalide anion stabilization resulting from subsequent reaction of X$^-$ with HX [55].

The absorptions attributed to ClHCl$^-$ by Milligan and Jacox were independent of the nature of the electron donor, which characterized these anions as "isolated" from their counterion in the matrix. In the alkali metal experiments with HCl and HBr, weak bands appeared before photolysis which were produced in great yield by the matrix reaction of the hydrogen halide with alkali halide molecules and identified by Ault and Andrews as the bound ion-pair $M^+HCl_2^-$ and $M^+HBr_2^-$ species [17,18,56]. The near agreement between the infrared absorptions for the $M^+HCl_2^-$ species and the 696 cm^{-1} absorption provides strong support for the isolated HCl_2^- anion identification.

A series of microwave discharge experiments performed in this laboratory showed that hydrogen atoms and chlorine codeposited without vacuum ultraviolet light produced HCl without any of the 696 cm^{-1} absorption. However, when the coaxial discharge was exposed to the matrix with the same reagents, the 696 cm^{-1} band was an intense component of the product spectrum. It was suggested that ionizing radiation from the coaxial discharge was required to produce the 696 cm^{-1} band which supported its charged identification [13].

The most stable member of this series, bifluoride ion, is well-known as a crystalline material with a centrosymmetric $(FHF)^-$ ion [57], and its characterization in matrices completes the case for bihalide ions in matrices. In studies by Ault, the matrix reaction of CsF and HF produced a very strong 1364 absorption and a sharp 1217 cm^{-1} band which were assigned to ν_3 and ν_2 of HF_2^- in the $Cs^+HF_2^-$ species; the 1364/969=1.41 ratio verified the centrosymmetric nature of the bifluoride ion in the ion-pair species [58].

Photoionization experiments in this laboratory using an open discharge tube like that shown in Figure 1 and Ar/HF=100/1 samples produced a strong new absorption at 1377.0 cm^{-1}; the 1377 cm^{-1} band intensity was increased in similar experiments with F_2 and NF_3 added to the argon-hydrogen fluoride sample. The strongest new product band appeared at 965.5 cm^{-1} in the corresponding DF experiment with no new absorption between this new band and the 1377 cm^{-1} feature observed due to HF remaining in the vacuum system. The 1377.0 and 965.0 cm^{-1} absorptions are assigned to the ν_3 modes of the isolated HF_2^- and DF_2^- ions, respectively. The 1377.0/965.5=1.426 ratio shows that HF_2^- is also centrosymmetric as the isolated or "gas-phase" species [59]. Finally, agreement between the 1377 cm^{-1} isolated HF_2^-, the 1364 cm^{-1} "bound" ion-pair $Cs^+HF_2^-$, and the 1450-1550 cm^{-1} crystalline bifluoride stretching fundamentals reinforces the comparison between HCl_2^- and $M^+HCl_2^-$ species [18] and confirms the anion identification of the 696 cm^{-1} absorption as HCl_2^-.

Ar_nH^+ and Ar_nD^+

Bondybey and Pimentel codeposited argon-hydrogen and argon-deuterium mixtures from a coaxial discharge tube and observed photosensitive new absorptions at 904 and 644 cm^{-1} which were assigned to interstitial H-atom and D-atom vibrations, respectively [60]. Although these workers considered the possibility of a charged species, it was rejected at least in part because of the absence of a counterion absorption. Milligan and Jacox observed these two absorptions in a number of hydrogen resonance photolysis

experiments, noted the increased yield when strong electron
acceptors were added, and reassigned the bands to Ar_nH^+ and Ar_nD^+
[61]. In radiolysis experiments performed in this laboratory, the
codeposition of deuterons with argon produced the 644 cm^{-1} band in
moderate yield; however, with added Cl_2 reagent the very strong
DCl_2^- band was observed along with a completely absorbing 644 cm^{-1}
band [39]. The observation of the 644 cm^{-1} band with the deuteron
beam, coupled with the absence of this absorption when D-atoms were
codeposited from an off-axis discharge tube strongly supports the
charged species identification [62].

Wight, et al. passed Ar/D_2 mixtures through a coaxial discharge and
observed both DO_2 and the 644 cm^{-1} band in contrast to the off-axis
discharge experiments which produced only DO_2 [13]. This suggested
that the most probable mechanism for the formation of Ar_nD^+ is
direct photoionization of a D atom in the argon matrix with a large
solvent shift in the ionization to lower energy owing to the large
proton affinity of argon. With this solvent shift, argon resonance
(11.6-11.8 eV) and hydrogen resonance (10.2 eV) radiation are
energetic enough to photoionize a D atom. The 644 cm^{-1} band has
been observed in numerous argon resonance photoionization
experiments in this laboratory where deuterium enriched precursors
were used, for example DF, DCl, CDF_3 and CD_2F_2. The requirements
for the production of the 644 cm^{-1} band are a D atom in solid argon
and ionizing radiation which contribute more evidence for the Ar_nD^+
charged species assignment.

In an interesting series of haloform experiments, which will be
described later, Jacox and Andrews and coworkers have observed
marked growth of the 904 and 644 cm^{-1} absorptions on mercury arc
photolysis of $(CHX_2^+)X$ and $(CDX_2^+)X$ species (X=Cl, Br) [12,63].
Similar photochemical results have been found for the CH_2Cl_2 and
CD_2Cl_2 precursors with the 904 cm^{-1} band reaching A = absorbance
units = 0.18 and the 644 cm^{-1} absorption reaching A=1.2 on 220-1000
nm photolysis [64]. In these examples, mercury arc photolysis
affects a transfer of the proton and deuteron from the halocarbon
cations to the argon matrix and provides a photochemical synthesis
of the Ar_nH^+ and Ar_nD^+ species.

The photochemical stability and yield of the Ar_nH^+ and Ar_nD^+ species
depend in general on the nature and abundance of the counteranion in
the matrix. In experiments depositing only deuterons or D-atoms
with vacum ultraviolet light, the counterion must come from common
impurities in the vacuum system such as O_2 and H_2O which could
produce anions like O^-, O_2^-, and OH^- by electron attachment to
molecular fragments. Counterions like O_2^-, O^- and OH^- with
relatively low energy detachment thresholds are expected to release
electrons at comparatively lower energies for neutralization of the
Ar_nD^+ ion isolated nearby in the matrix. On the other hand,
chlorine present in the sample traps electrons as Cl^- and allows
more Ar_nD^+ to be isolated during sample preparation. A photochemical reduction of the positive ion absorption in these
experiments requires detachment from Cl^- in the near ultraviolet
range, as has been observed in this laboratory. In experiments with

CDF_3, the CF_3^+ and Ar_nD^+ cations were each reduced about 10% on 220-1000 nm photolysis for 2 h by photoelectrons detached from fluoride ion electron traps in the argon matrix.

TRIHALOMETHYL CATIONS

Jacox and Milligan performed hydrogen resonance photolysis of argon/chloroform samples and observed a large yield of CCl_3 radical at 898 cm^{-1} and a new 1037 cm^{-1} absorption. The latter band showed the 3/1 doublet splitting characteristic of three equivalent chlorine atoms and a carbon-13 shift to 1003 cm^{-1} which supported its identification as CCl_3^+ isolated in the matrix [65]. Subsequent radiolysis studies of CCl_4, CCl_3Br, CCl_2Br_2, $CClBr_3$ and CBr_4 in this laboratory produced the 1037 cm^{-1} band, absorptions at 1019 and 957 cm^{-1} for CCl_2Br^+, absorptions at 978 and 894 cm^{-1} for $CClBr_2^+$ and a new band at 874 cm^{-1} for CBr_3^+, each from two of the above precursors, which confirmed the trichloromethyl cation identification of the 1037 cm^{-1} band. Absorptions for trihalomethyl cations are listed in Table II.

Table II. Carbon-halogen stretching vibrations (cm^{-1}) observed for trihalomethyl cations in solid argon.

Ion	C-F	C-Cl	C-Br
CF_3^+	1665		
CF_2Cl^+	1514,1414		
$CFCl_2^+$	1351	1142	
CCl_3^+		1037	
CCl_2Br^+		1019,957	
$CClBr_2^+$		978	894
CBr_3^+			874
CF_2Br^+	1483,1367		
CF_2I^+	1432,1320		

The radiolysis experiments with CCl_4 produced major new photosensitive bands at 927, 502 and 374 cm^{-1} in addition to the strong photochemically stable 1037 cm^{-1} band. In order to rationalize the different bulb filament photolysis behavior, the photosensitive bands were assigned to isolated cations and the stable 1037 cm^{-1} band to CCl_3^+ with a nearby chloride ion which would not attract electrons in the photobleaching process [10]. In a following argon resonance photoionization study of CCl_4 using filtered mercury arc photolysis of the deposited sample, the different photochemical behavior of each new absorption was attributed to the photochemical stability of the cation in question. The photosensitive 927 and 374 cm^{-1} bands were assigned to an asymmetric CCl_4^+ species and the 502 cm^{-1} absorption to Cl_3^+. The

1037 cm^{-1} band was decreased slightly by prolonged mercury arc
photolysis, which required photodetachment from chloride electron
traps in the matrix, since CCl_3^+ itself probably does not dissociate
in the mercury arc energy range [11]. A small shift of the 1037
cm^{-1} CCl_3^+ band in solid argon to 1035 cm^{-1} in solid krypton
suggests that CCl_3^+ does not interact strongly with either matrix.
An important conclusion from the mercury arc photolysis studies is
that isolated cations in matrices will be of two types: those which
photodissociate with mercury arc light and those photochemically
stable in the mercury arc range which must exhibit a decrease in
intensity from photoneutralization by detachment of electrons from
counteranions in the matrix. In this regard, appearance potential
data are useful for predicting the photochemical behavior of
cations. Clearly, photodetachment depends on the electron trapping
species, as will be discussed below for CF_3^+.

In order to bridge the gap from CCl_3^+ at 1037 cm^{-1} to CF_3^+,
photoionization and radiolysis studies were performed in this
laboratory on the Freon series $CFCl_3$, CF_2Cl_2 and CF_3Cl [66,15].
Sharp new 1352 and 1142 cm^{-1} bands in $CFCl_3$ experiments exhibited
appropriate carbon-13 shifts and photolysis behavior for assignment
to $CFCl_2^+$. Analogous bands in CF_3Cl experiments at 1415 and 1515
cm^{-1} showed large carbon-13 shifts and slight photolysis with the
full mercury arc which indicated assignment to CF_2Cl^+. These
vibrations for $CFCl_2^+$ and CF_2Cl^+ are 200-300 cm^{-1} above the
corresponding free radical values which predicts that CF_3^+ may
absorb above 1600 cm^{-1}.

Photoionization studies have been performed on the trifluoromethyl
compounds CF_3Cl, CF_3Br, CF_3I and CHF_3 [67]. The CF_2Cl^+ absorptions
at 1514 and 1414 cm^{-1}, which shifted to 1483 and 1367 cm^{-1} for
CF_2Br^+, as has also been observed by Jacox [68], and the CF_2I^+
absorptions were displaced to 1432 and 1320 cm^{-1}. The C-F
stretching modes in the CF_2X^+ ions exhibited a pronounced heavy
halogen effect. A weak 1665 cm^{-1} band in the CF_3Cl study was
produced with greater intensity in the CF_3Br, CF_3I and CHF_3
experiments at the same frequency [67]. This 1665 cm^{-1} absorption
was reduced 10% by full high-pressure mercury arc photolysis for 2 h
in fluoroform experiments. The production of the same 1665.2 cm^{-1}
band from four different trifluoromethyl precursors, and the
photolysis behavior indicate assignment of the 1665 cm^{-1} band to
CF_3^+. The trifluoromethyl cation is formed by photoionization of
CF_3 radicals produced in the matrix photolysis process [67].

Infrared spectra for CF_3^+ and $^{13}CF_3^+$ in fluoroform experiments are
compared in Figure 3. The strong absorption produced from $^{13}CHF_3$ at
1599.2 cm^{-1} (A=0.20) is appropriate for ν_3 of $^{13}CF_3^+$; note the weak
$^{12}CF_3^+$ absorption (A=0.02) at 1665.2 cm^{-1} due to 10% ^{12}C present in
the enriched fluoroform precursor. The ^{12}C-^{13}C shift, 66.0 cm^{-1}, is
however, greater than expected for the antisymmetric C-F vibration,
ν_3, of a planar centrosymmetric species. This unexpectedly large
carbon-13 shift can be explained by Fermi resonance between ν_3 and

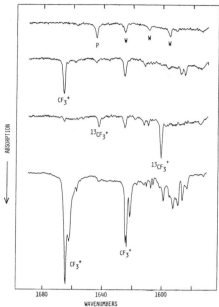

Figure 3. Infrared spectra in the 1580-1680 cm^{-1} region for fluoroform samples condensed with excess argon at 15 K. Spectrum (a) Ar/CHF_3=200/1 sample deposited with no discharge, P denotes precursor and W denotes water absorptions; trace (b) Ar/CHF_3=800/1 with concurrent argon resonance photoionization. Spectrum (c) Ar/$^{13}CHF_3$=800/1 with argon resonance photoionization; trace (d) Ar/CHF_3=700/1 with electric discharge on matrix during condensation.

the combination band ($\nu_1 + \nu_4$) for the $^{13}CF_3^+$ species, since the new 1641.7 cm^{-1} product band in the $^{13}CF_3^+$ spectrum can be assigned to the combination band. The calculated position of ν_3 for $^{13}CF_3^+$, 1620 cm^{-1}, appears to coincide with the apparent position of ($\nu_1 + \nu_4$), and these modes strongly interact and shift ν_3 down to 1599 cm^{-1} and ($\nu_1 + \nu_4$) up to 1642 cm^{-1}. In the absence of Fermi resonance, the ($\nu_1 + \nu_4$) combination band for $^{12}CF_3^+$ is expected to be 2-4 cm^{-1} above the $^{13}CF_3^+$ counterpart, in the region of the 1624 cm^{-1} water absorption; the spectrum in Figure 3(b) shows the 1624 cm^{-1} band, which may contain additional absorption. The markedly increased CF_3^+ yield in the electric discharge experiment, Figure 3(d), revealed a very strong 1665.2 cm^{-1} absorption (A=0.84) and 1624 cm^{-1} absorption (A=0.40) clearly in excess of the other water absorptions present [23]; the latter absorption is assigned to the combination band ($\nu_1 + \nu_4$) for $^{12}CF_3^+$. The combination ($\nu_1 + \nu_4$)

for $^{12}CF_3^+$ at 1624 cm^{-1} provides a basis for determining the infrared inactive symmetric C-F bond stretching mode ν_1 of $^{12}CF_3^+$. The intensity of the ν_2 and ν_4 modes of BF_3 are approximately a factor of ten weaker than that of ν_3, and the failure to observe these weaker fundamentals of CF_3^+ in the present study is not surprising. However, ν_4 may be estimated to be 500 ± 30 cm^{-1} since $\nu_4 = 512$ cm^{-1} for CF_3 [69] and 480 cm^{-1} for BF_3 [70], which predicts $\nu_1 = 1125 \pm 30$ cm^{-1} for $^{12}CF_3^+$.

It is interesting to consider the bonding in CF_3^+ in view of its substantially increased antisymmetric stretching frequency and the increased symmetric stretching frequency deduced from the $(\nu_1 + \nu_4)$ band. This increase is considerable relative to the pyramidal CF_3 species ($\nu_1 = 1086$ cm^{-1}, $\nu_3 = 1251$ cm^{-1}) [69] and the planar BF_3 molecule ($\nu_1 = 888$ cm^{-1}, $\nu_3 = 1454$ cm^{-1}) [70]. From the well-known back-donation of fluorine 2p electron density to the positive carbon, it is readily apparent that CF_3^+ should exhibit extensive pi bonding, and the markedly increased ν_3 and ν_1 modes are consistent with this bonding model.

The photolysis behavior of the 1665 cm^{-1} CF_3^+ absorption in CHF_3, CF_3Cl, CF_3Br and CF_3I experiments provides evidence for the halide counterion in these studies [67]. Since dissociation of CF_3^+ to $CF_2^+ + F$ requires at least 5 eV, which is above the mercury arc range, a photoneutralization mechanism is required for the slight decrease of CF_3^+ on mercury arc photolysis. The relative decrease in the CF_3^+ band on 290-1000 and 220-1000 nm photolysis in these experiments was more pronounced in the order $I^- > Br^- > Cl^- > F^-$ where these halide ions are the most likely counterions in photoionization studies with CF_3I, CF_3Br, CF_3Cl and CHF_3, respectively. This photolysis behavior parallels the expected photodetachment cross section in the halide series and supports the photoneutralization model for the photolysis of CF_3^+ isolated in a matrix containing halide counterions.

PARENT CATIONS

Tetrahalomethane parent cations are characterized by their instability with respect to halogen elimination owing in part to the unusual stability of the CX_3^+ daughter cations discussed above. A good example is CCl_4^+, which has escaped gas-phase detection due to the ease of chlorine atom elimination at thermal energies to give CCl_3^+. The strongest photosensitive band at 927 cm^{-1} and a weaker 374 cm^{-1} band in matrix photoionization studies of CCl_4 have been assigned to CCl_4^+ in an asymmetric C_{2v} structure; these absorptions were virtually destroyed by 500-1000 nm photolysis [11]. A subsequent optical absorption study of the matrix photoionization of CCl_4 produced a very strong 425 nm band, also destroyed by 500-1000 nm photolysis, which was assigned to CCl_4^+ in solid argon [71]. Since the difference between bands in the photoelectron spectrum [72] predicts a strong transition near 243 nm, substantial structural relaxation of CCl_4^+ must occur for the CCl_4^+ parent cation to be stabilized. Hence, the role of the matrix in quenching excess internal energy and allowing structural relaxation before dissociation would normally happen is apparent in these experiments.

The matrix photoionization studies with $CFCl_3$ produced infrared bands at 1214 and 1041 cm^{-1} and an optical band at 405 nm which were destroyed by 500-1000 nm photolysis; these absorptions were assigned to $CFCl_3^+$ [66,71]. Similar investigations of CF_2Cl_2 yielded infrared bands at 1234, 1067, 609 and 406 cm^{-1} and an optical band at 310 nm related by their partial photolysis with 290-1000 nm light, which were assigned to $CF_2Cl_2^+$ [15,71]. Similar spectroscopic observations have been made for the analogous CBr_4^+, $CFBr_3^+$ and $CF_2Br_2^+$ species [11,71,73].

In studies of the matrix photoionization of CF_3X precursors, strong bands at 1299 cm^{-1} in the CF_3Cl spectrum, 1293, and 1255 cm^{-1} in the CF_3Br scan and 1229 cm^{-1} in the CF_3I spectrum, which along with lower frequency absorptions at 455, 469, and 497 cm^{-1}, respectively, photolysed with 290-1000 nm light that had little effect on the CF_3^+ and CF_2X^+ absorptions. An optical band at 295 nm in the CF_3Cl study showed similar photolysis behavior. These absorptions were assigned to the CF_3X^+ parent cations [67,71]. It is interesting to note that the matrix photolysis behavior parallels the gas phase stability [74,75] of these parent cations with more energetic radiation required to dissociate the more stable ion in the order $CF_2Cl_2^+ > CF_3Cl^+ > CFCl_3^+ > CCl_4^+$.

The methylene halide cations provide another interesting series of parent cations for comparison to gas phase studies including photoelectron spectra. Matrix photoionization work on CH_2Cl_2 produced infrared absorptions at 1194 and 764 cm^{-1} which were destroyed by 650-1000 nm photolysis; analagous bands were observed at 1129 and 685 cm^{-1} in CH_2Br_2 studies. Since $CH_2Cl_2^+$ and $CH_2Br_2^+$ require only 0.8 eV for dissociation in the gas phase [74,76], the photosensitive matrix absorptions were assigned to the $CH_2Cl_2^+$ and $CH_2Br_2^+$ parent ions [64]. Further 290-1000 nm mercury arc photolysis in these experiments produced marked growth in CHX_2 radical, and CHX_2^+ and Ar_nH^+ cation absorptions. The $CHBr_2$ radical and cation absorptions have been observed in both argon and krypton matrices; $CHBr_2$ absorbs at 1166 and 786 cm^{-1} in argon and at 1162 and 782 cm^{-1} in krypton; $CHBr_2^+$ absorbs at 1229 and 897 cm^{-1} in argon and 1222 and 890 cm^{-1} in krypton. The argon-krypton matrix difference is 7 cm^{-1} for $CHBr_2^+$ and 4 cm^{-1} for $CHBr_2$ suggesting that the cation interacts only slightly more with the matrix than the free radical.

The optical spectra for a series of methylene halides subjected to argon resonance photoionization during sample deposition are illustrated in Figure 4. New absorptions were observed at 342, 362, and 375 nm in the series CH_2Cl_2, CH_2Br_2 and CH_2I_2. These bands were virtually destroyed by visible-near infrared photolysis (500-1000 nm), as shown in the Figure, which supports their assignments to the parent cations [64]. Photoelectron spectra for these compounds

exhibited a group of sharp bands from ionization of halogen
nonbonding electrons, followed by a broader band from ionization of
a C–X bonding electron [72]. The wavelength corresponding to the
energy difference between the first sharp lone–pair ionization and
the C–X bond ionization is noted with an arrow in each trace of the
Figure. The agreement between the energy of the observed electronic
transition for the parent ions in solid argon and the difference
between the two photoelectron bands is excellent, and it provides
strong support for the presence of $(CH_2X_2)^+$ parent ions in these
matrices. Small differences may arise from different Franck–Condon
factors between the electronic and ionizing transitions involved,
although very good agreement is expected where ionization causes no
major change in geometry of the ion as compared to that of the
neutral species.

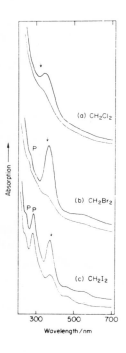

Figure 4. Absorption spectra from
200–700 nm for methylene halide
samples, $Ar/CH_2X_2=400/1$, deposited at
20 K with simultaneous exposure to
argon discharge radiation. (a) CH_2Cl_2,
(b) CH_2Br_2, (c) CH_2I_2. The trace
displaced below each scan was recorded
after 500–1000 nm photolysis for 30
min. The arrows denote the position
of transitions from the ground state
of the ions to the excited state as
determined from photoelectron spectra.
(Andrews, Prochaska and Ault, Ref.
[64])

Methylene fluoride provides an interesting contrast to methylene
chloride since first ionization in the former involved a C–H bonding
electron. Matrix photoionization of CH_2F_2 using the open discharge
tube shown in Figure 1 produced a rich infrared spectrum [77].
Sharp bands at 2854, 2744, 1408 and 1255 cm^{-1} exhibited large
carbon–13 displacements, in agreement with the ν_6, ν_1, ν_8 and ν_9
modes of the parent molecule, and photolysed with 420–1000 nm light,
which suggests their assignment to the parent cation. The infrared
spectrum of $CH_2F_2^+$ is consistent with SCF calculations on the

highest occupied orbital for CH_2F_2 which is strongly C-H$_2$ bonding and C-F$_2$ antibonding. The removal of an electron from this orbital reduces the C-H$_2$ stretching fundamentals by 214 and 178 cm^{-1} in the cation and increases the antisymmetric C-F$_2$ stretching mode by 176 cm^{-1} [77]. An 80-130 cm^{-1} increase in the C-F stretching fundamentals has also been observed for the CF_3Cl^+, $CF_2Cl_2^+$ and $CFCl_3^+$ parent ions [15,66].

The fluorohalomethane studies follow the example of methylene fluoride even though a halogen lone pair electron is removed in ionization. In the CH_2FCl experiments, a sharp band at 2902 cm^{-1} and a chlorine isotopic doublet at 874, 869 cm^{-1} exhibited carbon-13 and deuterium shifts similar to the symmetric C-H and C-Cl stretching modes of the neutral molecule and photolysed with 290-1000 nm light, which indicates their assignment to CH_2FCl^+ [78]. The C-H stretching mode for the ion again falls below the 2997 cm^{-1} parent value and the C-Cl stretching mode is above the 750 cm^{-1} neutral value. Similar observations were found for the CH_2FBr^+ and CH_2FI^+ parent ions. With each of the CH_2FX precursors, the $CHFX^+$ daughter ion was characterized by its C-F stretching mode near 1400 cm^{-1} and C-H deformation mode near 1290 cm^{-1} [78].

The boron trichloride and boron tribromide cations provide an interesting counterpoint to the carbon tetrachloride and carbon tetrabromide cations. In studies of the argon resonance photo-ionization of BCl_3 during condensation with excess argon at 15 K [79], new product absorptions at 1091 and 1133 cm^{-1} assigned to $^{11}BCl_3^+$ and $^{10}BCl_3^+$ were resolved into 9/1/3/1 quartets, as expected for a doubly degenerate vibration of three equivalent chlorine atoms, identical to that observed for $^{11}BCl_3$ and $^{10}BCl_3$ at 946 and 984 cm^{-1}, respectively. The $^{11}BBr_3^+$ and $^{10}BBr_3^+$ absorptions at 931 and 971 cm^{-1} also appeared above the analogous neutral precursor absorptions at 815 and 850 cm^{-1}. Optical absorption studies revealed new ultraviolet bands at 320 and 355 nm in solid argon which were assigned to BCl_3^+ and BBr_3^+, respectively, in excellent agreement with the difference between the a_2' and 2e' photoelectron bands corresponding to dipole allowed transitions of the ions at 318 ± 8 and 340 ± 10 nm in the gas phase [80]; this indicates no substantial structural difference between the BX_3 molecules and cations. The infrared and ultraviolet parent cation absorptions were substantially reduced by 340-600 nm photolysis and destroyed by 290-1000 nm irradiation. This behavior is consistent with the dissociation of BCl_3^+ to BCl_2^+ and Cl which is expected from appearance potential data. Boron trihalide and halomethane parent cations are photosensitive species, and their photolysis behavior has facilitated their identification from infrared matrix spectra.

CX_2^+ AND CHX_2^+ CATIONS

Very recent infrared studies of dihalocarbene cations are of interest as the first observation of second daughter cations and the largest frequency increase found on ionization. In the CH_2Cl_2

experiments described above, a weak 1197, 1194, 1191 cm^{-1} triplet remained after mercury arc photolysis [64]. This same resolved triplet was three-fold stronger in matrix photoionization studies with CHFCl$_2$ and exhibited the 9/6/1 relative intensities expected for a vibration of two equivalent chlorine atoms [81]. No deuterium shift was found in CD$_2$Cl$_2$ and CDFCl$_2$ studies. An experiment with a 90% carbon-13 enriched ^{13}CHFCl$_2$ sample produced weak carbon-12 absorptions at 1197 and 1194 cm^{-1} and strong new carbon-13 bands at 1158 and 1155 cm^{-1}, showing that this vibration involves a single carbon atom. The antisymmetric carbon-chlorine stretching absorption of CCl$_2$ has been well documented at 746 cm^{-1} [82]. The 1197, 1194, 1191 cm^{-1} triplet due to a new (CCl$_2$) species is higher than CHCl$_2$$^+$ at 1045 cm^{-1} and CFCl$_2$$^+$ at 1142 cm^{-1} [66,69], which indicates that the new triplet is due to CCl$_2$$^+$ [81]. Dichloro-carbene cation was produced in these experiments by photoionization of the CCl$_2$ intermediate photolysis product. The substantial increase in ν_3 from 746 cm^{-1} for CCl$_2$ to 1197 cm^{-1} for CCl$_2$$^+$ is consistent with increased pi bonding in the cation. Isotopic shifts show that the Cl–C–Cl angle increases 25-30° on ionization to CCl$_2$$^+$.

Similar matrix photoionization experiments with CH$_2$ClBr produced a 3/1 doublet at 1122 and 1118 cm^{-1} and studies with CHFBr$_2$, CDFBr$_2$, CH$_2$Br$_2$ and CD$_2$Br$_2$ gave a sharp 1019 cm^{-1} absorption. The observation of new products at 1197, 1122 and 1019 cm^{-1} in CH$_2$Cl$_2$, CH$_2$ClBr and CH$_2$Br$_2$ experiments verifies the infrared identification of the dihalocarbene cation species CCl$_2$$^+$, CClBr$^+$ and CBr$_2$$^+$ [81]. The CClBr$^+$ and CBr$_2$$^+$ absorptions are higher than CHClBr$^+$ at 993, 989 cm^{-1} and CHBr$_2$$^+$ at 897 cm^{-1} [64], and substantially higher than CClBr and CBr$_2$ at 744, 739 and 641 cm^{-1}, respectively [83], suggesting increased pi bonding in the dihalocarbene cations.

In recent CH$_2$Cl$_2$ matrix photoionization experiments with high resolution measurements using a Nicolet Fourier-Transform Infrared Spectrometer, a particularly large yield of CHCl$_2$$^+$ was produced upon 290-1000 nm photolysis of the CH$_2$Cl$_2$ sample subjected to argon resonance photoionization [84]. The intense triplet at 1044.7, 1041.9 and 1039.5 cm^{-1} (A=0.63, 0.43, 0.08, respectively) has been assigned to the antisymmetric C–Cl stretching mode in CHCl$_2$$^+$ [64]; these absorption bands are very sharp, with full-widths at half-maximum (FWHM) of 0.3 cm^{-1} using 0.1 cm^{-1} resolution. In addition a weaker triplet was observed at 845.4, 841.4, 837.7 cm^{-1} (A=0.54, 0.037, 0.007, respectively) which is assigned to the C–H stretching mode; these assignments are supported by ^{13}C substitution. It can be seen by comparison with similar molecules (for example, HBCl$_2$ has B–Cl stretching modes at 886 and 740 cm^{-1}) [79] that the 1044 and 845 cm^{-1} values for CHCl$_2$$^+$ are high for C–Cl stretching fundamentals, and substantial C–Cl pi bonding is indicated. On the other hand, the 3033 cm^{-1} value for the C–H mode in CHCl$_2$$^+$ is near the 3058 cm^{-1} value for CHCl$_3$; clearly the carbon-hydrogen bond is not affected by the adjacent positive ion center.

BENZYL AND TROPYLIUM CATIONS

Many investigators have been concerned with the structures of gas-phase $C_7H_7^+$ ions produced from different aromatic compounds. A number of workers have shown that the $C_7H_7^+$ ion population contains reactive and unreactive fractions and that the two populations probably have different structures [85-88], which have been identified as the benzyl 1 and tropylium 2 cations, respectively [87,88].

The first spectroscopic studies of $C_7H_7^+$ ions identified 2 in concentrated H_2SO_4 by strong absorptions at 274 and 217 nm [89,90]. Although several methyl-substituted derivatives of 1 have been observed in superacid solutions [91], 1 itself is sufficiently reactive to require pulse radiolysis and submicrosecond observation of a 363-nm absorption in dichloroethane solution [92]. The matrix photoionization study of benzyl bromide provides ultraviolet absorption spectra and information on the photochemical rearrangement of 1 and 2.

Using a new technique, neat benzyl bromide vapor was condensed with argon from a 3-mm orifice discharge tube [93]. A strong, broad, symmetrical band was observed beginning at 430 nm with a maximum and band center at 353 nm (A=0.60) and a minimum at 300 nm on a background with little slope; a weak shoulder was detected at 263 nm (A=0.06) on the side of the 225-nm precursor absorption. The spectrum after codepositon of benzyl bromide vapor with discharged argon for 1 h in another experiment is illustrated in Figure 5(a); the product bands were 353 nm (band center, A=0.36) and 263 (A=0.24). Photolysis with 380 nm cut-off radiation for 10 min markedly reduced the 353 nm band (to A=0.10) and produced a new maximum at 325 nm (A=0.10), and increased the 263 nm band (to A=0.33), Figure 5(b). Irradiation through the 340-600 nm band pass filter for 10 min further reduced the 353 and 325 nm bands (to A=0.02) and increased the 263 nm band (to A=0.37), Figure 5(c). Exposure of this sample to 290 nm light for 10 min destroyed the 353 nm band, reduced the 325 nm absorption (to A=0.01) and slightly reduced the 263 nm absorption (to A=0.35), Figure 5(d). A final full arc photolysis for 10 min produced new absorption centered at 353 nm (A=0.05), markedly reduced the 263 nm band (to A=0.06), and destroyed the 325 nm band, Figure 5(e).

The strong 353 nm absorption in solid argon is assigned to the benzyl cation, 1, based on agreement with the 363 nm band produced by pulse radiolysis of several benzyl compounds in 1,2-dichloro-ethane solution [92]. Typically, solution spectra are red shifted from argon matrix spectra which are in turn red shifted from gas

phase spectra.   Likewise, the 263 nm band produced on photolysis of
the 353 nm absorption is assigned to the tropylium cation, 2, owing
to agreement with the 274 nm absorption of tropylium salts in
concentrated acid solutions.   Since the ionization energies of
benzyl radical (7.2 eV) and tropyl radical (6.2 eV) [94,95] are
relatively low compared to argon (15.8 ev), 1 and 2 do not interact
significantly with the argon matrix, and the argon matrix spectra
should be representative of gas phase spectra.   Further evidence
supporting these spectroscopic observations of 1 and 2 is found in
their photochemical interconversion.

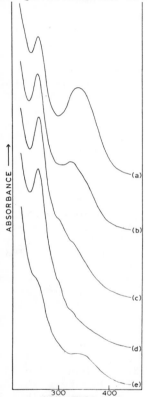

Figure 5.   Ultraviolet spectra of
products formed upon codeposition of
benzyl bromide vapor with argon from a
microwave discharge: (a) after 1 h
codeposition with 140 mtorr discharge;
(b) after 380 nm cutoff photolysis for
10 min; (c) after 340–600 nm
irradiation for 10 min; (d) after 290
nm photolysis for 10 min: (e) after
220–1000 nm photolysis for 10 min.
(Andrews and Keelan, Ref. [93]).

The figure shows five general photochemical observations that are of
interest here: (1) the 353 nm benzyl cation band photolysed at an
appreciable rate within the absorption band beginning at 420 nm (the
longest wavelength used here), (2) brief photolysis with 420 and 380
nm cut-off radiation produced a new species absorbing at 325 nm, (3)
photolysis with 340–600 nm and 290–1000 nm light reduced the benzyl
cation band symmetrically and produced the 263 nm tropylium cation
absorption, (4) photolysis with 220 nm radiation destroyed 2 and
regenerated a small yield of 1 without any 325 nm absorption, and
(5) continued ultraviolet photolysis decreased both 1 and 2
absorptions owing to neutralization by electrons photodetached from
bromide ion electron traps.   These observations will be described in
more detail.

(1)   The postulated equilibrium between 1 and 2 at relatively low internal energies in the gas phase suggests a ready rearrangement of 1 upon excitation. MINDO/3 calculations predict an activation energy of 33 kcal/mole [96] and experimental energy relationships give approximately the same barrier for the 1-2 isomerization, which is exceeded by absorption at 420 nm (68 kcal/mole). The least endothermic dissociation process, $C_7H_7^+$ to $C_5H_5^+$ and C_2H_2, however, requires 95 kcal/mole (300 nm) [86]. The near ultraviolet electronic absorption of 1 provides ample internal energy for rearrangement, but insufficient energy for photodissociation.

(2)   The interconversion between 1 and 2 may involve intermediate $C_7H_7^+$ isomeric species with different structures. Calculations have predicted local minima for norcaradienyl cation and 1-cyclo-heptatrienyl cation. The 325 nm absorption formed on 380 nm photolysis of 1 (Figure 5(b)) is attributed to an intermediate $C_7H_7^+$ structure different from 1 and 2; the weak absorption revealed at 300 nm on 340 and 290 nm photolysis (Figure 5(c), 5(d)) may be due to a still different structure.

(3) Photolysis with 290-1000 nm radiation gives essentially complete destruction of absorption in the 300 nm region, which results in a maximum yield of the 263 nm absorption due to 2. The 1 → 2 rearrangement is complete insofar as the competitive photo-neutralization process will allow.

(4)   The full mercury arc provides photon energies up to about 125 kcal/mole which clearly exceed the 95 kcal/mole activation energy for dissociation of 2 to $C_5H_5^+$. Nevertheless, a small amount of 1 is reformed from 2 although dissociaton clearly dominates (Figure 5(d)). The rearrangement process can be cycled back and forth [93], but both absorptions are reduced in the process owing to the favorably competitive dissociation pathway. The role of the matrix in quenching internal energy makes it possible for some 2 → 1 rearrangement to be observed.

(5)   The final point in these photolysis studies is that when the 369 nm photodetachment threshold [97] of the bromide ion electron trap is exceeded, photoneutralization of all isolated cations will proceed. Experiments with CF_3^+ and Br^- have shown that this is a relatively slow process (approximately 30% reduction in 125 min) [67], hence the dominant mechanism for 220-1000 nm photolysis is ion photodissociation.

STYRENE CATIONS

The spectroscopic properties of styrene and substituted styrene radical ions are of interest as conjugated aromatic cations. An extensive study of photodissociation spectra (PDS) [98] of these ions provides a basis for detailed comparison with optical spectra and photochemistry of styrene ions in solid argon [99].

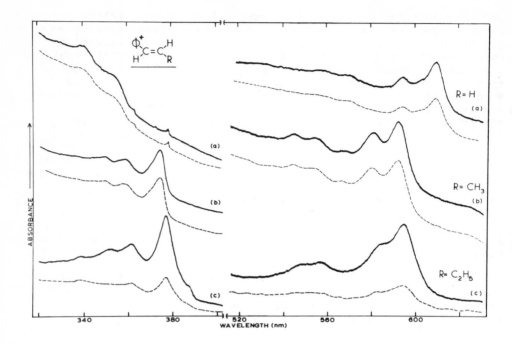

Figure 6.   Absorption spectra of styrene and substituted styrenes
subjected to argon resonance photoionization during condensation.
(a) Styrene, UV recorded on 0.2 A range, red on 0.1 A range, dashed
trace after 500-1000 nm photolysis. (b) β-Methylstyrene, UV recorded
on 0.5 A range, red on 0.1 A, dashed trace after 420 nm photolysis.
(c) β-Ethylstyrene, UV recorded on 0.5 A range, red on 0.2 A, dashed
trace after 500 nm photolysis of argon/precursor.   Sample
concentrations were approximately 200/1. (Andrews, Harvey, Kelsall
and Duffey, Ref. [99]).

Figure 6 compares the matrix absorption spectra of (a) styrene, (b)
β-methyl styrene and (c) β-ethyl styrene cations and shows marked
similarity in the electronic structure of the absorbing species.
The dashed traces, recorded after visible photolyses, demonstrate
that the red and near ultraviolet systems are due to the same
species since the two absorptions decrease in concert upon
irradiation into the red absorption.

Styrene cation exhibits a 630 nm absorption with a 590 nm shoulder
and a 340 nm absorption in s-butyl chloride glass [100]. Owing to
the 518 nm threshold for one-photon dissociation, the PDS shows only
a 330 nm peak with a 315 nm shoulder although some presumably
two-photon dissociation was observed at 579 nm [98].   The photo-

electron spectrum (PES) for styrene exhibits three sharp bands for π ionizations before the broad σ orbital ionization [98,101]; the excitation energy between the first onset and the third vertical ionization energies, 2.13 ± 0.02 eV, suggests a red absorption near 582 nm. The matrix spectrum reveals two ultraviolet bands at 353 and 339 nm and a structured red band system beginning at 608.5 nm, in agreement with the PDS and glassy matrix studies. The PDS for β-methyl styrene cation exhibits peaks at 354 and 579 nm with FWHM of about 3500 and 2000 cm^{-1}, respectively. The origins of the matrix absorptions at 373.6 and 591.1 nm are displaced 1464 and 343 cm^{-1}; however, the matrix bandwiths are less than 600 and 300 cm^{-1}, respectively, and vibrational structure is clearly resolved. The PDS for β-ethyl styrene is indistinguishable from the PDS of β-methyl styrene; however, Figure 6 reveals resolvable differences in the matrix band positions, although the vibrational intervals are almost indentical.

The matrix absorption spectrum of α-methyl styrene cation is substantially different; strong absorptions were observed at 686.3, 360.7 and 346.4 nm. The α-methyl substituent is expected to change the orientation between the phenyl and olefin skeletal planes and thus decrease interaction between the two pi systems. Photolysis of α-methyl styrene cations in matrices containing CH_2Cl_2 added to serve as an electron trap produced the more stable β-methyl styrene cation at the expense of the less stable α-isomer. This α-to-β substituent rearrangement has also been observed for the chlorine and bromine substituted styrene cations [102].

An interesting aspect of the matrix work on styrene cations is the vibrational structure. Two vibrational modes, one near 1150 cm^{-1} and one in the 330-420 cm^{-1} range, are active in the vibronic absorption bands of the styrene cations described here. The 1150 cm^{-1} intervals have been assigned to ϕ-C stretching vibrations. The lower frequency mode observed twice in each red absorption showed a decrease from about 420 cm^{-1} for styrene cation to 350 cm^{-1} for β-methyl styrene cation and 330 cm^{-1} for β-ethyl styrene cation. The observation of such a pronounced substituent effect indicates that this mode involves the alkene group [99]. It is reasonable to expect the ϕ-C stretching mode and ϕ-C=C bending mode to be active in the π(mostly ring) ← π(mostly olefin) transition, since conjugation between the two pi systems will be different for each electronic state.

NAPHTHALENE CATION

The naphthalene cation is a particularly interesting case for matrix study since a wealth of data are available from photoelectron spectroscopy [72,103,104], absorption spectra in glassy matrices [105] and multiphoton dissociation spectra [106]. Argon discharge photoionization of naphthalene (herafter N) vapor during condensation with excess argon produced new absorption band systems at 675.2 nm (A= absorbance = 0.17), 461 nm (A=0.01) and 381 nm

(A=0.03) [107].   The spectrum of naphthalene cation from matrix
two-photon ionization of naphthalene is illustrated in Figure 7:
trace (a) shows the spectrum of an Ar/CCl$_4$/N=3000/4/1 sample
deposited at 20 K, trace (b) illustrates the spectrum after
photolysis for 15 sec with 220-1000 nm high pressure mercury arc
radiation [25,108].   The above absorption bands were observed with a
3-fold intensity increase, and two strong new systems were produced
at 307.6 nm (A=0.78) and 274.6 nm (A=0.9) at the expense of N
precursor absorption systems begining at 313.3 and 272 nm, which
were reduced by 70%.

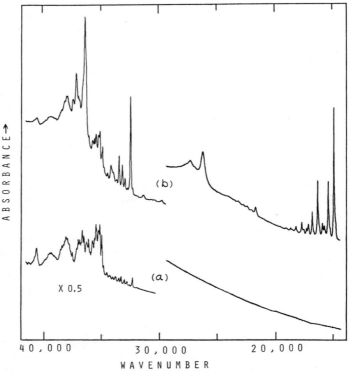

Figure 7.   Visible and ultraviolet absorption spectra of naphthalene
samples in solid argon at 20 K. (a) Ar/CCl$_4$/N=3000/4/1, (b) same
sample after 15 sec exposure to 220-1000 nm high-pressure mercury
arc radiation.

The five band systems are assigned to the naphthalene radical cation
(N$^+$) for the following reasons: (a) Three of these absorption
systems were observed with two different techniques capable of
forming positive ions.   Argon resonance radiation ionizes N
molecules during matrix condensation and the N$^+$ product is trapped
in solid argon; in these studies the electron removed in ionization
is probably trapped by a molecular fragment or another parent
molecule.   Intense mercury arc radiation performed two-color
resonance photoionization of N molecules isolated in solid argon
with the relatively long lived S$_1$ state serving as an intermediate

step in the photoionization process [24,25]. In the latter experiments CCl_4 was added as an electron trap, and positive ion yields increased by a factor of 3 over single photon ionization experiments, giving further support to the product identification as a positive ion. The two highest energy absorption systems were obscured by precursor absorption in the first experiments, but the latter studies used one-tenth as much N, and the N^+ absorptions were produced at the expense of N bands (Figure 7).

(b)   The two visible band origin positions are in excellent agreement with photoelectron spectra (PES). The red absorption provides an excellent basis for comparison between PES and the sharp matrix absorption spectrum since sharp vertical PES band origins, which can be measured to $\pm$ 0.01 eV, were observed for the first and third ionic states of N^+. The difference between the first and third PES band origins is 1.85 $\pm$ 0.02 eV, which predicts an absorption band origin at 14 920 $\pm$ 160 cm^{-1} in the gas phase, assuming no change in structure between the neutral molecule and the two ionic states. The argon matrix origin at 14 810 $\pm$ 3 cm^{-1} is in agreement within measurement error, which confirms the identification of N^+ and indicates relatively little perturbation of N^+ by the argon matrix. The blue absorption at 21 697 $\pm$ 10 cm^{-1} is also in excellent agreement with the 21 859 $\pm$ 160 cm^{-1} difference between sharp origins of the first and fourth PES bands. The three ultraviolet bands are attributed to electron-promotion transitions with upper states not reached by PES, hence a comparison cannot be made.

(c)   The first four transitions are in excellent agreement with the 690, 467, 387 and 308 nm absorptions assigned to N^+ produced by radiolysis of N in a Freon glass at 77 K [105]. The small red-shifts (42-400 cm^{-1}) are due to greater interaction with the more polarizable Freon medium. Vibronic structure is similar in the solid argon and solid Freon samples; the bands are sharper in solid argon which facilitates more accurate vibronic measurements.

(d)   The first 3 transitions are predicted reasonably well by simple Hückel molecular orbital theory with the molecular orbital energies calculated using β =2.77 eV from PES [103].

(e)   The sharp UV band origins at 32 510 and 36 417 cm^{-1} are in agreement with the broad 4-photon dissociation [106] bands at 33 200 and 38 200 cm^{-1} for N^+; the agreement would probably be better if origins were resolved for the broad PDS bands.

(f)   The vibronic structure in the absorption systems, particularly the red band, and the $N-d_8^+$ counterpart spacings correspond closely with N and $N-d_8$ vibrational intervals [109]. The small blue shifts in the origin bands upon deuteration range from 75 to 145 cm^{-1}; the weak S_1 origin of N at 313.2 nm exhibits a similar 102 cm^{-1} blue shift upon deuteration.

Two regularly repeating vibrational intervals were observed in the red N^+ absorptions. The first interval, 1422 $\pm$ 6 cm^{-1}, was

unchanged upon deuterium substitution. This is in agreement with
the 1420 ± 40 cm^{-1} interval in the first PES band [72], which is
probably due to the C(9)–C(10) stretching mode $\nu_4(a_g)$. The strong
fundamental at 1376 cm^{-1} in the Raman spectrum of $C_{10}H_8$ is
complicated by Fermi resonance, but the strong fundamental for $C_{10}D_8$
is at 1380 cm^{-1} [109]. The second repeated interval in the spectra
was approximately 505 cm^{-1} for N-h_8^+ and 485 cm^{-1} for N-d_8^+; this
corresponds to the $\nu_9(a_g)$ skeletal distortion observed in the Raman
spectra for N-h_8 at 512 cm^{-1} and N-d_8 at 491 cm^{-1}. The 754 cm^{-1}
interval corresponds to $\nu_8(a_g)$, the skeletal breathing mode observed
at 758 cm^{-1} in the Raman spectrum of N. It is seen that ionization
has a relatively small effect on the vibrational potential function
of the N molecule; this effect has been correlated with HMO pi bond
orders in the ionic and neutral states of N [108].

Similar experiments have been performed with substituted
naphthalenes. Deposition of 2-methylnaphthalene vapor with argon gas
and argon resonance radiation produced new band origins at 14 903,
21 137 and 25 994 cm^{-1}. Condensation of a dilute Ar/2-MeN>3000/1
sample with Ar/CCl$_4$=500/1 sample followed by a 5 min exposure to
mercury arc radiation produced the same absorptions with a 3-fold
increase in intensity and additional strong absorptions at 32 051
and 35 753 cm^{-1} [108]. The two repeating intervals observed in the
red 2-MeN$^+$ absorption were 1423 ± 10 cm^{-1} and 436 ± 10 cm^{-1}. The
C(9)–C(10) stretching mode is unchanged by 2-methyl substitution;
however, the skeletal distortion mode is altered substantially. The
absorption band origins for 2-MeN$^+$ are slightly displaced from the
N$^+$ band origins.

The photodissociation, argon matrix and Freon matrix absorption
spectra of 2-MeN$^+$ are contrasted in Figure 8; a number of important
differences are apparent. The large bandwidths in the photodissociation spectra (2500 cm^{-1} for the 15 600 cm^{-1} band) completely
obscure vibrational structure that is clearly observed in the solid
state spectra. This vibrational structure is characteristic of the
cation itself, as verified by both matrix studies, and is not caused
by the solid environment, as suggested to explain the lack of
vibrational structure in photodissociation spectra [110]. It is the
cold solid, however, that quenches excess internal energy, reduces
the bandwidth, and makes possible the observation of vibronic
structure in the matrix absorption bands. The present comparison of
argon matrix absorption and photodissociation bandwidths and the
earlier comparison with styrene cations [99] suggest that excess
internal energy contributes to the gas-phase bandwidth of ions
produced by electron impact. Comparison of peak positions in the
gas-phase and argon matrix is complicated by the great difference in
bandwidths and the lack of band origins in the photodissociation
spectra. These measurements for the two visible transitions agree
within experimental error. Although the photodissociation peaks in
the ultraviolet region are 1700,1700 and 2100 cm^{-1} higher than the
argon matrix band origins for the 2-Me-N$^+$ isomer, the onsets of the
photodissociation bands correspond closely with the matrix band
origins. Hence, part of this apparent difference is due to
comparing centers of very broad bands with sharp vibronic origins,

and part is due to matrix interaction and resulting red shifts of
the transitions, which is expected to be more pronounced with higher
excited states of cations as the electron affinity of the cation
state approaches the ionization energy of the solid argon host.

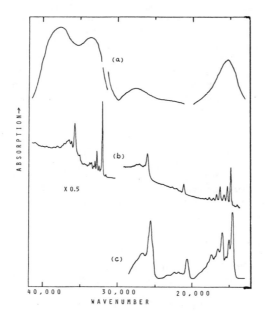

Figure 8.   Comparison of 2-methylnaphthalene cation absorption
spectra measured with different techniques.   (a) Gas-phase
photodissociation spectrum, replotted from Ref. [110], (b) argon
matrix absorption spectrum from Ref. [108], and (c) Freon matrix
absorption spectrum, replotted from Ref. [105].

Argon matrix absorption spectra for 1-XN$^+$ and 2-XN$^+$ (X=F, Cl, Br)
substituted naphthalene cations have been obtained in similar
experiments [108].   Substituent effects have been noted for the band
origins and the skeletal distortion vibronic intervals in the red
absorption bands for each substituted naphthalene cation.

HALOMETHANE PARENT ANIONS

Although cations have received more spectroscopic study in matrices
than anions, the latter are trapped in equal numbers.   The cations
have, in general, provided richer absorption spectra, since in many
cases the anion is a halide ion.   Several matrix systems give
strongly absorbing molecular anions, and this review will close with
a description of these systems.

The first infrared spectroscopic evidence for a parent radical anion

came from proton radiolysis studies on haloform molecules during
condensation [111]. New bands photosensitive to light bulb
photolysis were observed at 652, 621, 593 and 570 cm^{-1} in $CHCl_3$,
$CHCl_2Br$, $CHClBr_2$ and $CHBr_3$ experiments, respectively. The 652 cm^{-1}
$CHCl_3$ product band shifted to 645 cm^{-1} for $^{13}CHCl_3$ and 529 cm^{-1} for
$CDCl_3$. The vibrational mode is primarily a carbon-halogen stretch
although the large deuterium and small carbon-13 shifts indicate
considerable mixing with a hydrogen deformation mode. The
observation of four bands from the four different precursors
requires the presence of three halogen atoms in the species. These
carbon-halogen stretching modes fall 100-120 cm^{-1} below values for
the strongest neutral parent modes which, along with their extremely
photosensitive nature, suggests assignment to the parent radical
anion. The unusual mode mixing also infers that $CHCl_3^-$ is distorted
from C_{3v} symmetry.

Spectroscopic evidence has been presented for parent radical anions
in the matrix photoionization studies of CF_3Cl, CF_2Cl_2 and $CFCl_3$
systems [15,66], and in ESR studies following radiolysis [112].
Infrared absorptions have been assigned to the CF_3X^- species [67];
the C-F stretching mode near 900 cm^{-1} for these parent anions is
below the neutral parent fundamentals.

The first matrix photoionization studies on $CHCl_3$ by Jacox and
Milligan reported new bands at 2723, 2498, 1271 and 838 cm^{-1}, which
were also observed following mercury arc photolysis of $Ar/CHCl_3/Na$
samples and assigned to the molecular anion $CHCl_2^-$ [65]. An
investigation of the complete mixed chlorobromoform species in this
laboratory provided new spectroscopic information and assignments.
The $CHCl_2Br$ and $CHClBr_2$ studies each provided two new sets of four
product bands and a sixth new set was found with the $CHBr_3$,
precursor, which indicates the presence of a third inequivalent
halogen atom in the anion product species [63]. The carbon-13 and
deuterium isotopic data and the mixed bromochloroform spectra have
verified the $(CHX_2)(X)$ stoichiometry for the molecular anion
product. The heavy halogen substitution behavior requires the
intramolecular hydrogen-bonded arrangement X^---HCX_2 formed upon
electron capture by CHX_3 in the matrix photoionization experiments;
the gas phase products, however, would be X^- and the CHX_2 radical.
The matrix cage retains the photoproducts long enough for the stable
hydrogen bonding arrangement to be formed. The strong broad 2723
cm^{-1} C-H stretching absorption, ν_s, for Cl^---$HCCl_2$ exhibited a 9 ± 1
cm^{-1} carbon-13 shift and new counterparts absorbed at 2795 cm^{-1} for
Br^---$HCCl_2$ and at 2863 cm^{-1} for I^---$HCCl_2$ [113]. This is the
expected trend of larger displacement from a 3050 ± 50 cm^{-1} value
for the C-H stretching fundamental of $HCCl_2$ radical as the proton
affinity of the hydrogen bonding halide increases and the strength
of the hydrogen bond increases. The 2498, 1271 and 838 cm^{-1} bands
have been assigned to $2\nu_b$, ν_b (hydrogen bending) and ν_x
(antisymmetric C-Cl_2 stretching) modes, respectively, for
Cl^---$HCCl_2$. These modes demonstrate the expected spectroscopic
effect of hydrogen bonding in a type I species involving the
perturbation of an existing chemical bond, which exhibits the proper
changes as Cl^- is replaced by Br^- and by I^- [113-115]. The bending

mode, ν_b, is shifted up from the 1226 cm^{-1} CHCl$_2$ free radical value and the overtone $2\nu_b$ shows the effect of considerable anharmonicity in the bending vibration. The ν_x mode is shifted down from the 902 cm^{-1} CHCl$_2$, value [116] suggesting that hydrogen bonding a Cl$^-$ ion to CHCl$_2$ reduces pi bonding in the free radical [113].

Matrix photoionization studies on CHF$_3$ by Andrews and Prochaska revealed strong new bands at 3599, 1279, 1174 and 603 cm^{-1} in addition to CF$_3^+$ and free radical absorptions. The 3599 and 603 cm^{-1} bands exhibited 1 $\pm$ 1 cm^{-1} carbon-13 and large deuterium shifts, which characterize hydrogen vibrations not involving carbon, whereas the 1174 and 1279 cm^{-1} bands exhibited large carbon-13 and 1 cm^{-1} positive deuterium shifts, which are in accord with C-F vibrations. These new absorptions have been assigned to the fluoroform electron-capture photolysis product (F-H)--(CF$_2$)$^-$ produced by fluoride rearrangement and proton abstraction [117]. This intramolecular hydrogen bonded anion is a type III species which involves the formation of a new bond with residual perturbation. The similar species (F-H)--(CFCl)$^-$ and (F-H)--(CCl$_2$)$^-$ have been observed by Jacox and Millagan and in this laboratory and the corresponding (Cl-H)--(CF$_2$)$^-$, (Br-H)--(CF$_2$)$^-$ and (Cl-H)--(CFCl)$^-$ species have also been characterized [117-119].

The nature of the hydrogen bond formed in haloform electron-capture products, i.e., type I or type III, depends on the proton affinities of the halide ion eliminated and the CX$_2^-$ group that can be formed by proton abstraction. Carbon-13 isotopic studies are essential to characterize the hydrogen vibration and identify the new hydrogen bond. Haloform molecular anions rearrange to intramolecular hydrogen-bonded anions in matrix photoionization experiments, which provide interesting subjects for the study of hydrogen bonding.

CONCLUSIONS

Molecular ions in matrices may be classified as "chemically bound" or "isolated" with respect to the counterion. Alkali metal reactions necessarily produce bound ion pairs whereas ionizing radiation is required to produce cation and anions isolated from each other in the matrix. The isolated ion-matrix interaction is of interest as matrix spectra of ions are related to the gas phase. Although ionization exhibits a substantial shift to lower energy in solid argon, infrared absorption spectra of molecular cations are generally sharp (3 cm^{-1} FWHM) and matrix observations should correspond closely with gas phase vibrational fundamentals.

The effect that removing an electron from a molecule has on the vibrational potential function depends upon the location of the "hole" as can be seen from the examples described here. Halogens bonded to a carbocation center exhibit substantial pi bonding with carbon. The effect of a "hole" in an adjacent bond (C-H in CH$_2$F$_2^+$) or on an adjacent atom (X in CF$_3$X$^+$) is to increase the C-F vibrational frequencies presumably through polarization of the C-F bonding electrons and/or F lone pair electrons by the adjacent "hole". This effect is reduced when the "hole" is delocalized in a pi system as in the case of naphthalene cation. The bond stretching fundamental is, of course, reduced upon ionization of a bonding electron, as shown by the C-H fundamentals in $CH_2F_2^+$.

Mercury arc photolysis is an important diagnostic for isolated ions in noble gas matrices. Photolysis rearranges, dissociates or neutralizes cations; the latter requires electron detachment from anions and shows a dependence on the nature of the anion. For ions that do not photodissociate, such as naphthalene cation, a dynamic electron-transfer equilibrium is established between precursor (P) and electron traps (T):

$$P + T \; \underset{\text{photodetachment}}{\overset{\text{photoionization}}{\rightleftarrows}} \; P^+ + T^-$$

The important considerations are the photoionization efficiency of P and the photodetachment cross-section of T^- for the radiation employed. In the naphthalene system with CCl_4 added as an electron trap, high-pressure mecury arc photolysis is much more effective at two-color photoionization of N than photodetachment from Cl^-, and a large yield of isolated N^+ is produced in solid argon. The argon matrix also provides an effective medium for studying photochemical rearrangement of molecular cations. The cation is activated by irradiation into an absorption band; the matrix quenches internal energy and allows rearrangement to a non-absorbing structure to be competitive with dissociation. This rearrangement process can, in some cases, be reversed by irradiation in the absorption band of the second structural isomer.

Matrix studies of molecular ions are useful in their own right and, of perhaps more importance, matrix studies are complementary to gas phase investigations. The cold solid quenches internal energy, sharpens spectral bands and allows facile ions like CCl_4^+ to be stabilized for spectroscopic study.

ACKNOWLEDGEMENTS

The author gratefully acknowledges financial support for this research from the National Science Foundation and the contributions of many coworkers whose names appear in the following references.

REFERENCES

[1] L. Andrews, J. Am. Chem. Soc. 90 (1968) 7368.
[2] L. Andrews, J. Am. Chem. Soc. 50 (1969) 4288.
[3] P.H. Kasai and D. McLeod, Jr., J. Chem. Phys. 51 (1969) 1250.
[4] D.E. Milligan and M.E. Jacox, J. Chem. Phys. 51 (1969) 1952.
[5] L.E. Brus and V.E. Bondybey, J. Chem. Phys. 63 (1975) 3123.
[6] W.L.S. Andrews and G.C. Pimentel, J. Chem. Phys. 44 (1966) 2361.
[7] L. Andrews, J.T. Hwang and C. Trindle, J. Phys. Chem. 77 (1973) 1065.
[8] P.H. Kasai, Acct. Chem. Res. 4 (1977) 329.
[9] A. Gedanken, B. Raz and J. Jortner, J. Chem. Phys. 58 (1973) 1178.
[10] L. Andrews, J.M. Grzybowski and R.O. Allen, J. Phys. Chem. 79 (1975) 904
[11] F.T. Prochaska and L. Andrews, J. Chem. Phys. 67 (1977) 1091.
[12] M.E. Jacox, Chem. Phys. 12 (1976) 51.
[13] C.A. Wight, B.S. Ault and L. Andrews, J. Chem. Phys. 65 (1976) 1244.
[14] R.R. Smardzewski, Appl. Spectros. 4 (1977) 332.
[15] F.T. Prochaska and L. Andrews, J. Chem. Phys. 68 (1978) 5577.
[16] L. Andrews, D.E. Tevault, and R.R. Smardzewski, Appl. Spectrosc. 32 (1978) 157.
[17] B.S. Ault and L. Andrews, J. Am. Chem. Soc. 97 (1975) 3824.
[18] B.S. Ault and L. Andrews, J. Chem. Phys. 63 (1975) 2466.
[19] B.S. Ault and L. Andrews, J. Chem. Phys. 64 (1976) 4853.
[20] B.S. Ault and L. Andrews, J. Am. Chem. Soc. 98 (1976) 1591; Inorg. Chem. 16 (1977) 2024.
[21] J.H. Miller and L. Andrews, Inorg. Chem. 18 (1979) 988.
[22] B.S. Ault, Inorg. Chem. 18 (1979) 3339.
[23] B.J. Kelsall and L. Andrews, J. Phys. Chem. 85 (1981) 2938.
[24] M.A. Duncan, T.G. Dietz and R.E. Smalley, J. Chem. Phys. 75 (1981) 2118.
[25] B.J. Kelsall and L. Andrews, J. Chem. Phys. 76 (1982) 5005.
[26] D.A. Hatzenbühler and L. Andrews, J. Chem. Phys. 56 (1972) 3398.
[27] L. Andrews, Appl. Spectrosc. Rev. 11 (1976) 125.
[28] L. Andrews, J. Phys. Chem. 73 (1969) 3922; J. Chem. Phys. 54 (1971) 4935.
[29] L. Andrews and R.R. Smardzewski, J. Chem. Phys. 58 (1973) 2258.
[30] F.J. Adrian, E.L. Cochran and V.A. Bowers, J. Chem. Phys. 59 (1973) 56.
[31] D.M. Lindsay, D.R. Herschbach and A.L. Kwiram, Chem. Phys. Letts. 25 (1974) 175.
[32] L. Andrews, J. Mol. Spectrosc. 61 (1976) 337.
[33] M.H. Alexander, J. Chem. Phys. 69 (1978) 3502 and references therein.
[34] R.C. Spiker, Jr. and L. Andrews, J. Chem. Phys. 59 (1973) 1851.
[35] L. Andrews and R.C. Spiker, Jr., J. Chem. Phys. 59 (1973) 1863.

[36] F.J. Adrian, E.L. Cochran and V.A. Bowers, J. Chem. Phys. 61 (1974) 5463.
[37] M.E. Jacox and D.E. Milligan, J. Mol. Spectrosc. 43 (1972) 148.
[38] L. Andrews and D.E. Tevault, J. Mol. Spectrosc. 55 (1975) 452.
[39] L. Andrews, B.S. Ault, J.M. Grzybowski and R.O. Allen, J. Chem. Phys. 62 (1975) 2461.
[40] D.M. Thomas and L. Andrews, J. Mol. Spectrosc. 50 (1974) 220.
[41] L. Andrews, E.S. Prochaska and B.S. Ault, J. Chem. Phys. 69 (1978) 556.
[42] L. Andrews, J. Chem. Phys. 63 (1975) 4465.
[43] W.F. Howard, Jr. and L. Andrews, J. Am. Chem. Soc. 95 (1973) 3045; Inorg. Chem. 14 (1975) 409.
[44] L. Andrews, J. Am. Chem. Soc. 98 (1976) 2147.
[45] W.F. Howard, Jr. and L. Andrews, J. Am. Chem. Soc. 95 (1973) 2056; Inorg. Chem. 14 (1975) 767.
[46] L.C. Lee, G.P. Smith, J.T. Moseley, P.C. Cosby and J.A. Guest, J. Chem. Phys. 70 (1979) 3237.
[47] C.A. Wight, B.S. Ault and L. Andrews, Inorg. Chem. 15 (1976) 2147.
[48] W.F. Howard, Jr. and L. Andrews, J. Am. Chem. Soc. 97 (1975) 2956.
[49] L. Andrews, J. Am. Chem. Soc. 98 (1976) 2152.
[50] L. Andrews, E.S. Prochaska and A. Loewenschuss, Inorg. Chem. 19 (1980) 463.
[51] W. Kiefer and H.J. Bernstein, Chem. Phys. Letts. 16 (1972) 5.
[52] P.N. Noble and G.C. Pimentel, J. Chem. Phys. 49 (1968) 3165.
[53] D.E. Milligan and M.E. Jacox, J. Chem. Phys. 53 (1970) 2034.
[54] V.E. Bondybey, G.C. Pimentel and P.N. Noble, J. Chem. Phys. 55 (1971) 540.
[55] D.E. Milligan and M.E. Jacox, J. Chem. Phys. 55 (1971) 2250.
[56] B.S. Ault and L. Andrews, J. Chem. Phys. 64 (1976) 1986.
[57] J.A. Ibers, J. Chem. Phys. 40 (1964) 402.
[58] B.S. Ault, J. Chem. Phys. 82 (1978) 844; 83 (1979) 837.
[59] S.A. McDonald and L. Andrews, J. Chem. Phys. 70 (1979) 3134.
[60] V.E. Bondybey and G.C. Pimentel, J. Chem. Phys. 56 (1972) 3832.
[61] D.E. Milligan and M.E. Jacox, J. Mol. Spectrosc. 46 (1973) 460.
[62] D.W. Smith and L. Andrews, J. Chem. Phys. 60 (1974) 81.
[63] L. Andrews, C.A. Wight, F.T. Prochaska, S.A. McDonald and B.S. Ault, J. Mol. Spectrosc. 73 (1978) 120.
[64] L. Andrews, F.T. Prochaska, and B.S. Ault, J. Am. Chem. Soc. 101 (1979) 9.
[65] M.E. Jacox and D.E. Milligan, J. Chem. Phys. 54 (1971) 3935.
[66] F.T. Prochaska and L. Andrews, J. Chem. Phys. 68 (1978) 5568.
[67] F.T. Prochaska and L. Andrews, J. Am. Chem. Soc. 100 (1978) 2102.
[68] M.E. Jacox, Chem. Phys. Letts. 54 (1978) 176.
[69] D.E. Milligan and M.E. Jacox, J. Chem. Phys. 48 (1968) 2265.
[70] K. Nakamoto, Infrared Spectra of Inorganic and Coordination Compounds, John Wiley, New York, 1963.

[71] L. Andrews and F.T. Prochaska, J. Chem. Phys. 83 (1979) 368.
[72] D.W. Turner, C. Baker, C.W. Baker and C.R. Brundle, Molecular Photoelectron Spectroscopy, John Wiley, New York, 1970.
[73] F.T. Prochaska and L. Andrews, J. Phys. Chem. 82 (1978) 1731.
[74] A.S. Werner, B.P. Tasi and T. Baer, J. Chem. Phys. 60 (1974) 3650.
[75] H.W. Jochims, W. Lohr and H. Baumgartel, Ber. Bunsenges. Phys. Chem. 80 (1976) 130.
[76] T. Baer, L. Squires and A.S. Werner, Chem. Phys. 6 (1974) 325.
[77] L. Andrews and F.T. Prochaska, J. Chem. Phys. 70 (1979) 4714.
[78] F.T. Prochaska and L. Andrews, J. Chem. Phys. 73 (1980) 2651.
[79] J.H. Miller and L. Andrews, J. Am. Chem. Soc. 102 (1980) 4900.
[80] P.J. Bassett and D.R. Lloyd, J. Chem. Soc. A (1971) 1551.
[81] L. Andrews and B.W. Keelan, J. Am. Chem. Soc. 101 (1979) 3500.
[82] D.E. Milligan and M.E. Jacox, J. Chem. Phys. 47 (1967) 703; L. Andrews, J. Chem. Phys. 48 (1968) 979.
[83] L. Andrews and T.G. Carver, J. Chem. Phys. 49 (1969) 896; A.K. Maltsev, R.H. Hauge and J.L. Margrave, J. Phys. Chem. 75 (1971) 3984.
[84] B.J. Kelsall and L. Andrews, to be published.
[85] J.T. Bursey, M.M. Bursey and D.G.I. Kingston, Chem. Rev. 73 (1973) 191.
[86] F.W. McLafferty and J. Winkler, J. Am. Chem. Soc. 96 (1974) 5182; F.W. McLafferty and F.M. Bockhoff, J. Am. Chem. Soc. 101 (1979) 1783.
[87] J. Shen, R.C. Dunbar and G.A. Olah, J. Am. Chem. Soc. 96 (1974) 6227.
[88] J.M. Abboud, W.J. Hehre and R.W. Taft, J. Am. Chem. Soc. 98 (1976) 6072.
[89] W.E. Doering and L.H. Knox, J. Am. Chem. Soc. 76 (1954) 3203.
[90] G. Naville, H. Strauss and E. Heilbronner, Helv. Chem. Acta 43 (1952) 1221, 1243; E. Heilbronner and J.N. Murrell, Mol. Phys. 6 (1963) 1.
[91] G.A. Olah, C.U. Pittman, Jr., R. Waack and M. Dorna, J. Am. Chem. Soc. 68 (1966) 1488.
[92] R.L. Jones and L.M. Dorfman, J. Am. Chem. Soc. 96 (1974) 5715.
[93] L. Andrews and B.W. Keelan, J. Am. Chem. Soc. 103 (1981) 99.
[94] F.A. Houle and J.L. Beauchamp, J. Am. Chem. Soc. 100 (1978) 3290.
[95] B.A. Thrush and J.J. Zwolenik, Disc. Far. Soc. 35 (1963) 196.
[96] C. Cone, M.J.S. Dewar and D. Landman, J. Am. Chem. Soc. 99 (1977) 372.
[97] R.S. Berry and C.W. Reimann, J. Chem. Phys. 38 (1963) 1540.
[98] E.W. Fu and R.C. Dunbar, J. Am. Chem. Soc. 100 (1978) 2283.
[99] L. Andrews, J.A. Harvey, B.J. Kelsall and D.C. Duffey, J. Am. Chem. Soc. 103 (1981) 6415.
[100] T. Shida and W.H. Hamill, J. Chem. Phys. 44 (1966) 4372.
[101] J.W. Rabalais and R.J. Colton, J. Electron Spectrosc. and Rel. Phen. 1 (1972/73) 83.
[102] B.J. Kelsall, L. Andrews and H. Schwarz, to be published.
[103] J.H.D. Eland and C.J. Danby, Z. Naturforsch 23a (1968) 355.

[104] P.A. Clark, F. Brogli and E. Heilbronner, Helv. Chem. Acta 55
       (1972) 1415.
[105] T. Shida and S. Iwata, J. Am. Chem. Soc. 95 (1973) 3473.
[106] M.S. Kim and R.C. Dunbar, J. Chem. Phys. 72 (1980) 4405.
[107] L. Andrews and T.A. Blankenship, J. Am. Chem. Soc. 103 (1981)
       5977.
[108] L. Andrews, B.J. Kelsall and T.A. Blankenship, J. Phys. Chem.
       86 (1982) 2916.
[109] S.S. Mitra and H.J. Bernstein, Can. J. Chem. 37 (1959) 553.
[110] R.C. Dunbar and R. Klein, J. Am. Chem. Soc. 98 (1976) 7994.
[111] B.S. Ault and L. Andrews, J. Chem. Phys. 63 (1975) 1411.
[112] A. Hasegawa, M. Shiotani and F. Williams, Far. Disc. Chem.
       Soc. 63 (1977) 157.
[113] L. Andrews and F.T. Prochaska, J. Am. Chem. Soc. 101 (1979)
       1190.
[114] B.S. Ault, E. Steinback and G.C. Pimentel, J. Phys. Chem.
       79 (1975) 615.
[115] G.C. Pimentel and A.L. McClellan, The Hydrogen Bond, W.H.
       Freeman, San Francisco, 1960.
[116] T.G. Carver and L. Andrews, J. Chem. Phys. 50 (1969) 4235.
[117] L. Andrews and F.T. Prochaska, J. Phys. Chem. 83 (1979) 824.
[118] M.E. Jacox and D.E. Milligan, Chem. Phys. 16 (1979) 195, 381.
[119] B.W. Keelan and L. Andrews, J. Phys. Chem. 83 (1979) 2488.

Molecular Ions: Spectroscopy, Structure and Chemistry
Terry A. Miller and V.E. Bondybey (editors)
© North-Holland Publishing Company, 1983

Vibronic Spectroscopy and Photophysics of Molecular Ions in Low Temperature Matrices

by

V. E. Bondybey and Terry A. Miller

Bell Laboratories
Murray Hill, New Jersey 07974

ABSTRACT

Molecular ions in solid matrices are readily generated in adequate concentrations for spectroscopic studies by direct VUV photoionization of neutral parent compounds. The ions are investigated by electronic spectroscopy and, in particular, by time resolved laser induced fluorescence. The photophysics of the ions and their interactions with the matrix are discussed in some detail and solid neon is shown to be a particularly useful medium for ionic studies. Experimental techniques for matrix studies are described and experimental results summarized, tabulated and discussed.

I. INTRODUCTION

Matrix isolation techniques have been very successful in studies of free radicals and other transient species [1-3]. Highly reactive molecular species, which are extremely short lived under typical laboratory conditions are often indefinitely stable when prepared in a solid, chemically inert environment, and their properties and spectra can be conveniently studied. The use of rare gas solids specifically as a medium for spectroscopic studies, originally pioneered by Pimentel and coworkers [4-6], has contributed greatly to our knowledge of free radicals and their spectra. Of the numerous investigators active in this field, the early contributions of Milligan and Jacox and by Andrews and coworkers can perhaps be singled out [7-9].

Although the electronic spectra of a variety of free radicals were observed, overall the matrix studies of transient species have relied more heavily upon infrared spectroscopy. The major obstacles to exploiting the full potential of electronic spectroscopy are perhaps the various line broadening mechanisms operative in the solid matrix. If severe broadening of the individual vibronic transitions occurs, the vibrational structure may be lost and little useful information will be obtained. In many instances, the major contribution to the observed linewidths in the electronic spectrum comes from inhomogeneous broadening. This is due to the fact that different guest molecules in the glassy solid may be located in slightly different local environments or "sites," which may result in small differential shifts of the individual electronic states of the absorber. The overall absorption spectrum is then the result of the summation over all the "sites" present in the matrix.

When a broad band light source is used to excite sample fluorescence, the linewidths in emission are usually comparable to those observed in the absorption spectrum. On the other hand, when a monochromatic light source is used for sample excitation, one can often obtain sharp, well resolved spectra [10-12]. This is due to selective excitation of a particular "site" from the inhomogeneously broadened spectrum. Lasers, with their narrow bandwidths and high intensity are thus ideal sources for this type of experiments. The earliest studies of laser induced fluorescence from matrix isolated free radicals used fixed frequency, CW lasers, most often Ar ion lasers, and had thus to rely upon accidental overlap of the laser line with absorptions of the guest molecules [13-15]. The full potential of the laser induced fluorescence studies could only be realized by application of tunable dye lasers. In our laboratory, we have been for a number of years using excitation with tunable pulsed lasers coupled with time resolved detection of the resulting fluorescence for the study of the spectroscopy and dynamics of a large variety of transient species [16-18].

One class of transient species which are particularly elusive and difficult to study in the gas phase are molecular ions. Unlike the reactions of most molecules, even free radicals, that often have non-zero activation energies, most reactions of molecular ions proceed with large cross-sections and indicate the absence of any significant activation barrier. Another obstacle to preparing appreciable concentrations of ionic species in the gas-phase are the long-range coulombic interactions between charged species. Traditionally, discharges, flames and similar energy rich systems were the most common sources in which molecular ions were studied. Under the high temperatures which are characteristic of these types of systems, large numbers of molecular quantum states are typically thermally populated, and this increases drastically the complexity of the associated spectra particularly for larger molecular ions. The high temperatures thus lead to rotational and vibrational "dilution" of the signal, reducing the available signal to noise ratio for any given level and rendering observation of the spectrum more difficult. Even if a spectrum is observed, the increased complexity often makes its satisfactory interpretation impossible. In view of these difficulties it is not surprising that some 10-12 years ago the list of molecular ions observed by optical spectroscopy was restricted [19] to a variety of diatomics, 5-6 triatomic ions and a single polyatomic molecular ion [20], the cation of diacetylene, $C_4H_2^+$.

In the last few years there has been renewed interest in the spectroscopy of molecular ions and numerous important advances have occurred in this area [21-34]. The difficulties outlined above have indicated a clear need for the development of new low temperature techniques for studies of the optical properties of molecular ions. Matrix isolation in solid neon combined with time-resolved laser induced fluorescence spectroscopy was one of the techniques applied to this purpose in our laboratory. It has proved to be an extremely useful and efficient technique for generating intense and well resolved spectra and it has provided very detailed information about a large variety of molecular ions.

It is the purpose of this manuscript to review and summarize the matrix isolation studies of molecular ions, with particular emphasis on laser induced fluorescence. In Section II we will discuss the experimental techniques for generating and trapping transient species and molecular ions in low temperature matrices and for obtaining their laser induced fluorescence or absorption spectra. We will also review the advantages and drawbacks of the matrix techniques and consider the importance and magnitude of matrix perturbation and effects.

In Section III we will review in a more systematic way, the individual ionic species investigated. We will discuss separately several distinct groups of compounds whose molecular cations have been studied by the above matrix techniques and summarize the spectroscopic and photochemical information obtained.

II. EXPERIMENTAL TECHNIQUES AND GENERAL OBSERVATIONS

1. Matrix Technique and Spectroscopic Tools

The matrix technique itself has been described in numerous previous reviews [1-3]. We will therefore address it here only very briefly, and emphasize the particular systems and spectroscopic tools employed in our laboratory. A typical matrix spectroscopic setup is shown in Fig. 1. In the simplest experiment, the molecules of interest or a suitable precursor are diluted with a large excess of a matrix gas — usually neon in our experiments — and deposited on a cold solid substrate. A platinum or iridium plated copper mirror cooled to 4K by liquid helium is used in most of our work.

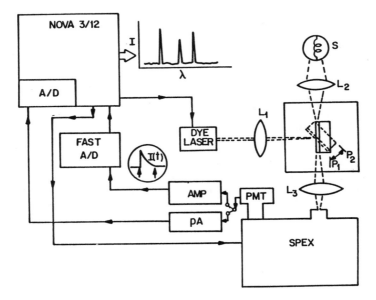

Fig. 1.   Schematic diagram of a typical apparatus for electronic fluorescence and absorption studies of matrix isolated molecules. The sample is deposited on a metal (Pt plated copper) substrate. The substrate in an evacuated shroud is rotatable, to permit optimum positioning for deposition, photolysis, absorption (position shown in solid lines) or LIF (position shown dashed) studies. The signal can be processed either in the DC mode (for absorption studies or very long-lived emissions) or in the time resolved, pulsed mode. Monochromator and lasers are controlled and the signal is processed by a minicomputer.

An important parameter is the dilution ratio of the host matrix to the solute guest, usually denoted M/R. A substitutional guest in a rare gas lattice has 12 nearest neighbor atoms. Thus, for instance, at M/R ~ 100, for an entirely statistical distribution, approximately one out of eight guest molecules will have another guest as a nearest neighbor. Typically guest-guest interactions are much stronger than those between guest and an inert gas host and the concentrations of guest dimers and clusters are much higher than expected purely on statistical considerations. Usually, M/R ratios of at least 1000 are desirable to assure effective isolation. In most of our work, even higher dilutions, in the neighborhood of $1:10^4$, are employed.

One of the ports of the cryostat (see Fig. 1) is fitted with a lithium fluoride window, permitting photolysis of the sample at wavelengths from near IR to vacuum UV. The sample substrate is rotatable within the vacuum shroud to permit optimum positioning of the sample for deposition, photolysis or for spectroscopic studies.

A variety of lasers are available for sample excitation in the fluorescence studies. Most commonly a Hänsch design dye laser pumped by a 300 kW N_2 laser is used for excitation. The laser gives pulses of ~10 ns duration and 15-30 GHz (0.5-1 cm^{-1}) spectral linewidth over the range of 3500-7000 Å. Typically, the laser is operated without amplifier and gives peak powers of 15-30 kW. In most fluorescence experiments it is convenient to lower the power by a factor of 10-100 to avoid saturating the transitions studied. This is conveniently done by the insertion of neutral density filters into the laser beam. The laser is usually loosely focused upon the sample using a 10 cm focal length cylindrical lens. In some experiments the laser frequency may be doubled using angle tuned KDP or KPB crystals. Alternatively, an excimer laser and a flash lamp pumped dye laser are also available for excitation.

The sample fluorescence is collected by a lens. It can be either viewed undispersed through a suitable combination of cutoff or interference bandpass filters, or be wavelength resolved in a SPEX 14018 double monochromator with holographic gratings. The high stray light rejection capability of this monochromator is not needed for most work with strongly emitting samples, but is quite useful in some experiments involving species or vibronic levels fluorescing with low quantum efficiency.

A disadvantage of techniques relying upon fluorescence is that they are restricted to species possessing excited electronic states with appreciable emission quantum yields. The ability to observe absorption spectra is therefore important for studies of nonfluorescent molecules and also facilitates the initial search for previously unknown or poorly characterized electronic transitions. Absorption studies are hampered by the inherently lower sensitivity of absorption spectroscopy coupled with the requirement of working with low concentrations. In order to enhance the sensitivity of the absorption experiments, a sample geometry shown schematically in Fig. 2 is used [33]. Two rectangular copper bars are mounted near each edge of the metal mirror substrate. These are held by suitable spacers above the mirror surface in such a way that ~50-200 μ slits are formed. The sample is then deposited to a thickness exceeding this slit-width, so that the space between the two slits is completely filled by a continuous sheet of the solid rare gas. The thickness of the deposit during the deposition is conveniently monitored in the usual fashion by counting the interference maxima in the intensity of a helium-neon 6328 Å laser beam reflected at an oblique angle from the sample surface.

Following deposition the sample is rotated into the position shown by the solid lines in Fig. 1. The light from the background source — usually either a xenon arc or a tungsten-halogen filament lamp — is focussed upon one of the slits in a direction parallel to the substrate surface [33,34]. The light injected into the sample is then prevented from exiting the sample by reflection on the mirror surface at one side and by total internal reflection at the rare gas-vacuum interface at the other, and propagates longitudinally in a waveguide fashion through the matrix. The light exiting from the slit at the opposite edge of the substrate is then focussed upon the entrance slit of the monochromator and analyzed. The advantage of this approach as opposed to conventional matrix absorption spectroscopy is that it increases the sensitivity by the ratio of matrix width (~20 mm) to its thickness (~0.1 mm), or, in the present case, a factor of approximately 200.

Application of this technique requires perfectly clear deposits without cracks and optical imperfections. Such deposits are readily grown at 4K with neon matrices. The heavier rare

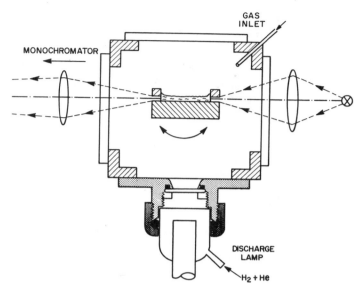

Fig. 2.   A detail of the vacuum shroud and sample substrate for matrix studies. Two copper bars ~3 × 2 × 40 mm are held 100 μ above the metal substrate, to form the entrance and exit slits. The light from the source is loosely focussed on the entrance slit by the lens (the lens is actually much further from the substrate than shown). The transmitted light is collected by a cylinderical lens and focussed upon the entrance slit of the monochromator. The substrate is rotatable to permit optimum positioning of the sample for deposition, photolysis or spectroscopic studies. A microwave powered atomic emission lamp is mounted on one of the faces of the cryostat to permit photolysis of the samples through a LiF window.

gases deposited at 4K usually give scattery deposits not suitable for the waveguide technique.

For all of the spectroscopic studies, the photomultiplier signals are amplified and digitized, usually in a Biomation 8100 waveform recorder with a 10 ns/channel resolution. The digitized signal is then averaged and further processed in a Nova III/12 minicomputer, which controls the entire experiment.

2. Generation of Transient Species

Numerous techniques have been employed to trap transient species in low temperature matrices. Perhaps the most straightforward approach involves fragmentation of suitable, stable precursors to obtain the desired reactive fragment. This fragmentation can occur either just prior to deposition in the gas phase, or following deposition in the solid solution. Early matrix studies commonly produced transient species to be studied in microwave or RF discharges through a mixture of the matrix gas with suitable precursors [35-38]. Other approaches, such as pyrolysis or gas phase photolysis of a larger precursor have also been used [39,40]. Another frequently used technique, developed among others by Milligan and Jacox and their coworkers involves direct photolysis of matrix deposits with high energy radiation. While mercury or xenon arc sources were used for this purpose in some instances [41,42], more often shorter wavelength vacuum UV emission from microwave powered atomic resonance lamps was needed to generate observable quantities of the transient products [43-45]. Also X-ray [46], electron beam and accelerated H^+ or D^+ ion irradiation have been successfully applied [47,48].

A useful alternative to photolysis or discharge techniques involves generation of the transient species by a bimolecular chemical reaction. Thus Andrews, Pimentel and coworkers [49,50] have extensively used and perfected techniques involving reactions of alkali metal atoms. In their experiments, the alkali metal vapors are trapped in the rare gas matrix together with a suitable reactant, for instance a methyl halide. Matrix reaction of the type

$$RX + M \rightarrow R + MX$$

produces alkali halide, MX, and a free radical R. Another bimolecular reaction frequently used to generate transient species involves abstraction of hydrogen by fluorine atoms,

$$RH + F \rightarrow HF + R .$$

This technique which is rather common in gas phase work has been successfully applied in numerous matrix studies mainly by Jacox [51]. A major drawback of these and many similar approaches is the formation of a stable molecule (in the above examples hydrogen or alkali metal halides) next to the radical or fragment of interest. This partially defeats the objective of the matrix isolation technique, whose primary purpose is to generate and study transient species in an inert and weakly interacting environment. The presence of a strongly polar nearest neighbor can result in rather strong perturbations of the guest spectra.

Alkali atoms are characterized by low ionization potentials, and when they are codeposited with molecules with high electron affinity such as O_2 or NO_2, spontaneous ionization can sometimes take place producing ion pairs [52,53] such as $M^+O_2^-$. In such ion pairs, the presence of the nearest neighbor alkali metal cation can again perturb rather strongly the spectrum of the molecular negative ion. Thus, for instance, in the above mentioned example of the $M^+O_2^-$ ion pairs, the O_2^- vibrational frequency exhibits a dependence upon the identity of the M^+ cation [54].

The disadvantage of the photolytic or discharge methods, on the other hand, is that they are usually nonselective and often a variety of different products are generated. Despite this drawback, we use in our laboratory almost exclusively UV photolysis to generate both free radical species and molecular ions. With good understanding of the processes and reactions occurring in the matrix, and of the effects of the solvent on the photophysics of the guest molecules, a fairly clean and selective production of the species of interest can often be achieved.

3. Solvent Cage Effect

The major mechanism by which the solid matrix affects and modifies the guest photochemistry is the so called "cage effect". For example, while Cl_2 or ICl dissociates in the gas phase in a single vibrational period when excited into the repulsive $^1\Pi_u$ state, in the matrix the solvent cage prevents diffusion of the fragments from the excitation site [55,56]. The atomic fragments quickly lose their excess energy in collisions with the solvent and eventually reform the molecular ground state. The quantum yield for permanent dissociation is $<10^{-5}$. Lighter atoms, on the other hand, and hydrogen in particular can diffuse through the rare gas matrices with relative ease. It is therefore usually convenient to choose a precursor which has the same heavy atom configuration as the desired fragment. The photolysis then removes one or several hydrogen atoms to generate the species of interest.

Unfortunately, to produce quantities of transient species observable by infrared spectroscopy, very long sample deposition and photolysis times are frequently required. Continuous photolysis for 10-20 hours appears to be no exception. The primary photolysis products can, of course, usually be further photolyzed and very extensive fragmentation of the precursor molecules can take place. This is evidenced by the long lived phosphorescence of atomic fragments $2p^3$ $^2D^o \rightarrow {}^4S^o$ N and $2p^4$ $^1D \rightarrow {}^3P$ O which is invariably observed following vacuum UV photolysis of precursors containing oxygen or nitrogen. Similarly, molecular fragments C_2 and C_2^- are usually readily detectable in photolyzed hydrocarbon samples. This extensive fragmentation can, of course, result in a large variety of products and may make unambiguous assignment of the infrared or optical spectra quite difficult.

A great advantage of electronic spectroscopy in general and laser induced fluorescence in particular is its much greater sensitivity. This permits the use of much more dilute matrices, typically $M/R = 1:10^4$ or more, and basically eliminates the possibilities of guest-guest interactions. Furthermore, only a rather short photolysis of the deposits is usually needed to generate observable amounts of the product of interest. In this way the influence of secondary photolytic processes may be greatly reduced.

4. The Photoionization Process and Formation of Molecular Ions

If the energy of the photolysis source is sufficient for ionizing the isolated guest molecule, an electron can be ejected and the corresponding cation formed. Microwave discharges through suitable low pressure gas mixtures are useful sources of such high energy radiation. Thus a discharge through a hydrogen-helium mixture produces the atomic H emission, while argon-helium is a source of Ar resonance fluorescence. The 1216 Å-Lyman α hydrogen line and the 1048 Å argon emission are efficiently transmitted through the LiF window usually employed to separate the lamp from the cryostat [57,58]. The corresponding energies of 10.2 and 11.8 eV, respectively, are adequate to ionize a large variety of both organic and inorganic molecules, particularly if one considers the red shifts which usually accompany ionizing transitions in a solid medium. In agreement with this expectation, photolysis of matrix samples with these sources readily generates numerous ions. Often even with highly diluted matrices clearly observable ionic spectra are easily produced. As an example, an absorption spectrum of the hexafluorobenzene cation, $C_6F_6^+$, obtained by the waveguide technique is shown in Fig. 3. Typically, only 5-30 seconds are required to produce intense ionic absorptions often with the stronger bands nearly 100% absorbing [33,58,59].

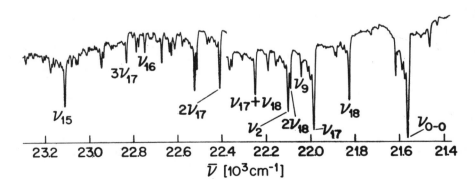

Fig. 3.   Part of a $C_6F_6^+$ absorption spectrum near the origin. A sample of C_6F_6 in neon ($1:10^4$) was photolyzed ~25s. The 0_0^0 origin is nearly fully absorbing, leading to some nonlinear distortion of the relative intensities.

In order to establish, whether much higher ionic concentrations could be produced by either extending the length of photolysis time, or by photolyzing during the deposition, numerous experiments were performed in our laboratory in which the ion concentration was monitored as a function of photolysis time. While in the first few seconds the concentration increases linearly, the rate of ion formation usually starts very quickly to level off. Little improvement in ion concentration is gained by photolyzing beyond 5-10 min.

The saturation of the ion concentration in most of these experiments could certainly not be attributed to the depletion of the parent molecules. It can easily be shown that even if each UV photon from our low intensity photolysis lamp were utilized to produce one cation, only <0.1% of the parent molecules originally present would be ionized in the available time. This was confirmed experimentally by examining IR spectra of the photolyzed samples which even after a very prolonged (several hours) irradiation, showed only marginal decreases in the parent compound absorptions.

Similarly, in a sample of $CS_2 + Ne$ (1:1000) exposed for 5 min. to Lyman 1216 Å radiation, using IR absorption only the strongest ν_3 fundamental of CS_2^+ was tentatively identified as an extremely weak band at 1211 cm^{-1}. This band, barely observable above the noise level, was at least a factor of 2000 weaker than the corresponding parent fundamental at 1523 cm^{-1}. On the other hand, the absorption of the CS fragment was clearly observable at 1280 cm^{-1} in the same sample, with about ½ the intensity of the $^{13}CS_2$ ν_3 fundamental in natural isotopic abundance. While the CS absorption grew in intensity approximately linearly with photolysis time, no further increase could be accomplished in the 1211 cm^{-1} absorption by prolonged photolysis. Samples prepared under similar conditions were after 5 min. photolysis fully absorbing in the visible in the region of the CS_2^+ $\tilde{A}^2\Pi_u \rightarrow \tilde{X}^2\Pi_g$ 0-0 band [60]. Similarly, no $C_6F_6^+$ infrared absorptions were observed in photolyzed deposits of hexafluorobenzene in neon, even after 30 min. photolysis, although the optical $\tilde{B}$-$\tilde{X}$ transition was fully absorbing after several seconds. Again, very little depletion of the parent could be detected.

These observations can be qualitatively explained with the aid of Fig. 4. The cations (e.g. CS_2^+, $C_6F_6^+$) are formed in the matrix following absorption of the 1216 Å (10.2 eV) photon by the parent molecule (C_6F_6) and ejection of an electron. It has been shown experimentally [61], that the electrons can readily propagate through the matrix over distances of $>10^2$-10^3 Å. They gradually lose their excess energy in collisions with the lattice atoms and eventually become attached to some guest or impurity molecule present in the matrix. At the beginning of the photolysis, the most abundant such "impurity" will be the parent molecule (C_6F_6) itself as shown in Fig. 4a, and thus the corresponding anions ($C_6F_6^-$) will undoubtedly be formed. At later times, represented schematically by Fig. 4b, a concentration of cations has built up in the matrix. Thus an electron ejected from one C_6F_6 molecule can, instead of attachment to another neutral C_6F_6 molecule, recombine with a $C_6F_6^+$ cation already present in the matrix. If this process occurs, one cation has been created and another destroyed, resulting in no net change in ionic concentrations. Since the cross-section for electron-cation recombination is obviously much larger than that for attachment to a neutral molecule, only a relatively small concentration of cations in the matrix is needed for this former process to dominate. Furthermore, the radiation used to ionize the neutral guest molecules can also destroy ions already present. This can occur either by electron detachment from an anion, followed again by electron-cation recombination, or by secondary photolysis and fragmentation of the cations. The net result of these processes is the observed levelling off in the ion production, and after very long photolysis times undoubtedly a gradual decrease in parent ion population.

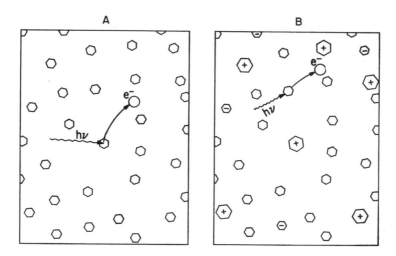

Fig. 4.   Schematic diagram explaining the photoprocesses occuring in the matrix. At short photolysis times — panel A — the prevalent process is photoionization. At longer times — panel B — when sizable concentrations of photoproducts have been formed in the matrix, competing processes hamper further build-up of ion concentration — see text.

With the above model it becomes quite understandable why attempts to enhance cation formation in the matrix by addition of "electron scavenger" molecules like SF_6 or NO_2 were not very successful. The electron affinities of molecules like C_6F_6 and of various impurities usually present in the matrix, e.g. O_2, N_2, or H_2O, are very much larger than kT at 4K, and they are therefore quite adequate as electron scavengers. Addition of deeper "trap" molecules with higher electron affinities is usually of little benefit, and makes the experiment more complex and interpretation of data potentially more difficult.

5. Vibrational Relaxation

As noted in the preceding sections, the cage effect preventing molecular dissociation is one of the important ways in which the matrix modifies the photophysical behavior of the guest molecules. Another important effect of the medium is vibrational relaxation. A small molecule in the gas phase when placed into a vibrationally excited level will, in the absence of collisions, remain in that level and can only return to the ground state by a radiative process. However, a solid matrix can act as an effective energy sink and vibrational relaxation can occur and, in fact, in most cases does occur even for small molecules. In diatomic molecules vibrational relaxation can often be relatively slow [17]; only in a handfull of triatomic species in low temperature matrices has emission from vibrationally unrelaxed levels of an excited electronic state ever been observed [62,63]. In very small molecules with large vibrational spacing, relaxation requires relatively inefficient high order multiphonon process. In larger, polyatomic molecules which are characterized by a high density of vibrational states and usually possess low frequency skeletal vibrations, relaxation can occur by low order phonon processes, and usually only vibrationally fully relaxed fluorescence is seen.

Because of this efficient vibrational relaxation, regardless of excitation wavelengths and of the initially populated level, the bulk of fluorescence typically occurs from the vibrationless level of the excited electronic state. An advantage of the fast relaxation is the simplicity of the resulting emission spectrum. It is also this fast vibrational relaxation, which permits one to obtain "site selected" fluorescence excitation spectra of the molecule of interest. As we have noted in the introduction, the rare gas matrices are usually imperfect, polycrystalline samples and individual guest molecules can be isolated in slightly differing local environments, which often produce slight shifts in the relative electronic energies. If one monitors the intensity of the undispersed fluorescence of a matrix sample as a function of laser wavelength, one obtains a so called "total" fluorescence excitation spectrum. This is usually quite analogous to the absorption spectrum and represents a weighted sum of the excitation spectra of molecules in all these "sites" or local environments.

The high signal/noise ratio usually available in matrix experiments makes it possible to disperse the sample fluorescence in a monochromator and monitor selectively, with a narrow bandpass (~ 1 cm^{-1} or even less) a single vibronic fluorescence transition from the vibrationless level of the excited state of the guest molecule in a particular "site". If one now scans the laser wavelength, one sees enhanced fluorescence only when the laser is in resonance with a vibronic transition of a guest molecule in that particular "site" and one obtains a greatly simplified, "site selected" excitation spectrum.

While the fact that vibrational relaxation is usually fast compared with the radiative rate is useful in simplifying the emission spectra, it can in some cases limit the amount of obtainable spectroscopic information. The vibrational wavefunction of the vibrationless level is totally symmetric and the level has a vibronic symmetry equivalent to the symmetry of the emitting electronic state. Assuming an allowed electronic transition, relaxed fluorescence will therefore terminate on, and provide only information about, the totally symmetric levels of the ground electronic state. While this may not be a serious problem in molecules of low symmetry, in highly symmetric species it limits severely the obtainable information.

However, spectroscopic work has shown that for many of the organic molecular cations studied in solid neon the individual vibronic transitions have linewidths of the order of 1 cm^{-1} or less and that therefore the vibrational relaxation rates must be relatively slow, i.e. $\tau > 1$ ps. Radiative lifetimes of many allowed electronic transitions in the visible and near UV range are in the low nanosecond range. This suggests that fluorescence originating from vibrationally unrelaxed levels might be observable, albeit with low quantum yields. Very recently, such vibrationally unrelaxed emission has indeed been observed for a variety of substituted benzene cations in solid neon.

Under usual circumstances one finds for the benzenoid cations that the highest energy band in the emission spectrum coincides with the lowest energy band in absorption and is readily identified as the 0-0 origin band. Other bands to lower frequencies correspond to the $O \rightarrow v''$ type transitions. This relaxed emission spectrum is independent of which of the upper state vibrational levels is excited. If special measures are taken to favor observation of the weak, vibrationally unrelaxed fluorescence, additional sharp structure can be observed to the blue of the origin band. This weak emission is, unlike the relaxed fluorescence, strongly dependent on the excitation wavelength and is readily identified as fluorescence originating from the directly excited level and from levels populated by the vibrational relaxation process [64,65].

The fact that vibrationally unrelaxed fluorescence is observable is useful in providing two different types of new information. In the first place, by comparing the relative intensities of the unrelaxed, partially relaxed, and fully vibrationally relaxed emissions for a variety of excitation wavelengths, one can deduce information about the relaxation rates of the individual vibronic levels of the guest molecules, and about the pathways and mechanisms of the relaxation process. A comparative study of a variety of halogenated benzene cations has shown that the relaxation rates are indeed unexpectedly slow, with many low vibrational levels having lifetimes in the 10-100 ps range and hence quantum yields for emission of order 10^{-3}. The rates exhibit a generally increasing trend with increasing vibrational energy

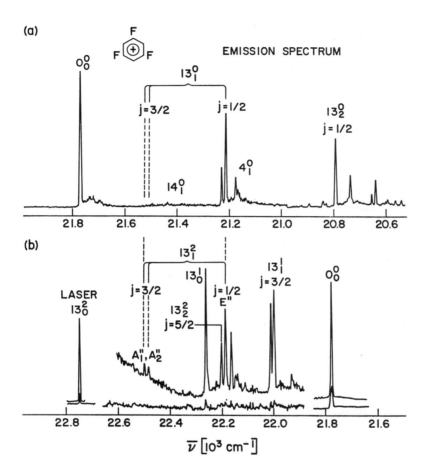

Fig. 5.   Parts of the fluorescence spectrum of sym-$C_6H_3F_3^+$ in solid neon excited by a laser at the 13_0^2 frequency. (a) Section near the 0_0^0 origin band obtained with low detection gain setting, showing a vibrationally relaxed fluorescence from the $v' = 0$, $\tilde{B}^2A$ into the vibrational levels of the ground state. Only transitions terminating in the $j=\frac{1}{2}$ components of the Jahn-Teller split ground state levels are permitted by the selection rules; thus transitions to the indicated $j=3/2$ levels are not observable for relaxed emission. (b) Spectrum showing the region *above* the 0_0^0 origin obtained (top trace) by increasing gain by 10^3 and using a narrow time gate to favor short lived unrelaxed fluorescence. Both the unrelaxed 13^2 and partially relaxed 13^1 levels are observed in fluorescence. Transitions terminating in the $j=3/2$ ground state levels, forbidden in relaxed fluorescence, are present in the unrelaxed emission as indicated.

content. Another interesting observation is that lower symmetry molecules generally relax faster than more symmetrical species, with the most symmetric molecule studied, $C_6F_6^+$ (D_{6h}) exhibiting the most extensive and intense vibrationally unrelaxed emission [65].

A second type of information obtained from these experiments is of a spectroscopic nature, and is particularly useful for high symmetry molecules. As we will discuss in a later section, benzenoid cations such as $C_6F_6^+$ or $C_6H_3F_3^+$ undergo a Jahn-Teller distortion [66,67] of their doubly degenerate ground states and as a result each of the ground state excited vibrational levels of the Jahn-Teller active modes will split into several components. Because of spectroscopic selection rules, only one of these sublevels will be observable in vibrationally relaxed fluorescence from the vibrationless level. On the other hand, in unrelaxed fluorescence from excited vibrational levels of the Jahn-Teller modes, all components can in principle be directly observed and thus provide a more complete description of the distorted ground state vibrational potential.

This is exemplified by Fig. 5 showing fluorescence of the sym $C_6H_3F_3^+$ cation with excitation of the v=2 level of the Jahn-Teller active e' mode ν_{13} at 22745 cm^{-1}. Trace 5a shows typical, vibrationally relaxed fluorescence which is observed under usual circumstances [68]. It shows a prominent 0_0^0 origin at 21776 cm^{-1} and a strong 13_0^1 (e') vibronic band 557 cm^{-1} lower in energy. In 5b, the detection sensitivity is increased (top trace compared with bottom) by $\sim 10^3$ and the detection system gated to detect preferentially emission occurring during the ~ 40 ns time interval centered around the laser pulse. This trace shows a number of additional sharp bands at energies higher than the 0_0^0 band. All three components of the Jahn-Teller split 13_1^2 band can be identified. In addition, the partially relaxed fluorescence from the 13^1 level is present, including again the 13_1^1 (a_1'', a_2'') sublevels inaccessible in the relaxed emission. By probing individual excited state levels in this way, very complete information about the ground state vibrational structure can be accumulated. As is discussed in a separate chapter of this book, this information can be inverted to provide a detailed description of the Jahn-Teller distorted ground state potential of these species.

6. Fluorescence Lifetimes and Nonradiative Processes

The density of states in small molecules is typically too low for nonradiative transitions to take place. A free molecule in a bound excited electronic state can only return into the ground state radiatively [17]. When such a molecule is placed in a condensed medium, the solvent provides the necessary density of states, and statistical limit, irreversible nonradiative transitions can take place. Because of this one often finds pronounced differences in behavior of small molecules in the gas phase and in low temperature matrices. Thus many molecules which emit with high quantum efficiencies in the gas phase have vanishingly small fluorescence quantum yields in rare gas solids. This is particularly likely to occur when already in the free molecule a strong quantum mechanical coupling exists between the emitting state and the ground state or another lower lying electronic state, as evidenced by perturbations in the spectrum. NO_2 and NH_2 free radicals, strong gas phase emitters which do not fluoresce in the matrix, are two typical examples.

From the ions studied in the matrix, CS_2^+ is an example of such situation [60]. The second excited electronic state, $B^2\Sigma_u^+$, of this cation is known to fluoresce strongly in the gas phase [69,70]. On the other hand, excitation of the $B^2\Sigma_u^+$ state in neon matrix produces only relaxed emission from the lower lying $\tilde{A}^2\Pi_u$ state [60,71]. An efficient $\tilde{B}^2\Sigma_u^+ \rightarrow \tilde{A}^2\Pi_u$ nonradiative transition apparently takes place in the matrix. Accordingly, mutual perturbations between the two states are present in the spectrum of the free ion, and the $\tilde{B}^2\Sigma_u^+ \rightarrow \tilde{A}^2\Pi_u$ interconversion seems to occur collisionally in the gas phase [72,73]. On the other hand, the $A^2\Pi$ state is characterized in solid neon by a long, 2.35 μs lifetime and probably unity quantum yield.

Unlike small molecules, the densities of states in larger, polyatomic molecules or ions are sufficiently great for nonradiative transitions to occur unimolecularly in the free species. Because of this the matrix usually has less effect upon the nonradiative processes in larger molecules. Large molecules or ions which fluoresce efficiently in the gas phase are quite likely to do so in solid neon and conversely molecules which relax nonradiatively in the gas phase will invariably do so also in the matrix.

The cations of substituted benzenes exemplify well this behavior. The radiative properties of the excited $\tilde{B}$ states of these species are governed by the relative positions and intramolecular coupling of several excited electronic states and of the ground state. Direct coupling between the ground state and the excited $\tilde{B}$ state of these species is usually inefficient, mainly because of very unfavorable Franck-Condon factors. Many of these cations, particularly with several fluorine substituents, where the $\tilde{B}$ state is the lowest excited electronic state, therefore fluoresce with unity or near unity quantum yields both in the gas phase [74] and in the matrix [75]. On the other hand, in many species with fewer fluorine substituents another electronic state, the "$\tilde{C}$ state", arising from σ ionization, drops below the first excited π ionization state, the "$\tilde{B}$ state". This state is usually strongly coupled with the $\tilde{B}$ state and provides an intermediary "gateway state" for efficient nonradiative relaxation. These species then exhibit severely shortened lifetimes and weak fluorescence, both in the matrix and in the gas phase. In many cases fluorescence is not detected at all and in the matrix one finds only broad, featureless absorption spectra.

Even in the absence of medium induced nonradiative processes the matrix modifies the fluorescence lifetimes of the isolated molecules. The major mechanism by which this occurs is by the medium providing an increased index of refraction into which the molecule radiates. (Other, usually smaller, lifetime modifying effects of the medium, are due to matrix shifts in emission wavelength caused by differential shifts of electronic states, minor changes in Franck-Condon factors due to small, medium induced changes in molecular structure and changes in molecular wavefunctions due to guest host interactions.) In a variety of systems it is found that the lifetimes scale approximately with the square of the index of refraction [76]. Several more sophisticated formulas, some taking into account the refractive indexes of both the quest and the host have been proposed [77,78].

In solid neon one finds empirically that for most species the lifetimes are shortened to ~85% of their gas phase values. We summarize the lifetimes of a variety of ions studied in neon matrices in Table I and show for comparison their gas phase values.

7. Effects of the Matrix on Ionic Spectra

Typical energies of chemical bonds in covalently bound molecules are of the order of 10 000-50 000 cm^{-1} (30-150 Kcal/mole). On the other hand, the pairwise interaction energies of rare gas atoms are much smaller, for instance only ~14 cm^{-1} for solid neon. The interaction energies between typical neutral guest molecule and a rare gas atom may be somewhat larger but of the same order of magnitude [79]. As a result, the entire solvation energy of a typical neutral guest will be at least 2 orders of magnitude lower than the strength of the internal bonds of the guest. This large disparity between the intramolecular bonding forces and the guest-host interactions results usually in a quite negligible perturbation of the guest's geometry and spectrum by the rare gas solvent and forms the basis of matrix isolation spectroscopy.

Unlike neutral guests, photoionization studies have shown that the solvation energies of molecular ions can, even in the inert gas solids, be quite considerable. The measured solvation energies for a variety of organic guests in inert gas matrices were found [80] to be of the order of 1-2 eV (~5-15000 cm^{-1}). In view of these much stronger interactions it seems necessary to inquire whether, or to what extent, can the spectra of the guest ions be perturbed by the solid medium.

The most straightforward way to test the magnitude of medium effects is clearly to investigate in matrices molecular cations whose spectra were previously studied in the gas phase. The only organic polyatomic cation which has been investigated in the gas phase with full rotational resolution is that of diacetylene, $C_4H_2^+$, and this ion was therefore also examined in some detail in the matrix [81]. In solid neon, both the laser excitation spectrum and the dispersed fluorescence spectrum show well resolved vibrational structure as shown in

TABLE I.

Comparison of the Electronic State Energies (ν_{00} in cm^{-1}) and Lifetimes (nsecs) of Molecular Ions in Neon Matrix and in the Gas Phase.

	Ne	Gas	$\Delta\nu$	τ_{Ne}	τ_{gas}	Ref.
Benzene						
F_6	21557	21617	-60	42	48	98,99,74,110
F_5	23014	23097	-83	40	47	74,75,99,110
1,2,3,4 F_4	23192	23298	-106	44	50	74,75,99,110
1,2,3,5 F_4	23232	23331	-99	43	50	74,75,99,110
1,2,4,5 F_4	24358	24440	-82	30	35	74,75,99,110
1,2,3 F_3	22288	(22460)	-	51	58	127
1,2,4 F_3	24177	24274	-97	~10	10	74,99,110,111
1,3,5 F_3	21774	21862	-88	53	58	68,74,112-114,83
1,3 F_2	22998	23229	-231	<0.1	<6	74,99,109,110
1,4 F_2				<0.1		
F_5Cl	19914	19990	-76	37	43	59,117
1,3,5 F_3 2,4,6 Cl_3	16785	16859	-74	29	34	117,118
1,3 F_2-5 Cl	19095	19220	-125	-	<6	111,117
1-F-3,5 Cl_2	17073	17205	-132	-	8	111,117
1,3,5 F_3-2,4,6-Br_3	13255	13510	-255	<10	-	121,122
1,2,4,5-F_4 3,6 Br_2	14680	14870	-190	-	-	121,116
1,3,5 Cl_3	15266	15411	-145	19	22	83,63
1,4 Cl_2	19452	19620	-168	-	-	116
1,3, Cl_2	18610	18770	-160	-	-	116
F_5-CH_3	21750	21900	-150	40	43	59
2,4,6 F_3-1-CH_3	22636	-	-	41	-	128
2,4,6 F_2 1,3(CH_3)$_2$	22086	-	-	27	-	128
2,4,6 F_3-(CH_3)$_3$	21140	(20900)	-	33	37	128,108
1- F-2,4,6 -(CH_3)$_3$	21468	-	-	<0.01	-	128
F_5CF_3	-	-	-	-	45	59

TABLE I. (Continued)

	Ne	Gas	$\Delta\nu$	τ_{Ne}	τ_{gas}	Ref.
OHF$_5$	22407	22410	-3	32	33	104-106
1-OH-2,3,5,6-F$_4$	21860	21922	-62	40	41	104,105,107
1-OH-3,5-F$_2$	21908	21994	-86	32	36	104,105,107
Pyridine						
F$_5$	27534	(27300)	-	-	-	119,120
2,6 F$_2$	27158	-	-	~5	-	120
2-F	(27500)	-	-	-	-	120
Diacetylene-C≡C-C≡C-						
H,H	19708	19723	-15	63	68	20,81,97
H,CH$_3$	20229	20374	-145	-	-	115,116
CH$_3$, CH$_3$	20499	20562	-63	-	-	115,83
trans hexatriene-(CH=CH-)$_3^-$						
H,H	15810	15868	-58	~15	17	123-126
H, CH$_3$	16081	16170	-89	≤10	12	126,116
CH$_3$, CH$_3$	16290	-	-	-	-	116
CS$_2^+$	21017	-	-	2.35^a	-	102,103,60

^{a} CS$_2^+$ τ in microseconds.

Fig. 6. The 0_0^0 origin band at 19708 cm^{-1} in emission coincides, within the experimental accuracy, with the corresponding absorption/excitation band, and is only insignificantly red shifted from its gas phase position [20] at 19723 cm^{-1}. The sharp (typically <2 cm^{-1}) zero phonon lines of the individual vibronic bands permit accurate measurement of the band positions and vibrational intervals. The vibrational frequencies which were reliably measured in the gas phase are compared with the corresponding matrix values in Table II, and the observed shifts can be seen to be extremely small.

Quite different are the spectra of the diacetylene cation in solid argon as shown in Fig. 7. The excitation spectrum in trace A is only broadly structured, with individual vibronic bands being more than 200 cm^{-1} wide, and the vibrational structure appears to be strongly perturbed. The 0_0^0 origin band is strongly red shifted from its gas phase position by some 400 cm^{-1}. The resolved emission spectrum in the bottom trace is completely structureless, and the onset of emission is extremely strongly red shifted to ~17500 cm^{-1}.

These results, and in particular the large difference between solid neon and argon can be explained qualitatively with the help of Fig. 8, where the energies of the relevant electrons are shown schematically, relative to the conduction band of the host. The situation in solid Ne is shown on the left-hand side. The electron affinity of C$_4$H$_2^+$ in the ground state is 10.17 eV, while that of the Ã state (12.62 eV) is higher by its excitation energy [82]. On the other hand, in view of the high 24.6 eV ionization energy of neon atoms, a much higher energy is needed to promote one of the Ne valence electrons into the conduction band of the solid. Accordingly, very little charge transfer interaction will take place between the C$_4$H$_2^+$ guest

TABLE II.

Comparison of Ne Matrix and Gas Phase Vibrational Frequencies
for Several Molecular Cations
(all frequencies in cm^{-1})

	Ne	Gas	$\Delta\bar{\nu}$	%	Ref.
$C_4H_2^+$ $\tilde{X}$					20,83,97
ν_3	865	860.5	+4.5	0.52	
$2\nu_7$	973	971.5	+1.5	0.15	
ν_2	2177	2176.6	+0.4	0.02	
ν_1	3143	3137	+6.0	0.19	
$C_6F_6^+$ $\tilde{B}$					98-100
ν_{18}	265	267	−2.0	−0.75	
ν_{17}	426	425	+1.0	0.24	
ν_{16}	1190	1195	−5.0	−0.42	
ν_{15}	1552	1550	+2.0	0.13	
ν_2	540	539	+1.0	0.19	
1,2,3,5 $C_6H_2F_4^+$ $\tilde{X}$					75,100,101
ν_{11}	305	305.7	−0.7	−0.23	
ν_{10}	426	426.6	−0.6	−0.14	
ν_5	1305	1314	−9.0	−0.68	
ν_3	1449	1453	−4.0	−0.31	
ν_2	1647	1656	−9.0	−0.55	
CS_2^+ $\tilde{X}$					60,70,71,101-103
ν_1	618	617	+1.0	0.16	
$2\nu_2$	698	694.5	+3.5	0.50	
$2\nu_3$	2416	2406	+10.0	0.42	

and the neon host. Furthermore, the interaction potential will not depend strongly upon the particular electronic or vibrational state of the ion. As a result, the observed unperturbed spectra consist, in most instances of sharp zero phonon lines (ZPLs) and show little indication of phonon sidebands.

The corresponding process in solid Ar will be much less endothermic as shown on the right hand side of Fig. 8 in view of the much lower, 15.76 eV ionization potential of argon. Because of this, a much stronger chemical, charge transfer, guest-host interaction will occur in the more polarizable argon solvent. The stronger interaction will result in rather strong perturbations in the spectrum. Furthermore, such perturbations will generally be ionic state specific, with usually stronger interaction in the excited electronic state which has the higher electron affinity. This differences in the guest-host interaction potentials between the ground

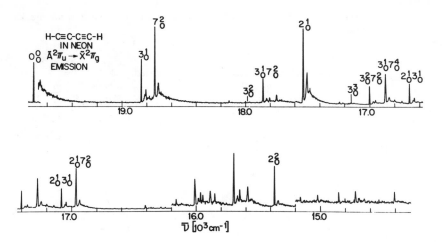

Fig. 6.   Laser induced fluorescence emission spectrum of the diacetylene radical cation in solid neon. Parent C_4H_2 concentration 1:5000. Photolysis time: 10s. The 0_0^0 band at 19708 cm^{-1} was excited by the laser. Assignments of some of the stronger emission bands are shown.

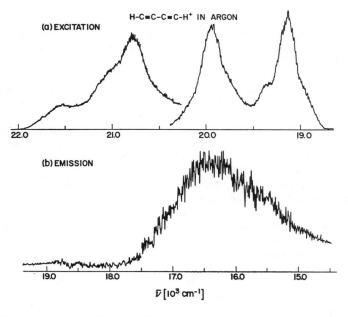

Fig. 7.   Spectra of diacetylene radical cation in solid argon. Parent concentration 1:5000. Photolysis 30s. (a) Excitation spectra over the ranges of two dyes are not corrected for variation of laser power. The emission around 16500 cm^{-1} was monitored. (b) Emission spectrum with excitation of the 0_0^0 origin at ~19150 cm^{-1}.

and excited electronic states of the guest ions will result in an intensity shift from the ZPLs into the phonon sidebands.

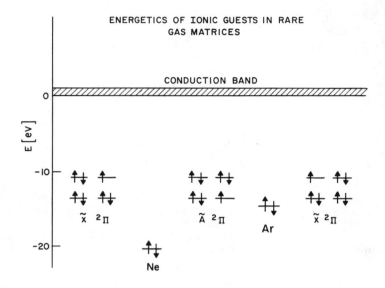

Fig. 8.   Schematic diagram showing the energetics of a matrix isolated ion, e.g. $C_6H_6^+$, dimethyldiacetylene cation, and its interaction with the host. The energies are relative to the rare gas conduction band. The diagram shows that solvent → solute charge transfer is highly endothermic in solid neon (left) but much closer to being thermoneutral in the more polarizable argon (right). See discussion in text.

The appearance of the spectrum in Fig. 7 can be explained using the potential energy diagram in Fig. 9. Here the curves shown represent the interaction potentials between the matrix atoms and the guest ion in the ground and excited electronic state, respectively. In the laser excitation process one will probe the relatively flat outside limb of the interaction potential. Following excitation, a subpicosecond "resolvation" of the excited state will take place forming a "tighter" neon cage around the more strongly interacting excited electronic state of $C_4H_2^+$. In the fluorescence process, the inner, steeply rising part of the ground state guest-host interaction potential will be probed, producing a much broader and consequently unresolved emission spectrum. In either case the ZPL's are very strongly Franck-Condon forbidden and remain unobserved.

The above arguments would seem to be of considerable generality. For instance, very similar are the observations for dimethyl diacetylene [83,84]. The ionization potentials of this molecule are somewhat lower (8.92 and 11.48 eV to produce the ions in the $\tilde{X}$ and $\tilde{A}$ state, respectively [85]) than those of unsubstituted diacetylene, and, accordingly, the argon matrix spectra of the cation are less strongly perturbed. The excitation spectrum of $C_6H_6^+$ in argon is much sharper than that observed for unsubstituted $C_4H_2^+$, and even the fluorescence spectrum exhibits hints of vibrational structure. Cations of some species with low ionization potentials, such as several of the halogenated benzenes, produce relatively unperturbed spectra in solid argon and, occasionally, even in krypton.

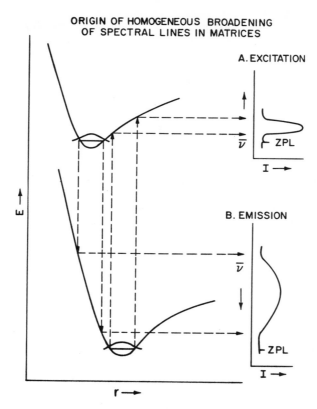

Fig. 9. Diagram explaining the homogeneous line broadening of spectral lines of matrix isolated ions. The curves represent, in one dimension, the potential surface of the interaction of the matrix isolated ion (e.g. $C_6H_6^+$) with the solvent atoms (e.g. Ar). The interaction will, in general, be stronger for ions in the excited electronic state (top) and the minimum of the potential energy surface will shift to shorter r values. Application of the Franck-Condon principle will predict broader lines in emission spectrum (bottom right) than in excitation-absorption spectrum (top right).

In some cases even other matrices can be used, as can be exemplified by the extensive studies of Shida [87-90] and of Haselbach [91,92] and coworkers. Their studies indicate that for a variety of organic cations, electronic spectra can be obtained in halocarbon glasses. In view of the stronger interactions in these solvents and the line-broadening mechanisms discussed above, usually only broad electronic absorptions, exhibiting little or no vibrational detail are observed. Such studies are therefore very useful in providing insights into the photochemical processes involving these molecules, but provide usually little structural information about the ionic species involved.

It may also be noted that since the solvent interactions are often weaker in the ground states of the molecular cations, the more polarizable argon and krypton matrices may in many cases be quite adequate for studies probing directly the ground states of the ions, such as IR or Raman spectroscopy. The guest-host interaction potentials are usually not strongly sensitive and specific to the vibrational state of the guest and the line broadening mechanisms discussed above will therefore usually be much less important in pure vibrational spectroscopy. Indeed, the observations of IR absorptions of a variety of cations have been reported, primarily by Milligan and Jacox [7,93,94] and by Andrews [9,95,96] and coworkers. Such studies are the subject of a separate chapter in this book. The major obstacles to IR studies would, of course, appear to be the much lower absorption cross-sections at infrared wavelengths, as well as the lower sensitivity of infrared detectors.

In general, however, one of the keys to obtaining sharp and unperturbed electronic spectra for ionic guests appears to be a large difference between the guest electron affinity and the host conduction band which is closely related to the bare atom's ionization potential. This condition will be well fulfilled for most organic and inorganic cations in solid neon. Neon is thus clearly the preferred medium for matrix molecular ion studies. In the last few years, a number of ionic species have been examined both in the gas phase and in solid neon and more extensive comparisons of their vibrational frequencies are therefore possible. Table II summarizes some of this vibrational data for several of the more thoroughly investigated molecules. Almost in all cases one finds excellent agreement between the gas phase and the rare gas matrix values, with the shifts rarely exceeding a few tenths of one percent.

In addition to gas/matrix vibrational shifts, one usually finds that the entire electronic transition is slightly shifted in the solid medium, reflecting small differences in the absolute depth of the guest-host interaction potentials between the ground and the excited electronic state of the ion. Here the comparisons between the gas phase and the matrix are somewhat more difficult to make in view of the inhomogeneous broadening. As discussed previously, one usually finds that the inhomogeneous distribution for each vibronic band spans a certain wavelength range, often with several sharp and well defined "sites" or maxima. While for some species this range is quite narrow, in some cases it may extend over 100 or more cm^{-1}. For the purpose of comparison, we have therefore used in Table I the most prominent "site," usually in samples which were previously annealed at 8-10K.

Table I, which comprises a rather wide range of chemically distinct cations, confirms the very unperturbed nature of the ionic spectra in solid neon. The electronic transitions usually exhibit small shifts towards lower energies, reflecting the slightly stronger guest-host interactions in the excited states as discussed above, and only occasionally exceeding 100 cm^{-1}.

Overall, both the electronic and vibrational data seem to indicate that most molecular ions in solid neon are only very weakly perturbed by the solid medium. The gas/matrix shifts experienced by a wide variety of molecular guests in neon are of comparable magnitude and in some cases even smaller than those experienced by the parent compounds and similar neutral guests. The valence electrons of cationic species are more tightly held because of the excess positive charge on the nuclei, resulting in more compact electronic wavefunctions. The guest solvation energy while much larger in the absolute sense, consists mainly in polarization of the lattice atoms by the ionic charge. Such an electrostatic interaction will not be very state specific, that is will not depend strongly on a particular state of the ion, and will not lead to significant phonon broadening and perturbation of the guest spectra. Very little chemical, charge transfer, interaction seems to be occurring in solid neon.

The above model is nicely documented by comparison of the matrix isolated Ca^+ cation with the isoelectronic neutral potassium atom. The highest electron of a K atom is relatively loosely bound and has, particularly in excited electronic states, partially a Rydberg character. This loosely bound, diffuse electron will interact rather strongly with the lattice, and the magnitude of this interaction will vary with the particular electronic state. As a result, the spectra of alkali atoms in the matrix are very broad and strongly perturbed [131,132]. The K atom 4s ^{2}S ← 4p ^{2}P fluorescence in solid argon is strongly red shifted and over 1000 cm^{-1} wide. While in potassium the outer electron sees an Ar-like core with overall charge +1, the corresponding electron of the isoelectronic Ca^+ species is in the field of a charge +2 core. Its electronic wavefunction will therefore be less diffuse, the radius of the orbital will be reduced and it will interact less strongly with the host. Accordingly, the spectrum of Ca^+ is much less perturbed. The linewidth is more than an order of magnitude narrower and the $4s\ ^2S \rightarrow 4p\ ^2P$ transition is almost unshifted from its gas phase position [133].

8. Negative Ions

Most of the arguments presented in the preceding section are only applicable to positive ions. In fact, by extrapolating the above arguments one might expect the spectra of negative ions with their more loosely bound valence electrons and more diffuse electronic wavefunctions to be more strongly perturbed in the matrix than their positively charged counterparts and neutral species.

It is interesting to consider in this connection the question of the negatively charged species maintaining the overall charge neutrality of the matrix in all of the studies of matrix isolated cations discussed in this chapter. The solvation energy of a free electron in solid neon has a negative value [80], and the electrons formed during the ionization process will undoubtedly attach to some guest or impurity molecule present in the matrix. Most of the compounds studied and in particular the halogenated hydrocarbons have appreciable electron affinities. Since in these experiments the parent compound is invariably the most abundant "impurity" molecule present in the matrix, there seems, as noted before, to be little doubt that, in addition to the cations, parent anions will be formed in the matrix.

Simple molecular orbital theory predicts for anions derived from alternant hydrocarbons spectra very similar to those of the corresponding cations. However, in spite of — in some cases — a thorough search, no fluorescence attributable to the anions was observed. The absence of observable fluorescence is usually due to nonradiative transitions. In neutral compounds, two types of nonradiative transitions are usually available: predissociation or nonradiative relaxation into a bound lower lying state — internal conversion or intersystem crossing. For negative ions at visible wavelengths a third channel opens up — electron detachment. Since ionization — electron tunneling — is intrinsically much more efficient than predissociation — heavy particle tunneling, one might expect the appearance of electronic fluorescence spectra for negative ions to be an exception, rather than a rule.

If the electron detachment would occur on a timescale of $10^{-11}-10^{-12}$s, one might even in the absence of fluorescence observe structured absorption spectra. On the other hand, if the detachment rate is in the $10^{14}-10^{16}s^{-1}$ range, the uncertainty principle limited bandwidths will be of the order of a vibrational spacing and completely featureless and continuous absorption spectra will result. In spite of rather thorough searches for several organic anions, no structured absorptions were observed, suggesting that the latter statement is closer to the truth and that the anionic spectra are broad. In most experiments involving photoionization, we observe in addition to the appearance of the structured absorption spectra of the cations upon photolysis, also a growth of continuous background. It is quite possible that negatively charged species are responsible for, or at least contribute to, this background. This is also consistent with the observation that the spectra of most matrix isolated ions are readily destroyed by continued irradiation with broadband near UV or visible sources. This can be easily interpreted in terms of photodetachment of electrons from negative ions and their subsequent recombination with the positively charged cations.

In view of the above considerations it appears unlikely that laser induced fluorescence and related techniques will be as useful and prolific in studies of negative ions as they are in molecular cation spectroscopy. Other techniques which probe directly the ground electronic state, such as infrared, Raman or EPR spectroscopy would appear to be more generally useful approaches to studies of negative ions. However, optical matrix spectroscopy and laser induced fluorescence may still make valuable contributions in studies of selected species, in particular of anions of species with high electron affinities.

One such anion would appear to be C_2^-, which has been studied rather extensively in solid rare gas matrices [12,46,134-136]. This ion is characterized by a rather high 3.54 eV photo detachment threshold [137], and its low lying excited electronic states are therefore stable, and in particular the $B^2\Sigma_u^+$ state has been well characterized in the matrix. In agreement with the arguments presented above, the spectrum is more strongly perturbed than are typically the spectra of molecular cations. The origin of the $B^2\Sigma_u^+ \rightarrow X^2\Sigma_g^+$ transition is shifted in solid neon to $T_e = 18839$ cm^{-1} from its gas phase position at 18281 cm^{-1}, a *blue* shift of ~560 cm^{-1} or more than 3%, and also the vibrational frequencies experience moderate gas/matrix shifts.

III. SPECTROSCOPIC STUDIES OF INDIVIDUAL IONIC SPECIES

The sum of the data presented in the preceding sections shows quite clearly that meaningful spectroscopic information can be obtained for many molecular ions using the experimental techniques discussed. In the next several sections we will attempt to summarize and review the types of interesting information that have been obtained for several groups of chemically related compounds that have been examined to date.

Most of the ionic species which were thus far studied by laser induced fluorescence in the matrix can be classified as derivatives of several important organic molecules. While there is no fundamental reason why inorganic ions could not be studied, in many cases the simple inorganic species can be more simply investigated by conventional gas phase spectroscopy, and the matrix techniques have less to offer. Also, the ionization potentials of many small inorganic species are too high to permit photoionization with the usual microwave powered atomic discharge lamps. A fundamental limitation of laser induced fluorescence studies, as opposed to, for instance, infrared spectroscopy, is the requirement that the ion must possess, in addition to a stable ground electronic state, also a sufficiently long lived, fluorescent excited electronic state.

The above requirements seem to eliminate, for instance, most saturated hydrocarbons either due to instability of their ionic ground state, or to the fact that their lowest excited electronic states lie high above the threshold for fragmentation. Most of the ionic species for which fluorescent decay has thus far been observed belong to several broad classes of unsaturated compounds. In the following we will consider, one at a time these groups of compounds and summarize the results obtained regarding the ions' structure and photophysics.

1. Benzenoid Radical Cations

One of the groups of ions which have been particularly extensively studied are the radical cations derived from substituted benzenes [21,22]. These species are characterized by a strong electronic transition in the visible or near ultraviolet spectral range and, in particular, many of the derivatives which possess several fluorine substituents, exhibit a high quantum yield for fluorescence. In order to understand the nature of this transition, one needs to consider bonding in the parent benzene. The predictions of the Hückel molecular orbital treatment of benzene are shown schematically in Fig. 10. This simplified treatment in which only π electrons are considered, predicts the existence of three strongly bonding π electronic orbitals. In neutral benzene, these bonding orbitals are just filled with the available six π electrons, as shown on the left-hand side of Fig. 10, giving benzene its extraordinary chemical stability. The lowest allowed electronic excitation of benzene corresponds to promotion of an electron from the higher energy, degenerate pair of orbitals into the corresponding anti-bonding, π^* orbital. Such excitation from the stable configuration of benzene requires

SCHEMATIC OF THE ELECTRONIC
STRUCTURE OF BENZENE AND ITS IONS.

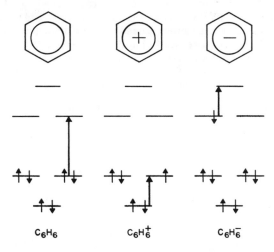

C_6H_6 $C_6H_6^+$ $C_6H_6^-$

Fig. 10.   Electronic structure and spectra of benzene and its ions in terms of simple Hückel molecular orbital theory. In neutral benzene (left) three π bonding orbitals are just filled by the six available electrons. The lowest excitation shown by the arrow involves promotion of an electron into the antibonding π^* orbital. In the cation (center) one electron is missing from the doubly degenerate highest occupied molecular orbital. A low energy $\pi-\pi$ transition involves electron exchange between the lower non-degenerate π orbital and the degenerate pair. In the anion (right) the extra electron is placed in the doubly degenerate π^* pair. A low energy transition interchanges electron between the antibonding π orbitals.

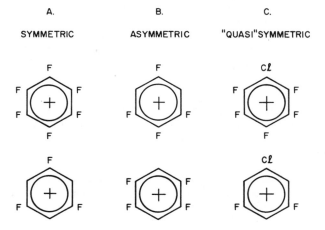

Fig. 11.   Examples of symmetric (A), asymmetric (B), and "quasisymmetric" (C) halogenobenzene cations, as discussed in the text.

considerable energy and the corresponding transition is located for most benzenoid compounds fairly far in the UV, near 2500 Å.

The situation in the corresponding cationic species is shown in the middle column of Fig. 10. Here, one electron is missing from the higher energy, doubly degenerate orbital of benzene, resulting in a doubly degenerate ground state of the cation. Furthermore, one can note that the cation will possess a relatively low-lying non-degenerate excited electronic state resulting from promotion of an electron from the non-degenerate π orbital into the doubly degenerate pair. The corresponding transition should occur at approximately ½ of the energy needed for excitation of parent benzene. This places the transition in the visible range near 5000 Å, far removed from all transitions of the parent and, perhaps even more importantly, in an experimentally very convenient energy range well accessible with available tunable dye lasers.

In the above simple Hückel treatment only π electrons are considered. In actuality the presence of σ electrons complicates the situation somewhat. While it does not affect significantly the energetics of the π orbitals discussed above, in numerous compounds the lowest unoccupied σ orbital lies between the degenerate and non-degenerate π bonding orbitals of Fig. 10 and, consequently, the lowest excitation corresponds to a π-σ and not a π-π process. As we have discussed in one of the earlier sections, the presence of a state arising from σ ionization (usually denoted the $\tilde{C}$ state in these compounds) below the excited non-degenerate π state ($\tilde{B}$ state), leads invariably to efficient non-radiative relaxation and strong lifetime broadening of the $\tilde{B} \rightarrow \tilde{X}$ spectrum. This situation arises for instance in unsubstituted benzene itself. Presence of some substituents such as, for instance, halogens on the benzene ring is known to stabilize preferentially the σ bonding orbitals [138,139]. Thus for many benzenes with several halogen, especially fluorine, substituents the second photoelectron band corresponds to a π ionization and the "$\tilde{B}$ state" of the ions lies below the state arising from σ ionization, the $\tilde{C}$ state. For these compounds sharp, well-resolved absorption and laser induced fluorescence spectra are usually observed, and these will be discussed in the following sections.

Depending on the position and nature of the benzene ring substitution the benzenoid radical cations can be divided into three clearly distinct groups, symmetric (A), asymmetric (B), and quasi-symmetric (C), each of them characterized by a different spectroscopic behavior. Some representative examples of these three groups are shown in Fig. 11. The physical reasons underlying these differences are explained in Fig. 12. Group (B) contains asymmetrically substituted compounds with D_{2h} or lower symmetry. In these compounds the originally doubly degenerate ground state of benzene is split into two components so that they effectively possess a non-degenerate ground state and a rather low-lying excited electronic state, usually referred to as the "A" state. In this case one observes two independent transitions terminating in the two ground state components.

Into group (A) belongs the benzene cation itself as well as species substituted in symmetric fashion so that they possess D_{6h} or at least D_{3h} symmetry and therefore retain a doubly degenerate ground state. In these species, as we will discuss later, the ground state is distorted due to Jahn-Teller interactions [66], and a single, strongly perturbed spectrum is seen. Finally, in group (C) are several molecules which can be referred to as quasi-symmetric. Here belong mainly species which are substituted in symmetric positions but by unlike, albeit similar substituents [111]. In these species the electronic ground state splitting is usually very small, often commensurate with vibrational frequencies. Also several asymmetric species which are substituted in such a way that the effects of the substituents are fortuitously cancelled belong into this group. Each of the three groups mentioned above is characterized by a distinctive spectroscopic behavior and they will be discussed in the following sections.

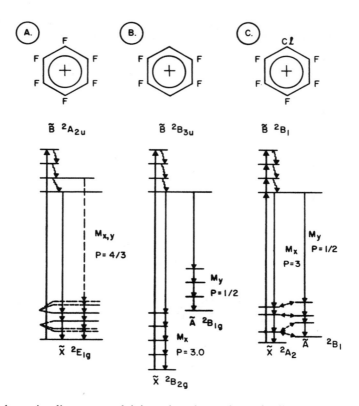

Fig. 12.   Schematic diagram explaining the electronic and vibrational structures of symmetric (A), asymmetric (B) and quasisymmetric (C) substituted benzene cations. In symmetric species (A) the ground state vibrational structure is distorted due to the Jahn-Teller effect. The transition dipole is oriented in the molecular plane with x,y axes equivalent. Photoselection gives a fluorescence polarization p=4/3. In asymmtric species (B) the ground state is split into two components which are far apart. The transition moments for transitions into the two components are orthogonal, giving fluorescence polarization p=3 and ½, respectively. In quasisymmetric molecules (C) the $\tilde{X}$-$\tilde{A}$ splitting is small. The two vibrational manifolds overlap and perturb each other, with the polarization a particular level being of dependent upon its parentage.

a. Asymmetrically Substituted Cations

In asymmetrically substituted cations the two components of the ground state are split far apart as shown in Fig. 12b, often by 0.5-1 eV and this is reflected by a corresponding splitting of the first photoelectron band of the parent compound [140]. The spectroscopic behavior of the asymmetric species is exemplified in Fig. 13 by the 1, 2, 4, 5 tetrafluorobenzene cation. Trace (a) of that figure shows the laser excitation spectrum, $\tilde{B} \leftarrow \tilde{X}$. This corresponds to transitions from the vibrationless level of the ground state into vibronic levels of the non-degenerate $\tilde{B}$ state. A common feature of the spectra of most of the benzenoid cations studied is a rather simple vibrational structure, consisting of a regular array involving a relatively small number of vibrational frequencies [75]. These vibrational frequencies are readily identified as the totally symmetric vibrational modes of the cations based on their similarity to the corresponding modes of the parent compounds.

The corresponding $\tilde{B}$ state emission spectrum produced by exciting one of the upper state levels and resolving the fluorescence is presented in Fig. 13b and 13c. As we noted previously, in the asymmetric compounds the degenerate ground electronic state is split into two separate components. This is, as we have noted, evidenced by the appearance of two bands in the photoelectron spectrum and should also lead to the appearance of two separate band systems in the emission spectrum, corresponding to transitions from the $\tilde{B}$ state into the two ground state components. This is indeed confirmed by the laser induced fluorescence studies. As shown for the 1,2,4,5 tetrafluorobenzene cation in Figs. 13b and 13c. The $\tilde{B} \rightarrow \tilde{X}$

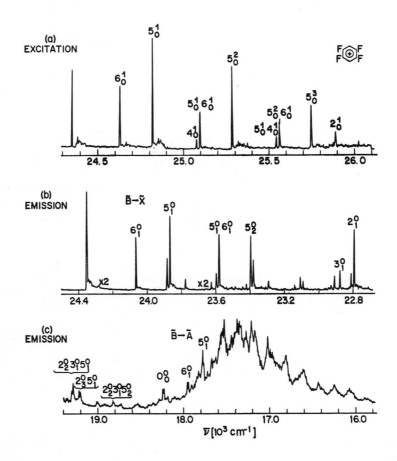

Fig. 13.   Spectra of 1,2,4,5 $C_6H_2F_4^+$ in solid neon. (a) Low energy section of the fluorescence excitation spectrum. (b) $\tilde{B}^2B_{3u} \rightarrow \tilde{X}^2B_{2g}$ fluorescence near the 0_0^0 origin, which exhibits a uniform parallel polarization p=3.0. (c) Fluorescence further in the red showing a region of the $\tilde{B}^2B_{3u} \rightarrow \tilde{A}^2B_{1g}$ spectrum. Starting with the sharp 0_0^0 origin, the polarization changes abruptly to p$\cong$0.5. Note that the frequency scale of panel (c) is compressed 2× compared with a and b.

emission in Fig. 13b exhibits extensive vibronic structure. The most intense band in the spectrum is the 0-0 band at 24358 cm^{-1} and intensities gradually decrease towards the red, reflecting the diminishing Franck-Condon factors for transitions from the vibrationless level of the $\tilde{B}$ state into higher vibrational levels of the ground state. The vibrational structure in the emission spectrum is for most of the asymmetric cations again easily analyzed in terms of a small number of vibrational modes which are very similar to the frequencies observed in the excitation spectra and to the totally symmetric vibrational modes of the parent compounds.

This similarity between the ionic frequencies, both in the $\tilde{X}$ state and the $\tilde{B}$ state, and the frequencies of the parent compounds is particularly apparent for low frequency vibrational modes below ~1000 cm and can be clearly seen in Table III, where the data for some of the cations studied are summarized. Based on the dominance of totally symmetric, in-plane, vibrational modes in the electronic spectra one can readily conclude that the ions remain planar in both states involved in the transition. Furthermore, the similarities of the vibrational modes of the ion to those of the parent suggest that removal of one of the π electrons does not alter significantly the in-plane vibrational potential.

TABLE III.

Comparison of Vibrational Fundamentals of Several Asymmetric Fluorobenzene Cations in the $\tilde{X}$ and $\tilde{B}$ States with the Corresponding Modes of the Parent Compounds (cm^{-1})

		X^+	B^+	P
$C_6HF_5^+$	ν_{11}	278	268	272
	ν_{10}	438	429	436
	ν_9	460	456	470
	ν_8	575	566	578
1,2,3,4-$C_6H_2F_4^+$	ν_{11}	273	277	280
	ν_{10}	340	324	324
	ν_9	443	442	459
	ν_8	680	668	684
1,3,4,5-$C_6H_2F_4^+$	ν_{11}	303	305	305
	ν_{10}	426	428	443
	ν_9	581	573	580
	ν_8	785	783	789
1,2,4,5-$C_6H_2F_4^+$	ν_6	287	276	280
	ν_5	485	465	487
	ν_4	726	725	748

b. $\tilde{B} \rightarrow \tilde{A}$ Emission and Fluorescence Polarization

If the $\tilde{B}$ state fluorescence of the asymmetric benzenoid cations is examined far enough into the red, a second maximum in the intensity distribution is invariably observed [111]. This second maximum which is clearly visible in Fig. 13c, is attributed to the $\tilde{B} \rightarrow \tilde{A}$

fluorescence terminating in the upper component of the split ground electronic state. The correctness of the assignment can be confirmed by studying fluorescence polarization. It is well known that the fluorescence of non-rotating matrix isolated molecules is often polarized. Even though the glassy rare gas matrices are generally isotropic and the molecules randomly oriented, the polarized exciting laser beam selects an oriented, anisotropic subset [56,141]. It excites preferentially the molecules oriented in such a way that their transition dipole is aligned with the electric vector of the incoming light. This then results in a non-random polarization of the fluorescence.

In the case when both the absorption and reemission occur on the same transition between two non-degenerate electronic states, that is when the transition dipole for the absorption and reemission process are oriented along the same molecular axis as is the case of the $\tilde{B} \rightarrow \tilde{X}$ emission of asymmetric benzenoid cations, one expects [141] a depolarization ratio $p = I_{||}/I_{\perp} = 3.0$. The heavier rare gases form at 4K highly scattering deposits which reduce the depolarization ratio and frequently almost completely depolarized emission is seen. On the other hand, neon usually forms clear, optically nearly perfect layers, and depolarization ratios very close to the theoretically expected values are often measured. Thus the $\tilde{B} \rightarrow \tilde{A}$ emission in Fig. 13b shows the expected value p=3.0. For the $\tilde{B} \rightarrow \tilde{A}$ emission on the other hand, the transition dipole is oriented perpendicular to that of the $\tilde{X} \rightarrow \tilde{B}$ absorption process. For this circumstance theory predicts p=0.5. Accordingly, the polarization properties in emission change abruptly starting with the sharp band denoted as O_0^0, at 18242 cm^{-1} in Fig. 13c, and prevailing perpendicular polarization with p $\sim$ 0.5 is seen. Clearly, this sharp band 6116 cm^{-1} from the O_0^0 $\tilde{B} \rightarrow \tilde{X}$ origin band is the origin of the $\tilde{B} \rightarrow \tilde{A}$ transition.

Similarly, in the other asymmetric benzenoid cations the first perpendicularly polarized band determines the energy of the $\tilde{A}$ state and hence the magnitude of the ground state splitting. In view of the sharpness of the bands the splitting can be determined with an accuracy of a few wavenumbers, 2-3 orders of magnitude more accurately than from the incompletely resolved broad bands in photoelectron spectra. While, in principle, one has to contend with the possibility of environmental effects, the very small gas-matrix shifts experienced by the $\tilde{B} \rightarrow \tilde{X}$ origin bands in solid neon suggest that these effects will be minor, and that the matrix values will be quite representative of the ground state splitting in the free ions.

The origin and magnitudes of this splitting can again be satisfactorily explained using the simplest Hückel molecular orbital theory. As noted previously, the highest occupied molecular orbital in benzene is doubly degenerate, and contains four electrons. In this representation, each of the two wavefunctions ψ_a and ψ_b are characterized by one nodal plane perpendicular to the plane of the benzene ring. These are represented schematically at the left side of Fig. 14, where the numbers shown represent the coefficients, c_i, of the atomic orbitals to the molecular wavefunctions. In benzene or in symmetrically substituted benzenes the two orbitals are degenerate. In asymmetrically substituted species the two wavefunctions will place differing amounts of electron density upon the substituted carbon atoms, resulting in splitting of their degeneracy.

The magnitude of this splitting can be estimated for the variously substituted fluorobenzenes by adding up the squares of the coefficients on the carbon atoms bonded to the fluorine substituents and computing the difference $\Sigma c_{ia}^2 - \Sigma c_{ib}^2$. As shown in Fig. 14, this difference is, of course, zero for the sym $C_6H_3F_3^+$. For the 1,2,3,5 and 1,2,4,5 isomers of tetrafluorobenzene cations the respective values are 1/3 and 2/3. Indeed, the ground state splitting in the latter isomer (6115 cm^{-1}) is found to be somewhat more than twice as large as in the former, 1,2,3,5 species (2442 cm^{-1}). Interestingly, this simple treatment predicts (see Fig. 14) for the very asymmetrically substituted vicinal 1,2,3 trifluorobenzene also zero difference. Again in excellent agreement with this prediction one finds experimentally that in

**GROUND STATE SPLITTING IN
SUBSTITUTED BENZENE CATIONS**

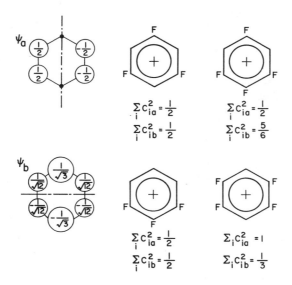

Fig. 14.   Diagram explaining how a rough estimate of the relative magnitude of the ground state splitting in variously substituted benzene radical cations (or anions) can be made based on a simple Hückel molecular orbital theory. If one adds up the squares of the atomic coefficients c_i on the *substituted* carbon atoms for the two molecular orbitals ψ_a and ψ_b, the difference of the two sums should be roughly proportional to the splitting. See text.

this cation the splitting is extremely small, with the $\tilde{A}$ state being merely 197 cm^{-1} above the ground state.

The studies of polarization in the matrix emission spectra thus provide a rather accurate measure of the ground state splitting and of the energies of the $\tilde{A}$ states. In all cases studied the measured values are in good qualitative agreement with the estimates provided by the simple theoretical treatment outlined above. The experimental $\tilde{A}$ state energies for a variety of fluorobenzene cations, including some cations containing other substituents, are summarized in Table IV. In view of the insignificant differential shifts experienced by the electronic states of most of the cations studied in solid neon, as evidenced by Table I, these energies should in most cases be very close to their values in the free ions.

It is interesting to note that the $\tilde{B} \rightarrow \tilde{A}$ emission in the gas phase is in all the ions studied thus far completely featureless and no hints of vibrational structure have been detected. It was suggested [74] that this is due to lifetime broadening and a fast non-radiative transition into the ground state. The observation of sharp vibrational structure in the matrix spectra indicates that this is probably not the case and that at least the lowest vibrational levels of the $\tilde{A}$ state are relatively long lived. The lack of resolved structure in the gas phase emission is probably due to the complexity and congestion in the spectrum.

c. Symmetric Benzenoid Cations

The cations derived from symmetrically substituted benzenes with D_{3h} or D_{6h} symmetry exhibit a distinctly different spectroscopic behavior. In these species the ground state is

TABLE IV.

Ground State Splitting in Several
Substituted Benzene Cations in Ne
$[cm^{-1}]$

Substituent Pattern	T_e (Ã)	Substituent Pattern	T_e (Ã)
F_5	2950	ClF_5	185
1,2,4,5-F_4	6115	1,3-Cl_2, 5-F	159
1,2,3,5-F_4	2442		
1,2,3,4-F_4	2767	1,2,3-F_3	197
1,2,4-F_3	5089		
		1-OH,2,3,5,6-F_4	207
1-OH,3,5-F_2	1334	CH_3F_5	220

doubly degenerate and, accordingly, no doubling of the first photoelectron band is observed in the parent photoelectron spectrum. It has long been appreciated that in degenerate electronic states of non-linear molecules the potential energy function of certain non-symmetric modes will, as a result of the interaction of electronic and nuclear vibrational motion, be distorted in such a way that the potential minimum is no longer at the symmetric position. This has been first pointed out by Jahn and Teller [66] and examined since in numerous theoretical works [142-144]. The ground state of the benzene cation and its symmetrically substituted derivatives are classical examples of this situation. Experimental studies of the spectroscopy of these species thus provide very detailed information about the Jahn-Teller effect [145,146].

The presence of ground state distortion is apparent from the representative excitation spectrum of $C_6F_6^+$ shown in Fig. 15. In an allowed electronic transition only progressions in totally symmetric modes should appear, in the case of $C_6F_6^+$ the a_{1g} C-F stretch, ν_1, and C-C stretch, ν_2. Actually, these modes appear only weakly and the most prominent structure in the spectrum is associated with the e_{2g} in-plane modes, ν_{12}-ν_{16}, based on their close similarity to the corresponding modes of the parent. The appearance of the e_{2g} modes suggests that the effective symmetry is lowered to D_{2h}, in agreement with Jahn-Teller theory.

Quite different is the appearance of the fluorescence spectrum. In the case of asymmetric cations we have noted that two distinct systems exhibiting depolarization ratios p=3.0 and 0.5, respectively, were observed and that the $\tilde{B} \rightarrow \tilde{X}$ emission represents to a good approximation a "mirror image" of the excitation spectrum. In contrast to that a single system with uniform polarization p~1.3 is observed for $C_6F_6^+$ and other symmetric species. Furthermore, quite unlike the regular and harmonic vibrational structure observed in the excitation spectrum, the fluorescence seems quite erratic with no apparent regularity.

The reasons for the apparent irregularity of the emission spectrum were shown in Fig. 12. The Jahn-Teller effect will result in splitting each of the excited levels of the degenerate vibrational modes into several components. In emission originating from the vibrationless level of the excited electronic state typically only transitions into one of the components of the Jahn-Teller split ground state levels are symmetry allowed and this leads to the seemingly irregular appearance of the emission spectrum.

As we have noted previously, in emission from vibrationally unrelaxed upper state levels all components of the ground state are spectrally accessible and one can directly observe the

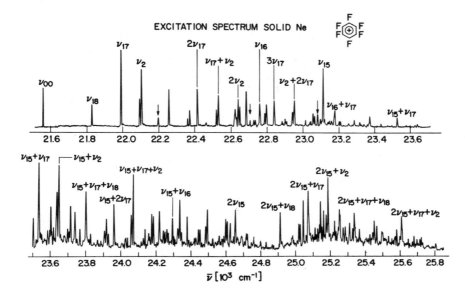

Fig. 15.   Laser excitation spectrum of $C_6F_6^+$, hexafluorobenzene cation in solid neon.

splitting of ground state vibrational levels [64,65]. This was exemplified by Fig. 5 for the sym-$C_6H_3F_3^+$ cation. By combining the data from the relaxed fluorescence spectra with the additional information contained in the unrelaxed emission and gas phase spectrum, one can get a very complete description of the ground state vibrational structure. As is discussed in a different chapter of this book, using sophisticated theoretical models of the Jahn-Teller effect, this extensive data can be inverted to provide information about the distorted ground state potential of the symmetrical benzene cations, and ultimately about their equilibrium geometry.

d. Quasisymmetric Cations

An interesting situation arises in cases where the electronic D_{6h} (or D_{3h}) symmetry is only slightly distorted and when the ground state electronic splitting is small, i.e. commensurate with vibrational frequencies. Such a situation prevails, for instance, in $C_6F_5Cl^+$ and similar species that are substituted in symmetric positions but with different halogens [59,111]. In these cases the vibrational manifolds of the $\tilde{X}$ and $\tilde{A}$ states overlap and one observes in the fluorescence spectrum interspersed parallel and perpendicularly polarized bands. One example of such an emission spectrum is shown in Fig. 16 for 1-fluoro 3,5-dichlorobenzene cation [111]. The vibrational manifolds of both electronic states are strongly perturbed as a result of vibronic interactions and the spectra appear quite irregular. A detailed theoretical analysis of these spectra would require simultaneous consideration of the vibronic, Jahn-Teller-like interactions and of the intrinsic electronic splitting.

As we have mentioned previously, based on simple molecular orbital theory, one could expect the 1,2,3-trifluorobenzene cation also to exhibit a very small ground state splitting and fall into this category. This is indeed confirmed by the neon matrix studies [127,148]. Several other asymmetrically substituted cations, in which the substituent effects are fortuitously cancelled, were found to have a very small separation between the $\tilde{X}$ and $\tilde{A}$

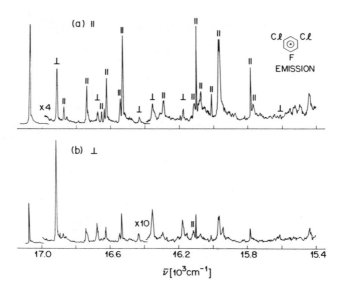

Fig. 16.   Fluorescence spectrum of a "quasisymmetric" cation, the 1-F, 2,3-Cl_2-$C_6H_3^+$.
               Trace (a) is taken with parallel polarized fluorescence being detected while (b)
               is with perpendicularly polarized fluorescence being detected. In the bottom
               trace the $\perp$ bands are enhanced approximately 2×, while the $\parallel$ bands are
               reduced by a factor of 3×.

states [59,107]. The neon matrix data for several benzenoid cations with small ground state
splitting is summarized on the right hand side of Table IV.

2. Pyridine Derivatives

While, as discussed in the previous section, a large variety of substituted benzene cations
have been studied in fluorescence, no gas phase optical spectra of the electronically similar
heterocyclic compounds, pyridines and other aza-benzenes, have as yet been reported. Very
recently four cations of the pyridine group have been examined in solid neon [119,120]. The
four closely related pyridine cations studied exhibited an amazing variety in their
spectroscopic behavior. However, it was shown that the observed differences and trends in
their spectroscopic properties can be well understood in terms of their electronic structure.
Studies of the cation spectra did, in fact, provide useful new insights into the photophysics
and electronic structures of the parent pyridines.

The spectroscopic observations can be rather easily explained with the help of Fig. 17,
which shows semiquantitatively the electronic structures of pyridine and some of its
fluorinated derivatives. In drawing this diagram, the orbital energies used for pyridine and
pentafluoropyridine were derived from photoelectron spectra [139]. For the partially
fluorinated species it was assumed that the electronic orbital energies are a linear function of
the number of substituents. While this surely is an oversimplification, it explains quite
adequately the observed spectra.

For the parent pyridine, for which the ground state arises from ionization from the
nitrogen lone pair "n" orbital [139], the only transition accessible in the near ultraviolet range

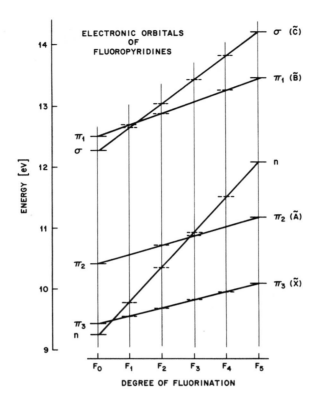

Fig. 17.   Diagram showing the electronic structure and orbital ordering in fluorinated
pyridines. The ionization energies of the electronic orbitals of pyridine (F_0)
and pentafluoropyridine (F_5) are taken from the photoelectron spectra (see
text). For the intermediate fluorination it is assumed that the orbital energies
are a linear function of the number of fluorine substituents.

is of the n-π type; such transitions are typically at least an order of magnitude weaker than
allowed π-π transitions. Furthermore, the excited π_1 state ($\tilde{B}$ state) is located close above the
state arising from σ-ionization ($\tilde{C}$ state) and this situation leads invariably to very efficient
nonradiative relaxation and severe lifetime broadening of the $\tilde{B}$ state levels [23].
Accordingly, no absorption or fluorescence were observed for photolyzed pyridine samples in
solid neon.

Introduction of fluorine substituents into the aromatic ring has a pronounced effect upon
the electronic structure as can be seen in Fig. 17. In particular, the σ (and n) type electronic
orbitals are strongly stablized [138,139], while the π type orbital ionization energies are much
less strongly affected. As a result of such differential shifts, the ordering of orbitals is
changed in 2-fluoropyridine, and the ground state of the ion arises from π ionization. As a
result, a strong π-π transition in the near UV absorption spectrum would be expected for the
ground state cation, and indeed a strong absorption is observed in the expected region in
photolyzed neon matrix. The "$\tilde{B}$ state" is, however, very near resonance with the lower lying
"$\tilde{C}$ state" formed from σ ionization; the absorption is broad and featureless, and no
fluorescence is observed.

In 2,6-difluoropyridine with two substituents, the σ orbitals are further stabilized (i.e. the corresponding ionization energies in Fig. 17 increased) and the $\tilde{B}$ state is now located below the perturbing $\tilde{C}$ state. Accordingly, the $\tilde{X} \rightarrow \tilde{B}$ absorption in the cation is now sharp and structured. Excitation within the region of absorption produces an intense fluorescence (see Fig. 18), albeit with less than unity quantum yields and a very short lifetime. The resulting fluorescence spectrum is shown in Fig. 18. The low quantum efficiency, which we crudely estimate at 10-15% is again due to nonradiative relaxation. Possibly the state arising from the "lone pair", n, ionization now takes over the role of the "gateway state" for the relaxation process.

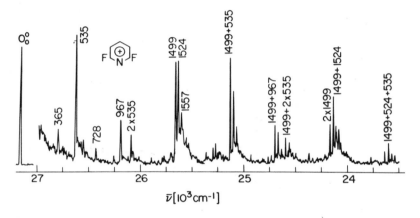

Fig. 18.   Fluorescence spectrum of 2,6 difluoropyridine cation in solid neon excited by a
laser tuned to the 0^0_0 origin. The individual vibrational modes are labeled by
their respective frequencies in cm^{-1}.

Consistent with this picture is also the situation in the pentafluoropyridine, where the σ and n type orbitals are further stabilized and the "lone pair n state" of the cation moves relatively up in energy, closer to the $\tilde{B}$ state. This enhances the nonradiative relaxation. No fluorescence is seen for this cation, and individual vibronic bands of the $\tilde{X} \rightarrow \tilde{B}$ absorption exhibit a moderate lifetime broadening [119].

The photophysical behavior of the cations studied can thus in a very self consistent way be understood in terms of their electronic structure and of the substituent induced shifts of the individual electronic states. Conversely, the studies of the ions lend strong support for the n, π, π ordering of the first three photoelectron bands in the parent pyridine from the several suggested assignments of its photoelectron spectrum [139].

Examination of the vibrational frequencies of the pyridines for the two cations for which vibrational fine structure is observed leads to similar conclusions as in the related substituted benzene studies. The compounds are apparently planar in both of the ionic states and the in-plane vibrational potential is not strongly altered by removal of an electron from either of the two low-lying π orbitals, since there is a very close similarity between the vibrational frequencies derived from the ionic spectra and the totally symmetric vibrational modes of the parent.

Insertion of the heteroatom directly into the aromatic ring constitutes a much stronger perturbation of the benzene electronic structure than hydrogen substitution and accordingly, the splitting of the two components of the benzene cation ground state is considerably larger in the pyridines than in the asymmetrically substituted benzene cations. This again results in observation of two clearly resolved bands separated by more than 1 eV in the photoelectron spectrum and by the appearance of the second, $\tilde{B} \rightarrow \tilde{A}$ emission maximum in the fluorescence of difluoropyridine cation.

3. Acetylenic Cations

Another group of ions for which fluorescence has been observed and which have been rather extensively studied both in the gas phase and in low temperature matrices are various derivatives of acetylene. Considering the π electron alone, the nature of the transition observed in these species is elucidated in Fig. 19. The four π electrons of acetylene itself are located in a doubly degenerate, bonding orbital, contributing to the strong carbon-carbon triple bond. Electronic excitation of acetylene requires promotion of one of these electrons to the corresponding antibonding orbital as shown schematically in column A of Fig. 19. This necessitates a considerable amount of energy, and as a result, acetylene is colorless, with its electronic absorption spectra beginning rather far in UV. A similar situation prevails also in

ELECTRONIC TRANSITIONS IN ACETYLENE,
DIACETYLENE AND THEIR CATIONS

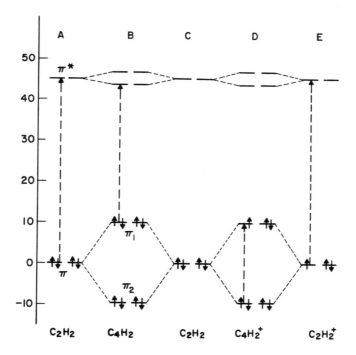

Fig. 19. Diagram explaining the electronic structures of acetylene, diacetylene and their cations. The HOMO of acetylene is a doubly degenerate bonding π orbital and is filled with 4 electrons (A and C). In diacetylene the two doubly degenerate orbitals π_1 and π_2 are split more than 2 eV apart (B). In the corresponding cation (D) only three electrons occupy the higher lying π_1 orbital. A low energy allowed electronic transition in the visible interchanges an electron between π_1 and π_2.

the acetylene radical cation (shown in column E), except in this case only 3 electrons occupy the highest π orbital. Excited π states of the ion are rather high in energy, considerably above the fragmentation threshold and consequently no radiative decay is detected for the parent acetylene cation.

In compounds with more than one triple bond, the doubly degenerate π orbitals associated with the individual acetylene units will interact, resulting in formation of two or more bonding molecular orbitals split apart in energy. In diacetylene (Fig. 19B), this splitting is over 2 eV and its UV absorption spectrum is considerably red-shifted compared with acetylene. Furthermore, in the diacetylene cation where only seven π electrons are available, there is now a low lying electronic transition in the visible which interchanges an electron between the two pairs of the doubly degenerate π orbitals as shown in column D of Fig. 19. The orbital situation will be similar and similar transitions will be present in most other polyacetylene compounds.

The diacetylene cation was in fact the first polyatomic molecular cation to be observed by optical spectroscopy, and thus far the only one to be studied with full rotational resolution [20]. Diacetylene and its isotopic derivatives $C_4D_2^+$ and C_4HD^+ have also been studied in solid neon matrix [81], as was the ^{13}C substituted molecule. This ion thus afforded the first possibility of detailed comparison of gas-phase and matrix-isolated ion. As noted previously, such a comparison has shown that the ionic species are in solid neon quite negligibly perturbed.

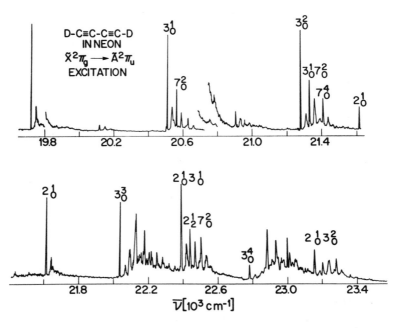

Fig. 20.   Laser excitation spectrum of deuterated diacetylene cation, $C_4D_2^+$ in solid neon. The spectrum is a composite of scans over the ranges of several dyes and is not corrected for laser power variation. Assignments of some of the stronger vibronic transitions are shown.

The gas phase fluorescence spectra consist mainly of emission originating from the vibrationless level and several low-lying vibrational levels, and contain therefore mainly information about the ground state vibrational potential. Laser excitation spectra, on the other hand, provide also the complementary information about the vibrational structure of the excited electronic state. A typical excitation spectrum for the $C_4D_2^+$ isotopic species is shown in Fig. 20. Similarly well defined and detailed structure is observed in the resolved emission spectrum and in the spectra of the other isotopic species.

The basic vibrational structure and the strongest features in the spectrum are readily assigned and they are quite consistent with an ion linear in both electronic states. The most prominent features in the spectrum are due to the symmetric C≡C and C-C stretching modes ν_2 and ν_3. Even overtones of the bending ν_7 mode also appear strongly due to Fermi resonances with the ν_3 vibration. On the other hand, finer details of the structure of $C_4H_2^+$ and the assignments of numerous weaker bands are not clearly understood at this time. Both the $\tilde{X}^2\Pi_g$ and the $\tilde{A}^2\Pi_u$ states are of course subject to Renner-Teller splitting, and this probably accounts for at least some of the weaker extra bands appearing in the spectrum. A full understanding of the nature and magnitude of these effects will require additional studies.

TABLE V.

Summary of the Vibrational Frequencies of Diacetylene Radical Cation and its Derivatives in Solid Neon
$[cm^{-1}]$

		≡C-C≡	C≡C	≡C-C-
H,H	X^+	865,973[a]	2177	-
	A^+	807,861	1971	-
D,D	X^+	829,932	2067	-
	A^+	782,836	1893	-
CH_3,H[c]	X^+	654,695	-,2215	1209
	A^+	613,669	2010,2138[b]	1132
CH_3,CH_3[c]	X^+	560	2246	1323
	A^+	529	2130	1223

[a] In all the compounds except dimethyl-diacetylene cation, the central C-C stretching vibration is in Fermi resonance with the overtone of a skeletal bending mode ($2\nu_7$ in diacetylene).

[b] In methyldiacetylene both the C≡C vibrational modes are active in the spectrum.

[c] In the methyl substituted compound which are not strictly speaking linear, several other modes appear weakly in the spectrum. These are not listed in the table.

In addition to diacetylene itself, ions of several of its alkyl derivatives were also studied, most of them both in the gas phase and by the matrix technique [83,116]. As an example, part of the laser excitation spectrum of the 1,3 pentadiyne cation is presented in Fig. 21. The spectroscopy of these species is quite similar to that of the unsubstituted diacetylene. In each case, the spectrum is dominated by the symmetric stretching vibrations of the carbon skeleton, with little involvement of the C-H bonds. Some of the frequencies of the dominant carbon-carbon stretching modes in three of these compounds are summarized in Table V. Again, numerous weaker bands are observed in each case which do not fit into the regular pattern of the prominent totally symmetric vibrations. These are probably best explained in terms of distortions of the degenerate electronic states. It should be noted, that strictly speaking the alkyl-substituted species are no longer linear molecules and therefore their distortions should be discussed in terms of the Jahn-Teller, rather than the Renner-Teller effect.

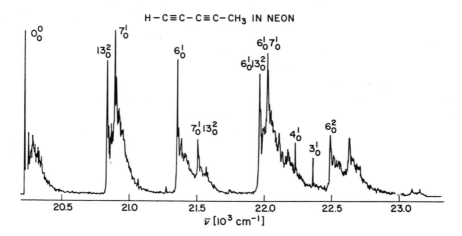

Fig. 21.   Laser excitation spectrum of methyldiacetylene (1,3 pentadiyne) radical cation in neon matrix.

4. Olefinic Cations

Numerous doubly bonded, olefinic cations have been investigated by a variety of techniques. The photoelectron spectra of the two simplest compounds from this class, ethylene and butadiene show no vibrational fine structure in their second photoelectron band. While other effects could result in broad photoelectron bands, the lack of vibrational structure is usually the consequence of extremely fast nonradiative relaxation. When the rate gets into the 10^{13}-$10^{14}\,s^{-1}$ range, the linewidths, limited by the uncertainty principle, become comparable to the vibrational spacings and preclude the observation of a resolved structure. Indeed, thus far in all cases where fluorescence of a particular ionic state has been observed, the corresponding photoelectron band is structured. Accordingly, no fluorescence has been as yet observed either for ethylene or for butadiene cations.

Conversely, the photoelectron spectra of hexatriene show sharp vibrational structure in the first two photoelectron bands [148]. Accordingly, also the absorption spectra of the hexatriene cation exhibit sharp vibrational structure [125] and apparently both the cis- and trans- isomers were also detected in fluorescence [149]. Shida and coworkers [150] have studied the spectra of the hexatriene cation and several related hydrocarbons in frozen halocarbon matrices. Following gamma irradiation, the frozen solutions of hexatriene in halocarbon mixtures exhibited strong absorptions which were assigned to several isomers of the hexatriene radical cation. The absorptions were all rather broad but showed some evidence of partially resolved vibrational structure.

These results and the nature of the electronic states of the ions can be explained with the help of Fig. 22, which shows the orbital energies of several simple olefins, as determined from their photoelectron [148] and electronic spectra. In ethylene (left) the highest occupied molecular orbital (HOMO) is a π orbital, which is only half filled in the cation. The lowest transition promotes an electron from this orbital into the corresponding antibonding π^* orbital. A lower energy transition, possible in the cation, which promotes an electron from a lower lying σ orbital into the half filled π orbital is symmetry forbidden. Butadiene and hexatriene have two and three bonding olefinic π orbitals, respectively, whose energies are split several electron volt apart. This is evidenced by the shifts of the π-π^* electronic transitions of the parent compounds to lower energies, as well as by direct observation of the splitting in the photoelectron spectrum. In the cations, the highest of the π bonding orbitals is only half filled, and this gives rise to low lying allowed electronic transitions indicated in Fig. 22, interchanging an electron between the π orbitals. While both in butadiene and in

ELECTRONIC TRANSITIONS IN OLEFINIC
COMPOUNDS AND THEIR CATIONS

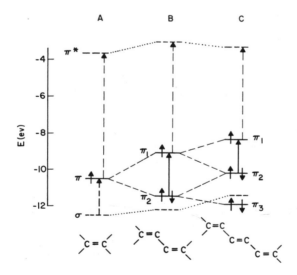

Fig. 22. Electronic structure and electronic transitions in olefinic compounds and their cations. Ethylene has a single π orbital which is filled with 2 electrons. The cation ground state is formed by removal of one of these electrons. the allowed π-π^* transition is located fairly far in UV. In the conjugated olefins the π orbitals associated with the individual double bonds split apart. As a result, low energy transitions are available in their cations corresponding to interchange of an electron between the π orbitals as shown by arrows.

hexatriene cations the upper state of the shown transition is apparently still above the threshold for ion fragmentation, in the latter case it is sufficiently long lived to give a sharp band in the photoelectron spectrum and to permit observation of the transition in emission.

In a solid neon matrix a rather intense fluorescence attributable to trans 1,3,5 hexatriene cation is observed. Both the emission and the laser excitation spectrum show sharp, well resolved zero phonon lines with rather extensive vibrational structure. The emission spectra give extensive information about the vibrational structure of the cation in the ground electronic state, and they provide information similar to, but somewhat more extensive than, the gas phase emission spectra. There is quite good agreement between the gas phase and matrix data, suggesting that the ions are not strongly perturbed in solid neon. An advantage of the matrix experiments is that they provide, via the fluorescence excitation and absorption spectra, similarly detailed information about the excited electronic state.

In addition to hexatriene, also two of its methyl substituted derivatives, trans-heptatriene and 2,4,6 octatetraene cations were investigated and similarly extensive information about these species was obtained. The data obtained are exemplified in Fig. 23 by the low energy section of the excitation spectrum of the latter compound, the trans 1,3,5 heptatriene radical cation.

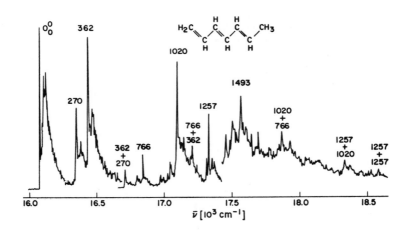

Fig. 23.   Fluorescence excitation spectrum of heptatriene radical cation in solid neon. The individual vibrational modes are labeled by their frequencies in cm^{-1}.

Parent hexatriene, as well as its methyl-substituted derivatives have low ionization potentials [148] and the corresponding cations are therefore very efficiently produced in solid neon matrix by in-situ Lyman-α photolysis. Somewhat more unexpectedly, photolysis of the cyclic C_6H_8 isomers, 1,3- or 1,4-cyclohexadienes, also produces the open chain hexatriene cations with comparable ease. This demonstrates that following the ionization process, the neon offers little resistance to the $C_6H_8^+$ cations in rearranging to the most stable straight chain trans conformer.

While some cations, in particular some of the fluorobenzenes, are apparently isolated in well-defined sites and their absorption spectra exhibit relatively sharp, narrow lines, hexatriene and related cations are characterized by rather broad inhomogeneous distributions.

Each vibronic band extends over ~150-250 cm^{-1} with several well defined maxima, some of which get very sharp (<1 cm^{-1}) in annealed samples. The long, extended molecule must replace a number of lattice atoms in solid neon and can apparently be isolated in several well-defined local environments. The inhomogeneous absorption band profile and the site distribution in the spectrum of trans hexatriene cation are strongly dependent on the C_6H_8 precursor. The nature of the cationic site is apparently dependent on the original matrix configuration around the neutral parent, and this is clearly quite different for the cyclic C_6H_8 compounds than for the extended hexatriene compounds. While we have not studied each of the cations in each of the possible sites, it was found that except for the overall shift of the electronic transition, the spectra of the cations in the different sites exhibit practically identical vibrational structure.

In view of the rather large inhomogeneous linewidths it is difficult to compare the electronic transition energies with the gas phase, but the most abundant sites again appear to

TABLE VI.

Vibrational Frequencies in the Spectra of trans 1,3,5-hexatriene
and trans 1,3,5-heptatriene in Solid Neon.

trans 1,3,5-hexatriene			trans-1,3,5-heptatriene	
$\tilde{X}^1A_g$	$\tilde{X}^2A_u$	$\tilde{A}^2B_g$	$\tilde{X}$	$\tilde{A}$
347[a]	350[b]	343	281	270
444	442	443	374	362
897	951			766
1187	1115	1071		120
1238	1239	1245	1236	1257
1280	1293		1299	
1394	1376	1373	1414	
1573	1513	1434	1519	
1623	1622	1486	1629	1493

[a] The a_g fundamentals (ν_{13}-ν_5) of trans-hexatriene are taken from ref. 151. The CH stretching vibrations (ν_1-ν_4) were omitted.

[b] The most prominent vibrational modes appearing in the cation spectra in neon matrix. The shown associations with particular vibrational modes of the parent are in many cases tentative.

[c] The symmetry of heptatriene cation is C_s or lower. Analysis of the parent vibrational spectrum is not available.

undergo only small (<100 cm^{-1}) shifts in solid neon. Some of the vibrational data for the related cations of 1,3,5 hexatriene and heptatriene in both states are summarized in Table VI. Rather prominent in each spectrum appear to be two low frequency modes which are quite similar in the emission and excitation spectra, and which by comparison with the parent compounds are identified as the lowest frequency, in-plane skeletal bending modes. These, for instance, occur at 343 and 443 cm^{-1} in the excitation spectrum of $C_6H_8^+$ and they shift to 270 and 362 cm^{-1} in $C_7H_{10}^+$.

The stretching frequencies in the $\sim$800-1700 cm^{-1} range show much less clear correlation with the vibrational modes of the parent and they change quite appreciably in the excited electronic states of the cations. Apparently the bond strengths and the conjugation effects between the three double bonds are a very sensitive function of the number of available electrons and their distribution between the bonding π orbitals.

The trans-hexatriene and heptatriene cations have in the gas phase lifetimes of 17 and 12 ns respectively [126]; in solid neon the lifetimes are less than 10 ns. These values are considerably shorter than the expected radiative lifetimes and it was estimated that the trans $C_6H_8^+$ fluorescence yield is $\sim$7%. On the other hand, the survival of a substantial fraction of the ions in the $\tilde{A}$ state in the photoelectron-phonon coincidence experiment of a $\sim$40 μs time of flight indicates that the rate of fragmentation is slow ($<10^6$ sec^{-1}). The observed lifetime must thus be controlled by nonradiative relaxation into high vibrational levels of the ground state. The ion decomposition in the gas phase thus proceeds in two steps, nonradiative relaxation, followed by a much slower fragmentation. In the matrix, the slow fragmentation can not compete with the extremely fast vibrational relaxation of the highly excited ground state levels. Accordingly, no permanent dissociation is observed in solid neon, and even prolonged excitation of the sample produces no bleaching of the radical cations.

5. CS_2^+ And Other Small Inorganic Cations

As noted previously, small inorganic ionic species are usually more readily amenable to studies by means of classical spectroscopic techniques than large organic ions and one might therefore anticipate less need for low temperature matrix studies. Furthermore, for some of the simplest small molecules the ionization potentials are relatively high and in these cases one would have to consider the possibility of the effects of the environment — even a Ne matrix, upon the spectroscopy of the ionic guest. In spite of this, in many specific cases matrix studies can provide very useful information not readily obtainable by other means, and indeed several small ions and ionic fragments have been investigated in this way.

A variety of species belonging in this category were examined by means of IR absorption spectroscopy, in particular by the groups of Andrews and Jacox [8,9]. Such investigations are particularly useful for species which do not posses fluorescent excited electronic states. As mentioned previously a drawback is that in the absence of the extensive and redundant information built into electronic spectroscopy and laser induced fluorescence studies, making unambiguous assignments is often difficult and in many cases the effects of medium perturbations are hard to assess.

CS_2^+ is an example of a molecular cation which has been examined in solid matrices by laser induced fluorescence in some detail. This cation has previously been investigated in the gas phase by a number of workers [70,102,103] using a variety of techniques. In spite of this extensive attention, the spectroscopy of CS_2^+ was incompletely understood. There was rather poor agreement between the individual studies, and even the location of the origin of the visible transition was not firmly established. The matrix spectra of CS_2^+ are extremely simple [60,71], almost self-explanatory and they provide rather complete information about the vibrational structure of CS_2^+.

CS_2^+ is a molecule with 15 valence electrons which, based on Walsh's rules [152] and on comparison with isovalent systems [153,154] like BO_2, CO_2^+ or NCO, should be linear in its ground state and in other low lying electronic states. The spectroscopy is quite consistent with this expectation. For a triatomic $D_{\infty h}$ molecule there is only one totally symmetric vibration, the symmetric C-S stretch, ν_1, which can appear as a progression forming mode in an allowed electronic transition. In CS_2^+, as in a number of similar linear molecules, the spectra are complicated by Fermi resonances between ν_1 and even overtones of the bending vibration ν_2. This can be seen in Fig. 24, which shows the CS_2^+ emission spectrum. The spectrum is characterized by a progression of "Fermi polyads" involving ν_1 and $2\nu_2$. In CS_2^+ the two modes are almost perfectly mixed as was shown [60] by examination of the isotopically substituted $^{13}CS_2^+$. This study has found for all the levels involved in the Fermi resonance, isotopic shifts intermediate between the shifts expected for the ν_2 mode and the zero shift expected for ν_1. On the other hand, the third vibrational mode ν_3, the asymmetric CS_2^+ stretch, which is observed in the spectrum as the even overtone $2\nu_3$ and $4\nu_3$ levels, does not seem to mix with the other modes and exhibits the theoretically expected ^{13}C isotopic shift $\rho = 1.034$.

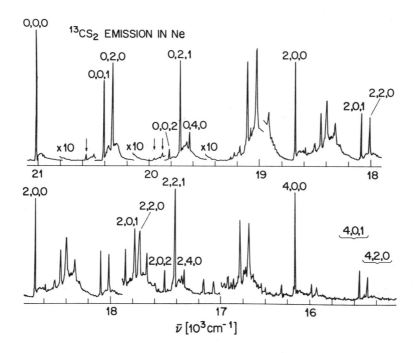

Fig. 24.   Resolved fluorescence spectrum of $^{13}CS_2$ cation in neon matrix. The vibronic bands are labeled by their v_1'', v_2'', v_3'' values.

With the insights provided by the matrix isolation studies, the spectroscopy of the CS_2^+ cation is now well understood and all three vibrational fundamentals are now known, both in the $\tilde{X}^2\Pi_g$ ground state and in the $\tilde{A}^2\Pi_u$ lowest excited state.

IV. SUMMARY

In this chapter we have examined the spectroscopy and photophysics of molecular ions in low temperature matrices. In the first part (Section II) the general techniques for producing, detecting and analyzing the electronic spectra of ionic species are discussed. It is shown that direct photoionization provides a very efficient and satisfactory technique for cation generation. The question of guest-solvent interactions is discussed in some detail. An important parameter is found to be the difference between the electron affinity of the ionic state studied and the ionization potential of the solvent. This difference is large for most ions in solid neon and accordingly, ionic spectra in neon matrix are shown to be largely unperturbed by the medium. Stronger charge-transfer interactions are found to occur frequently in solid argon and the heavier, more polarizable rare gases. Neon is therefore shown to be the preferred medium for ion studies.

The second part of the manuscript (Section III) then discusses the spectroscopy and photophysics of several chemically distinct classes of molecular cations, whose electronic spectra and properties were investigated using matrix techniques. The major categories involve derivatives of benzene, pyridine, acetylene, olefins as well as miscellaneous other ions. In each case the matrix provides useful new information not readily obtainable by other means and conveniently complements gas phase studies.

REFERENCES

1. B. Meyer, Low Temperature Spectroscopy, Elsevier, New York 1971.

2. H. E. Hallam (ed), Vibrational Spectroscopy of Transient Species, Wiley, New York (1973).

3. M. Moskovits and G. A. Ozin, (ed), Cryochemistry, Wiley, New York 1976.

4. E. Whittle, D. A. Dows and G. C. Pimentel, J. Chem. Phys. **22**, 1943 (1954).

5. G. C. Pimentel, Spectrochim. Acta, **12**, 94 (1958).

6. E. D. Becker, G. C. Pimentel and M. Van Thiel, J. Chem. Phys. **26**, 195 (1957).

7. L. Andrews, Annu. Rev. Phys. Chem. **22**, 109 (1971).

8. M. E. Jacox, Rev. Chem. Intermed. **1**, 1 (1978).

9. L. Andrews, Ann. Rev. Phys. Chem. **30**, 79 (1979).

10. R. I. Personov, E. I. Al'Shits and L. A. Bykovskaya, Optics Commun. **6**, 169 (1972).

11. K. Rebane, Impurity Spectra of Solids, (Plenum, New York, 1970).

12. V. E. Bondybey and L. E. Brus, J. Chem. Phys. **63**, 2223 (1975).

13. J. S. Shirk, J. Chem. Phys. **55**, 3608 (1971).

14. V. E. Bondybey and J. W. Nibler, J. Chem. Phys. **56**, 4719 (1972).

15. D. E. Tevault and L. Andrews, J. Mol. Spectrosc. **54**, 54 (1975).

16. V. E. Bondybey, J. Chem. Phys. **66**, 995 (1977).

17. V. E. Bondybey and L. E. Brus, Adv. Chem. Phys. **41**, 269 (1980).

18. V. E. Bondybey and J. H. English, J. Chem. Phys. **67**, 3405 (1977).

19. G. Herzberg, Rev. Chem. Soc. **25**, 201 (1971).

20. J. H. Callomon, Can. J. Phys. **34**, 1046 (1956).

21. J. Daintith, R. Dinsdale, J. P. Maier, D. A. Sweigart and D. W. Turner, Molecular Spectroscopy (Institute of Petroleum 1972), p. 16.

22. M. Allan and J. P. Maier, Chem. Phys. Lett. **34**, 442 (1975).

23. J. P. Maier, Chemia **34**, 219 (1980).

24. P. C. Engelking and A. L. Smith, Chem. Phys. Lett. **36**, 21 (1975).

25. V. E. Bondybey and T. A. Miller, J. Chem. Phys. **67**, 1790 (1977).

26. C. Cossart-Magos, D. Cossart and S. Leach, J. Chem. Phys. **69**, 4313 (1978).

27. C. Cossart-Magos, D. Cossart, S. Leach, J. P. Maier and L. Misev, J. Chem. Phys., in press.

28. R. C. Woods, T. A. Dixon, R. J. Saykally and P. G. Szanto, Phys. Rev. Lett. **35**, 1269 (1975).

29. R. J. Saykally and R. C. Woods, Ann. Rev. Phys. Chem. **32**, 403 (1981).

30. A. Carrington, D. R. J. Milverton and P. J. Sarre, Mol. Phys. **35**, 1505 (1978).

31. R. C. Dunbar, Spec. Periodicals Report: Mass Spectroscopy VI, ed. R. J. W. Johnstone, The Chemical Society, London (1981).

32. J. Moseley, Ann. Rev. Phys. Chem. **32**, 53 (1981).

33. V. E. Bondybey, T. A. Miller and J. H. English, J. Chem. Phys. **72**, 2193 (1980).

34. R. Rosetti and L. E. Brus, Chem. Phys. Lett. **69**, 447 (1980).

35. F. O. Rice and M. Freamo, J. Am. Chem. Soc. **75**, 548 (1953).

36. M. Mc Carty, Jr. and G. W. Robinson, J. Am. Chem. Soc. **81**, 4472 (1959).

37. P. N. Noble and G. C. Pimentel, J. Chem. Phys. **49**, 3165 (1968).

38. V. E. Bondybey and J. W. Nibler, J. Chem. Phys. **58**, 2125 (1973).

39. C. B. Moore, G. C. Pimentel and T. D. Goldfarb, J. Chem. Phys. **43**, 63 (1965).

40. A. Snelson, J. Phys. Chem. **74**, 537 (1970).

41. J. H. Current and J. K. Burdett, J. Phys. Chem. **74**, 537 (1970).

42. K. Rosengren and G. C. Pimentel, J. Chem. Phys. **43**, 507 (1965).

43. D. E. Milligan and M. E. Jacox, J. Chem. Phys. **43**, 4487 (1965).

44. D. E. Milligan and M. E. Jacox, J. Chem. Phys. **43**, 4487 (1965).

45. M. E. Jacox and D. E. Milligan, J. Chem. Phys. **50**, 3252 (1969).

46. R. P. Frosch, J. Chem. Phys. **54**, 2660 (1971).

47. L. Andrews, J. M. Grzybowski and R. O. Allen, J. Phys. Chem. **79**, 904 (1975).

48. F. T. Prochaska and L. Andrews, J. Chem. Phys. **67**, 1091 (1977).

49. L. Andrews and G. C. Pimentel, J. Chem. Phys. **44**, 2361 (1966).

50. L. Andrews, J. Phys. Chem. **71**, 2761 (1967).

51. M. Jacox, Chem. Phys. **42**, 133 (1979).

52. L. Andrews, J. Am. Chem. Soc. **90**, 7368 (1968).

53. D. E. Milligan, M. E. Jacox and W. A. Guillory, J. Chem. Phys. **52**, 3864 (1970).

54. L. Andrews, J. T. Hwang and C. Trindle, J. Phys. Chem. **77**, 1065 (1973).

55. V. E. Bondybey and C. Fletcher, J. Chem. Phys. **64**, 3615 (1976).

56. V. E. Bondybey and L. E. Brus, J. Chem. Phys. **64**, 3724 (1976).

57. D. E. Milligan and M. E. Jacox, J. Chem. Phys. **43**, 4487 (1965).

58. H. Okabe, J. Opt. Soc. Am. **54**, 478 (1964).

59. V. E. Bondybey, T. A. Miller and J. H. English, J. Chim. Phys. **77**, 667 (1980).

60. V. E. Bondybey and J. H. English, J. Chem. Phys. **73**, 3098 (1980).

61. Z. Ophir, B. Raz, J. Jortner, V. Saile, N. Schwentner, E. Koch, M. Skibowski, W. Steinmann, J. Chem. Phys. **62**, 650 (1975).

62. V. E. Bondybey, J. Mol. Spectrosc. **63**, 164 (1976).

63. V. E. Bondybey and J. H. English, J. Chem. Phys. **67**, 2868 (1977).

64. V. E. Bondybey, T. A. Miller and J. H. English, Phys. Rev. Lett. **44**, 1344 (1980).

65. V. E. Bondybey, J. H. English and T. A. Miller, J. Phys. Chem., April, 1983.

66. H. A. Jahn and E. Teller, Proc. R. Soc. London, **A161**, 220 (1937).

67. A. D. Liehr, Z. Phys. Chem., Neue Folge **9**, 338 (1956).

68. V. E. Bondybey, T. A. Miller and J. H. English, J. Chem. Phys. **71**, 1088 (1979).

69. J. H. Callomon, Proc. R. Soc. London, **A244**, 220 (1958).

70. W. J. Balfour, Can. J. Phys. **54**, 1969 (1976).

71. V. E. Bondybey, J. H. English and T. A. Miller, J. Chem. Phys. **70**, 1621 (1979).

72. J. H. D. Eland, M. Devoret and S. Leach, Chem. Phys. Lett. **43**, 97 (1976).

73. S. Leach, M. Devoret and J. H. D. Eland, Chem. Phys. **33**, 113 (1978).

74. M. Allan, J. P. Maier and O. Marthaler, Chem. Phys. **26**, 131 (1977).

75. V. E. Bondybey, J. H. English and T. A. Miller, J. Mol. Spectrosc. **81**, 455 (1980).

76. J. B. Birks, "Photophysics of Aromatic Molecules," Wiley, New York (1970).

77. R. L. Fulton, J. Chem. Phys. **61**, 4141 (1974).

78. W. B. Person, J. Chem. Phys. **28**, 319 (1958).

79. M. L. Klein and J. A. Venables, eds., Rare Gas Solids, Vol. 1, Academic Press, New York (1976).

80. A. Gedanken, B. Raz and J. Jortner, J. Chem. Phys. **58**, 1178 (1973).

81. V. E. Bondybey and J. H. English, J. Chem. Phys. **71**, 777 (1979).

82. C. Baker and D. W. Turner, Proc. Roy. Soc. **A308**, 19 (1968).

83. V. E. Bondybey, J. H. English and T. A. Miller, J. Chem. Phys. **70**, 1765 (1979).

84. T. A. Miller and V. E. Bondybey, J. Chim. Phys. **77**, 695 (1980).

85. J. Dannacher, J. P. Stadelman and J. Vogt, Chem. Phys. (1980).

86. V. E. Bondybey, J. H. English and T. A. Miller, J. Am. Chem. Soc. **101**, 1248 (1979).

87. T. Kato and T. Shida, J. Am. Chem. Soc. **101**, 6869 (1979).

88. T. Shida and T. Kato, Chem. Phys. Lett. **68**, 196 (1979).

89. T. Shida, H. Kubodera and Y. Egawa, Chem. Phys. Lett. **79**, 179 (1981).

90. T. Shida, Y. Egawa, H. Kubodera and T. Kato, J. Chem. Phys. **73**, 5963 (1980).

91. E. Haselbach, T. Bally, R. Gschwind, V. Klemm and Z. Lanyiova, Chimia **33**, 405 (1979).

92. T. Bally, E. Haselbach, Z. Lanyiova and P. Baertschi, Helv. Chim. Acta, **61**, 2488 (1978).

93. M. E. Jacox and D. E. Milligan, J. Chem. Phys. **54**, 3935 (1971).

94. D. E. Milligan and M. E. Jacox, J. Chem. Phys. **48**, 2265 (1968).

95. F. T. Prochaska and L. Andrews, J. Am. Chem. Soc. **100**, 2102 (1978).

96. L. Andrews and B. W. Keelan, J. Am. Chem. Soc. **101**, 3500 (1979).

97. M. Allan, E. Kloster-Jensen and J. P. Maier, Chem. Phys. **17**, 11 (1976).

98. V. E. Bondybey and T. A. Miller, J. Chem. Phys. **73**, 3053 (1980).

99. C. Cossart-Magos, D. Cossart and S. Leach, Mol. Phys. **37**, 793 (1979).

100. T. A. Miller, B. R. Zegarski, T. J. Sears and V. E. Bondybey, J. Phys. Chem. **84**, 3154 (1980).

101. V. E. Bondybey and T. A. Miller, J. Chem. Phys. **70**, 138 (1979).

102. S. Leach, J. Chim. Phys. **67**, 74 (1970).

103. L. C. Lee, D. L. Judge and M. Ogawa, Can. J. Phys. **53**, 186 (1975).

104. J. P. Maier, O. Marthaler, M. Mohraz and R. H. Shiley, J. El. Spectrosc. Rel. Phen. **19**, 11 (1980).

105. J. P. Maier, L. Misev and R. H. Shiley, Helv. Chim. Acta **63**, 1920 (1980).

106. V. E. Bondybey, J. H. English, T. A. Miller and C. B. Vaughn, J. Phys. Chem. **85**, 1667 (1981).

107. V. E. Bondybey, J. H. English, T. A. Miller and R. H. Shiley, J. Phys. Chem., accepted.

108. J. P. Maier, O. Marthaler and M. Mohraz, J. Chim. Phys. **77**, 661 (1980).

109. V. E. Bondybey, J. H. English and T. A. Miller, Chem. Phys. Lett. **66**, 165 (1979).

110. G. Dujardin, S. Leach and G. Taieb, Chem. Phys. **46**, 407 (1980).

111. V. E. Bondybey, C. R. Vaughn, T. A. Miller, J. H. English and R. H. Shiley, J. Chem. Phys. **74**, 6584 (1981).

112. C. Cossart-Magos, D. Cossart, and S. Leach, J. Chem. Phys. **69**, 4313 (1978).

113. C. Cossart-Magos and S. Leach, Chem. Phys. **48**, 329 (1980).

114. V. E. Bondybey, T. J. Sears, J. H. English and T. A. Miller, J. Chem. Phys. **73**, 2063 (1981).

115. J. P. Maier, O. Marthaler, and Kloster-Jensen, J. Chem. Phys. **72**, 701 (1980).

116. V. E. Bondybey, unpublished results.

117. J. P. Maier, O. Marthaler, M. Mohraz and R. H. Shiley, Chem. Phys. **47**, 295 (1980).

118. V. E. Bondybey, J. Chem. Phys. **71**, 3586 (1979).

119. V. E. Bondybey, J. H. English and T. A. Miller, Chem. Phys. Lett. **90**, 394 (1982).

120. V. E. Bondybey, J. H. English and R. H. Shiley, J. Chem. Phys. **77**, 4826 (1982).

121. J. P. Maier, O. Marthaler, M. Mohraz and R. H. Shiley, Chem. Phys. **47**, 307 (1980).

122. V. E. Bondybey, T. J. Sears, T. A. Miller, C. Vaughn, J. H. English and R. H. Shiley, Chem. Phys. **61**, 9 (1981).

123. V. E. Bondybey, J. H. English and T. A. Miller, J. Mol. Spectrosc. **80**, 200 (1980).

124. R. C. Dunbar, J. Am. Chem. Soc. **98**, 4671 (1976).

125. R. C. Dunbar and H. Teng, J. Am. Chem. Soc. **100**, 2279 (1978).

126. M. Allan, J. Dannacher and J. P. Maier, J. Chem. Phys. **73**, 3114 (1980).

127. V. E. Bondybey, J. H. English, T. A. Miller and R. H. Shiley, J. Mol. Spectrosc. **84**, 124 (1980).

128. V. E. Bondybey, C. B. Vaughn, T. A. Miller, J. H. English and R. H. Shiley, J. Am. Chem. Soc. **103**, 6303 (1981).

129. J. P. Maier and O. Marthaler, Chem. Phys. Lett. **32**, 419 (1978).

130. J. P. Maier, O. Marthaler and M. Mohraz, Chem. Phys. **47**, 307 (1980).

131. A. A. Belyaeva, Y. B. Predtechenskii and L. D. Shcherba, Opt. Spectrosc. **34**, 21 (1973).

132. L. C. Balling, M. D. Havey and J. J. Wright, J. Chem. Phys. **70**, 2404 (1979).

133. V. E. Bondybey, J. Chem. Phys. **75**, 492 (1981).

134. D. E. Milligan and M. E. Jacox, J. Chem. Phys. **51**, 1952 (1969).

135. L. E. Brus and V. E. Bondybey, J. Chem. Phys. **63**, 3123 (1975).

136. L. J. Allamandola, A. M. Rohjantalab, J. W. Nibler and T. Chappel, J. Chem. Phys. **67**, 99 (1977).

137. D. Feldman, Z. Naturforsch. **25a**, 621 (1970).

138. C. R. Brundle, M. B. Robin, N. A. Kuebler and H. Bash, J. Am. Chem. Soc. **94**, 1451 (1972).

139. C. R. Brundle, M. B. Robin and N. A. Kuebler, J. Am. Chem. Soc. **94**, 1466 (1972).

140. A. W. Potts, W. C. Price, D. G. Streets and T. A. Williams, Disc. Farad. Soc. **54**, 168 (1972).

141. A. C. Albrecht, J. Mol. Spectrosc. **6**, 84 (1961).

142. H. C. Longuet-Higgins, Adv. Spectrosc. **2**, 429 (1961).

143. M. S. Child and H. C. Longuet-Higgins, Proc. Roy. Soc. **A245**, 33 (1961).

144. C. S. Sloane and R. Silbey, J. Chem. Phys. **56**, 6031 (1972).

145. T. Sears, T. A. Miller and V. E. Bondybey, J. Chem. Phys. **72**, 6070 (1980).

146. T. J. Sears, T. A. Miller and V. E. Bondybey, J. Chem. Phys. **74**, 3240 (1981).

147. V. E. Bondybey, T. A. Miller and J. H. English, J. Mol. Spectrosc. **90**, 592 (1981).

148. M. Beez, G. Bieri, H. Bock and E. Heilbronner, Helv. Chim. Acta, **56**, 1028 (1973).

149. M. Allan and J. P. Maier, Chem. Phys. Lett. **43**, 94 (1976).

150. T. Shida, T. Kato and Y. Nosaka, J. Phys. Chem. **81**, 1095 (1977).

151. E. M. Popov and G. A. Kogan, Opt. Spectrosc. **17**, 362 (1964).

152. A. D. Walsh, J. Chem. Soc. 2296 (1953).

153. J. W. C. Johns, Can. J. Phys. **42**, 1004 (1964).

154. R. N. Dixon, Can. J. Phys. **38**, 10 (1960).

Molecular Ions: Spectroscopy, Structure and Chemistry
Terry A. Miller and V.E. Bondybey (editors)
© North-Holland Publishing Company, 1983

EMISSION AND EXCITATION SPECTROSCOPY OF
OPEN-SHELL ORGANIC CATIONS

Dieter Klapstein, John P. Maier and Liubomir Misev

Physikalisch-Chemisches Institut, Universität Basel,
Klingelbergstrasse 80, CH-4056 Basel, Switzerland

The application of emission and laser excitation techniques
for the spectral characterisation of rotationally cooled
open-shell organic cations in the gas phase are illustrated
by means of selected examples. The principles of the techniques and the information forthcoming from such, and time-
resolved, studies are discussed. Vibrational frequencies of
the cations in their ground and lowest excited states can
be obtained to an accuracy of ± 1 cm^{-1}, or better. These can
thus serve as reference data for other methods and for the
identification of the cations in various environments.

1. INTRODUCTION

Molecular ions play an important role in terrestrial and extra-
terrestrial chemistry [1]. The characterization of the species is thus
of considerable importance and the usual means is by spectroscopy.
Important applications of the spectroscopic techniques are not only
the identification of such species in various environments but also
the spectral features can be used to provide information about the
physical conditions of the surroundings and on energy transfer
mechanisms [2].

This article is concerned with emission and laser excitation
spectroscopic studies on polyatomic organic open-shell cations in the
gas phase. Both these methods are dependent on the radiative decay of
the electronically excited cations to their ground state, and this
enables these states to be spectrally investigated.

Among organic cations only the emission spectrum of diacetylene
cation, $H\{C\equiv C\}_2H^+$ $\tilde{A}^2\Pi_u \rightarrow \tilde{X}^2\Pi_g$, was known [3] until the advent of
photoelectron spectroscopy [4]. The latter technique provided a ge-
neral method to measure the ionisation energies to the accessible
doublet states ($^2\tilde{X}$, $^2\tilde{A}$, ..., $^2\tilde{J}$) of open-shell cations. This mapping
of the lowest lying doublet states paved the way for an extensive
search for the radiative decay of organic cations, with the result
that this decay mechanism has been detected for over one hundred such
species [5]. Some examples of the types of cations studied are listed
in table 1.

Table 1
Typical open-shell organic cations for which radiative
decay has been detected from their lowest excited states.[a]

$X-C\equiv N^+$, X= Cl,Br,I
$H-C\equiv P^+$
$H\{C\equiv C\}_n H^+$, n= 2,3,4
$H\{C\equiv C\}_n X^+$, n= 1,2   X= Cl,Br,I   n= 2   X= CH_3
$X\{C\equiv C\}_n X^+$, n= 1,2   X= Cl,Br,I   n= 2   X= F,CH_3,CF_3
$N\equiv C\{C\equiv C\}_n C\equiv N^+$, n= 1,2
$N\equiv C\{C\equiv C\}_n X^+$, n= 1   X= Cl,Br   n= 1,2   X= CH_3
B-trifluoroborazine$^+$
fluoro-substituted benzenes$^+$
fluoro-substituted phenols$^+$
mixed halo-substituted benzenes$^+$, phenols$^+$, toluenes$^+$
<u>cis</u>- and <u>trans</u>-1,3,5-hexatriene$^+$
3,5-octadiyne$^+$
t-Bu$\{C\equiv C\}_3$-t-Bu $^+$

[a] for a more complete list of species studied and the individual references see ref. [6].

The emission spectra of such organic cations have been obtained by
electron impact excitation of the neutral species in its ground
state, $M^1 X$:

$$M\ ^1X \xrightarrow{e(E_i)} M^{\oplus}\ ^2\tilde{X},^2\tilde{A},^2\tilde{B}\cdots\cdots^2\tilde{J}(I \leqslant E_i)$$

$$\downarrow \rightsquigarrow \boxed{h\nu}$$

$$M^{\oplus}\ ^2\tilde{X}$$

Although the electron-impact excitation is nonspecific, i.e. all
accessible electronic states with ionisation energies (I_i) less than
the electron energy (E_i) may be reached, it was chosen for intensity
reasons and the fact that it is reasonably easy to vary the electron
energy as well as to pulse the beam for time-dependent studies. Any
resultant emission from populated excited states is wavelength dispersed.

In the laser excitation studies, the open-shell cations are prepared
by Penning ionisation, which is known to yield a sufficient concentration of thermalized ground state ions in the gas phase for the
detection of the laser-excited fluorescence [7], i.e.:

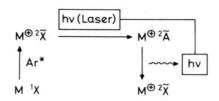

The ionisation is induced by energy transfer through collisional
complexes between the metastable rare gas atoms and the sample
molecules. An electron removal results in the population of the
ground or an excited state of the molecular open-shell cation, depending on the energy of the metastable. However, an excited state
is rapidly depleted by spontaneous emission and/or intramolecular
processes. Vibrational energy is removed by collisions with the rare
gas carrier since the pressure in the interaction zone is of the
order of 1 mbar. Thus a Boltzmann-like ground state distribution is
encountered for the cations reflecting the temperature of the
carrier gas.

.While both emission and excitation spectroscopies rely on the same
property, the radiative decay of the electronically excited cationic
state, the information obtained from such measurements is by no
means duplicate but, rather, complementary [5]. Figure 1 illustrates
the different excitation and monitoring methods of the two techniques.

In subsequent sections of this article these features and the information forthcoming from such studies on open-shell organic cations
will be illustrated by means of selected examples.

2. RADIATIVE AND NON-RADIATIVE DECAY

The presence or absence of radiative decay from an electronically excited state of a cation may be predicted using several types of information [5]. Most important is the photoelectron spectrum and the
assignment of the symmetry of the electronic states. The transition
between upper and lower states must be electric dipole allowed and
from the experimental point of view the energy separation of the two
electronic states should lie in the 200-900 nm wavelength range. If
these conditions are satisfied, the detection probability depends on
the relative rates of the radiative k_r and non-radiative k_{nr} (i.e.
internal conversion, fragmentation or isomerisation) pathways depleting the excited state. The respective rates are related to the lifetime, τ, of the excited state by $1/\tau = k_r + \Sigma k_{nr}$ where the summation
is over the accessible channels. To determine the individual rates,
the fluorescence quantum yields, ϕ_F, are required (e.g. $k_r = \phi_F/\tau$);
these can be absolutely determined by a photoelectron-photon coincidence technique, as discussed elsewhere [8].

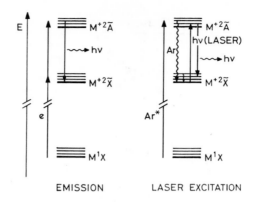

Figure 1
Complementary nature of emission and laser
excitation fluorescence spectroscopies

The present limit for the detection of the radiative decay by emission spectroscopy, i.e. instrumental sensitivity, is $\phi_F > 10^{-5}$. Since the measured radiative decay rates of polyatomic organic cations are typically of the order 10^7 s^{-1} [5][8], the undetectability of a radiative decay process implies that $\Sigma\ k_{nr} > 10^{11}$ s^{-1}. This value provides a further link to photoelectron spectroscopy, since the absence of vibrational structure on the photoelectron band corresponding to formation of the specified cationic state implies $\Sigma\ k_{nr} > 10^{12}$ s^{-1}. Such high rate constants are observed for fragmentation, internal conversion and isomerisation decay channels, especially for highly excited states. For the lowest excited electronic states (i.e. $^2\tilde{A}$ or $^2\tilde{B}$) usually the two competing relaxation pathways are internal conversion, $^2\tilde{A} \rightsquigarrow {}^2\tilde{X}$, and the radiative decay, $^2\tilde{A} \rightarrow {}^2\tilde{X}$. Also, for a small number of cases it has been found that the radiative decay is detectable even though the non-radiative decay leads ultimately to fragment ions [5]. These and related aspects of the radiationless decay of open-shell organic cations are the subject of a recent review [8].

3. SPECTRA OF COOLED IONS

The precision of the spectral data (e.g. vibrational frequencies) of
organic cations obtained by emission and laser excitation spectroscopy
is limited not by the optical resolution attainable but by the width
of the observed bands. This broadening arises from the rotational
transition envelope, which in most cases can not be resolved, and
leads to an uncertainty of ±5 to 10 cm^{-1}.

This limitation can be surmounted by carrying out the measurements at
reduced sample temperatures. This can be achieved by several approaches. At one extreme is the generation and study of the cations imbedded in a neon matrix at 4-5 K [7]. While the emission or laser
excitation bands in such cases can be quite sharp, the gain may be
offset by additional broadening due to matrix interactions, matrix
shifts with respect to gas-phase data and the presence of phonon side
bands. Another approach, which is also the one used in our studies,
is the electron-impact excitation of a supersonic molecular beam to
obtain the emission spectra of radical cations [9][10]. This is accomplished by expansion of the sample mixed with a rare gas through a
nozzle to produce a supersonic free jet which then intersects
an electron beam. The conditions within the jet are such that the
rapid expansion results in a cooling of the internal degrees of
freedom. With the apparatus described in the next section [11], rotational temperatures of less than 30 K are reached for the organic
cations; the vibrational temperatures are expected to be somewhat
higher (e.g. ≈50 K) [12]. Under these conditions the vibronic bands
of polyatomic cations become much sharper, enabling the vibrational
frequencies to be determined to ±1 cm^{-1}, or better, as well as
simplifying heavily congested spectra. Examples will be considered in
sections 5.2 and 5.5. In the laser excitation studies, rotationally
and vibrationally cooled cations can be produced by collisional relaxation with the rare gas carrier which has been precooled by a liquid
nitrogen bath [7]. As with the emission spectra, the narrowing of the
bands in the excitation spectra of cooled ions is apparent. This will
be illustrated in sections 5.3 and 5.5.

4. EXPERIMENTAL

4.1. Emission Spectroscopy

The main elements of the apparatus used to record the emission spectra
of open-shell radical cations in the gas phase are shown schematically
in figure 2. In the early work [13] the samples were admitted into
the apparatus through a 1 mm diameter effusive nozzle from glass
vessels held in constant temperature baths which gave sample vapour
pressures of ca. 1-100 mbar. The effusive nozzle has been replaced
with a supersonic free jet configuration in which the sample vapour
is expanded with a large excess of a rare gas, usually helium,
through a small aperture of 30-100 μm diameter [11]. To give efficient mixing of the sample vapour and helium carrier gas the sample
is first transferred into a stainless-steel bubbler through which the
helium is then passed. The vapour pressure of the sample can be controlled by a cryostated cooling jacket surrounding the bubbler which
can maintain a temperature as low as -120°C. Sample vapour and mixture pressures are read with an absolute pressure gauge. Gas mixtures
of ca. 0.1-10% sample are used; higher sample concentrations may give

rise to clustering phenomena. The total stagnation pressure is ca.
0.5-5 bar, depending on the nozzle diameter, while the pressure in
the ionisation region is < 10^{-3} mbar.

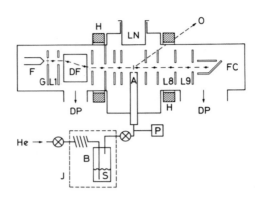

Figure 2
Crossed sample-electron beam apparatus.
F - filament, G - grid, L1-L9 - circular electron lenses,
DF - parallel deflecting plates (2), FC - Faraday cage,
H - Helmholtz coils, A - 30-100 μm diameter nozzle,
P - absolute pressure gauge, S - sample, B - bubbler,
J - cooling jacket, LN - liquid nitrogen-cooled trap,
DP - diffusion pump, O - optics, monochromator,
photon counting electronics.

Excitation is achieved with a collimated beam of electrons from a
differentially-pumped electron gun incorporating a directly heated,
thoriated tungsten filament and a trochoidal monochromator. The
electron energy can be varied over the range 10-200 eV. External
Helmholtz coils, besides providing the magnetic field for the trochoidal monochromator, help to keep the electron beam well collimated through the ionisation region. Electron currents of typically
3-6 mA at 200 eV energy are used. To compensate for variations in
alignment the nozzle is mounted on a 3-dimensional translational
stage which allows the nozzle position to be adjusted to give maximum
intersection of the free jet with the electron beam; the intersection
center is 4-8 mm downstream from the nozzle. It is important to have
the sample-electron beam interaction region far enough downstream so

that the maximum translational cooling is achieved, yet not near the
region where the supersonic free jet terminates at the Mach disc.
A limited range of translational temperatures is, however, attainable
by varying the nozzle-electron beam distance, as well as by varying
the nozzle diameter and stagnation pressure.

Emitted photons are gathered at right angles to both the electron
beam and free jet axes and focused onto the entrance slit of a
Spex f/9.5, 1.26 m monochromator which has a reciprocal dispersion of
0.43 nm/mm. Photons are detected by a cooled GaAs photomultiplier
tube (RCA 31034-02) coupled to single-photon-counting electronics;
the data acquisition system is outlined in figure 3. Three modes of
operation are possible and are indicated in the figure. In the normal
mode the photon pulses are counted in a clock interface, the spectrum
being recorded digitally with a microcomputer. For time resolved
spectroscopy the electron beam is pulsed at 500 kHz and only those
photons which are detected within a predetermined time interval are
processed further. For lifetime measurements the electron beam is
pulsed and the fluorescence signal as a function of time is monitored.
Calibration of the spectra is achieved with the numerous rare gas
emission lines which are excited concomitantly by the electron impact.

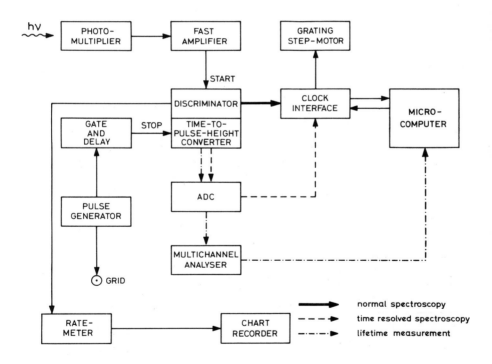

Figure 3
Block diagram of the signal acquisition system
for emission spectroscopy

4.2. Laser Excitation Spectroscopy

The essential features of the apparatus used to record laser excitation spectra of open-shell cations are represented schematically in
figure 4 [14]. The metastable states of He or Ar are generated between
two rolled tantalum electrodes of a dc discharge flow system. Ideal
discharge conditions are found at 250 V, 0.5-1 mA and rare gas
pressures of 0.3-1 mbar for Ar and 1-3 mbar for He. Beneath the exit
of the discharge tube the sample gas is admitted through an annular
inlet and mixed with the rare gas stream. The complete discharge
system is surrounded by liquid nitrogen to reduce the temperature of
the carrier gas. The reaction chamber is connected to a 1680 ℓ/min
rotary pump.
The thermalised ground state ions are optically pumped by means of a
repeatedly applied (20 Hz) 4 ns pulse from a narrow-bandwidth (1-0.05
cm^{-1}) dye laser pumped by a 1.4 MW nitrogen laser. Before entering
the interaction zone, however, the laser beam is passed through a
system of optical baffles of varying diameter designed to reduce
scattering of laser light into the reaction chamber. Undispersed fluorescence from the probed cation is detected by a photomultiplier tube
after passing through a biconvex lens and a cut-off filter discrimi-
nating against residual scattered laser light. The photomultiplier
signal and the output pulses of two photodiodes sampling laser intensity and external etalon fringes are acquired and digitized on a

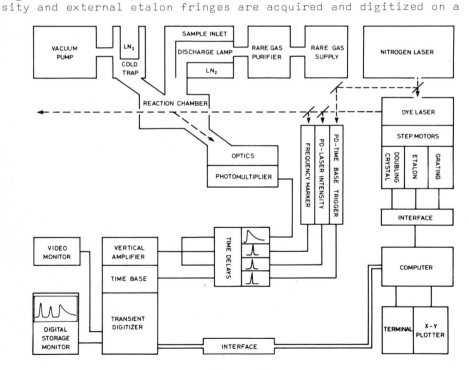

Figure 4
Schematic representation of the laser excitation
spectrometer and data acquisition system.

common time base by a transient digitizer which is controlled by a
microcomputer. In figure 4 the arrangement for the measurements is
depicted.

Signal statistics are improved by repetitive accumulation of typically
32 shots per sampling point. The resulting waveform is transferred to
the interfaced microcomputer where the data are reduced to fluorescence intensity corrected for laser power, referenced to the frequency
markers and stored. The grating of the dye laser is advanced to the
next wavelength position and the cycle is repeated. The scanning speed
for spectral resolutions around 1 cm^{-1} is about 30 nm/h. Spectral
sections resulting from the use of different dye solutions are norma-
lized by means of bands apparent in their overlapping regions.

The laser excitation spectra are calibrated using a single atomic line
apparent in the spectrum (Ta, Ar or He) as an absolute wavelength
reference and a dense manifold of equidistant frequency markers from
an extracavity Fabry-Perot etalon, the thickness of which has been
spectroscopically determined. The resulting precision of energy calibration for any spectral position is better than ±0.5 cm^{-1}, which can
be easily verified by means of the very large number of excitation
lines from tantalum atoms, the source of which are the tantalum
electrodes of the discharge system.

Higher resolution recordings of the excitation spectra are attained
using an intracavity Fabry-Perot etalon as an active filter between
the grating and the beam-expanding telescope of the dye laser oscillator. The solid state etalon is angle-tuned and synchronized to the
grating-selected wavelength to allow continuous scanning. This process
is entirely computer-controlled and interactive in so far that asyn-
chronism is automatically detected and countermeasures are taken.

5. EXAMPLES

5.1. General Remarks

Figure 5 represents a simplified energy level diagram for the ground
and first excited doublet states of an open-shell cation. If ν_1 and
ν_2 are the totally symmetric fundamentals (with respect to the common
symmetry elements of the $^2\tilde{X}$ and $^2\tilde{A}$ states), then the indicated levels
may be populated in the ionisation process. Their relative popula-
tions are determined by the population distribution in the molecular
ground state, M^1X, and the associated Franck-Condon factors for the
ionisation process, $M^{+2}\tilde{A}(\nu') \leftarrow M^1X(\nu'')$. The former is governed by the
vibrational temperature of the sample while the latter by the relative
geometries of the two states. Thus when the molecule is vibrationally
cooled as in the supersonic free jet, by and large only the zeroth
vibrational level of the ^1X state is populated but the various totally
symmetric levels of the ionic states are accessible in the ionisation
process. The resulting emission spectrum may then take on an appearance
as depicted in the figure; progressions in the totally symmetric vibrations and their combinations and sequence transitions are apparent.
In the spectra obtained with cations characterized by a room temperature rotational distribution, the sequence bands lying near the progressions are often obscured due to the inhomogeneous broadening of
the bands. In contrast, in the emission spectra of rotationally, and

vibrationally, cooled cations, the bands are narrow but the overall
appearance of the vibrational structure is similar. Sometimes the
excitation of degenerate modes in double quanta is apparent; the resulting totally symmetric component often gains intensity by Fermi-
resonance with energetically near-lying transitions involving the totally symmetric fundamentals. As far as the vibrational frequencies
are concerned, the emission spectra yield these values for many of
the fundamentals in the $^2\tilde{X}$ state and to a lesser extent in the $^2\tilde{A}$
state. To extend the information on the excited state, the laser excitation technique is utilized.

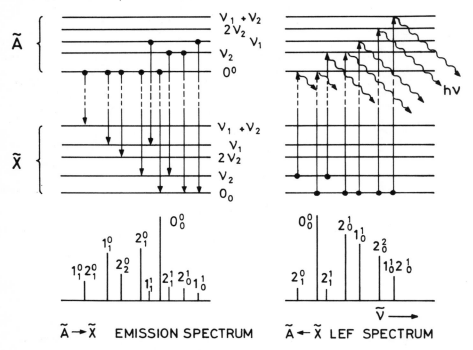

Figure 5
Schematic representation of the vibrational energy levels of the
$^2\tilde{X}$ and $^2\tilde{A}$ electronic states of a cation showing the
emission and laser excitation processes.

In the laser excitation experiment the cations are prepared in the
ground state $^2\tilde{X}$ by Penning ionisation and collisional deactivation by
the inert gas. Thus the vibrational distribution is governed by the
temperature of the bath; at liquid nitrogen temperature the population is concentrated in the lowest vibrational level. A consequence
of this is that, in contrast to the emission spectrum, the only sequence transitions apparent are from the residual population of vibrationally excited levels in the $^2\tilde{X}$ state. In the resonance excitation the totally symmetric levels of the excited state $^2\tilde{A}$ are populated
depending on the relative Franck-Condon factors. This again illustrates
the complementary nature of the emission and excitation spectroscopies;
in the former the Franck-Condon factors link the molecular ground state
with the excited ionic state and in the latter the $^2\tilde{X}$ and $^2\tilde{A}$ ionic
states. Comparison of the two spectra therefore leads to the identi-

fication of the 0_0^0 transition, which in cases of large geometry changes may not be evident in the emission spectrum alone. The extension
of this is the identification of spin-orbit sub-systems in an emission
spectrum, since in the laser excitation experiment the energetically
higher lying component is attenuated. Examples of this will be given
subsequently.

From figure 5 and the above discussion it is clear that emission and
laser excitation spectroscopies are advantageously applied simultaneously for the spectral characterisation of the cations. The emission
spectra yield the vibrational frequencies mainly for the ground state,
the excitation spectra mainly for the excited state although some
common frequencies for the other state are obtained by both. In addition the two sets of data are often required to interpret all the
features of the respective spectra.

5.2. Emission Spectroscopy

The emission spectrum of 2,4-hexadiyne cation illustrates the reasons
for the study of rotationally cooled ions. The emission system which
is shown in figure 6 has been identified as the $\tilde{A}^2E_u \rightarrow \tilde{X}^2E_g$ transition
(with assumed D_{3d} symmetry). The top trace is the spectrum obtained
initially using an effusive source [15], corresponding to a room temperature rotational distribution. The width of the bands (0.8 nm) is,
however, greater than the optical resolution (0.16 nm) due to the
spread of the rotational envelope. Consequently the vibrational frequencies could be inferred to an accuracy of only ± 10 cm^{-1}.

The bottom trace of figure 6 shows the same emission system obtained
by electron impact excitation of 2,4-hexadiyne seeded in a helium supersonic free jet [11]. The narrowing of the vibronic bands, relative
to the room temperature spectrum, due to the restricted rotational
level population distribution, is clearly evident. Bandwidths are
typically 0.1 nm and the vibrational frequencies can be inferred to
± 1 cm^{-1}. This degree of accuracy is comparable to, and often better
than, that to which the molecular vibrational frequencies have been
determined. The bands lying to lower wavelength (higher energy) of the
0_0^0 band correspond to transitions from vibrationally excited levels of
the $\tilde{A}$ state. These "hot" bands are a consequence of favourable Franck-
Condon factors during the ionisation process $\tilde{A}^2E_u \leftarrow X^1A_{1g}$ in which most
molecules are all in the zeroth vibrational level of the ground state
due to the cooling effect of the supersonic expansion.

The extent of rotational cooling can be deduced directly from the
emission spectra if the rotational transitions are resolved. An exam-
ple of this is the $\tilde{A}^2\Sigma^+ \rightarrow \tilde{X}^2\Pi_\Omega$, with $\Omega = 3/2, 1/2$, emission system of
N_2O^+ which has been rotationally analysed earlier [16]. In figure 7
the rotational structure of the 0_0^0 emission band is shown including
the rotational assignments. The top spectrum was obtained by electron
impact excitation on a pure effusive jet and corresponds to a rotational temperature of about 300 K.

The bottom trace of figure 7 shows the same band recorded using 10%
N_2O seeded in a helium supersonic jet. The P_1 band head is now completely absent since at the rotational temperature of ca. 10 K the
$J > 17\frac{1}{2}$ rotational levels of the excited $\tilde{A}$ state have an insignificant population. The shift in the J values of the branch maxima is seen.

Dieter Klapstein, John P. Maier & Liubomir Misev

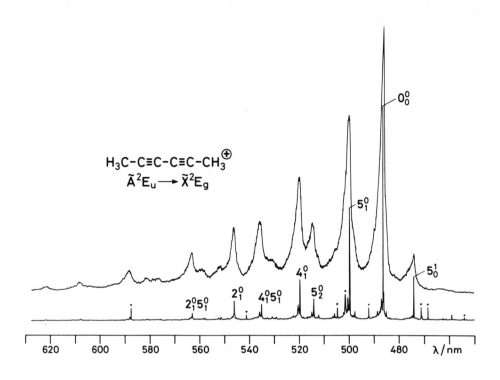

Figure 6

The $\tilde{A}^2E_u \rightarrow \tilde{X}^2E_g$ emission band system of 2,4-hexadiyne cation. The top trace, resolution 0.16 nm, was from electron impact on an effusive jet and the bottom trace, resolution 0.025 nm, on a seeded helium supersonic free jet (helium emission lines are marked with a dot).

In the case of the larger organic cations with smaller rotational constants, even lower rotational temperatures can be expected. The individual rotational transitions are not usually resolved but the bandwidth of a vibronic band in a cooled spectrum is at least three times smaller than for the uncooled species; thus close-lying bands may be more easily discernible.

The rotational temperature of a species, as determined by the rovibronic transition line intensity distribution, is important for inferring the temperature of the environment of the species. Well known examples of this are spectroscopic temperature determinations of the interstellar medium, stars and comets [17]. An interesting example of importance to astro- and geophysics is the visible emission system of H_2O^+ [18]. This was first identified in the laboratory and later analysed as the $\tilde{A}^2A_1 \rightarrow \tilde{X}^2B_1$ transition [19]. Some lines of this emission system were subsequently identified in the tail of comet Kohoutek (and earlier comets) [20] and in the twilight sky [18]. As supporting evidence in the former case a prediction was made of the intensity distribution within certain rovibronic bands at rotational tempera-

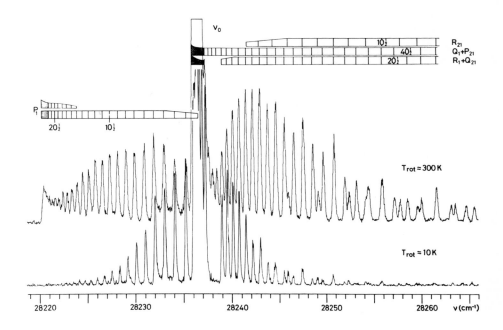

Figure 7

The 0^0_0 band of the $\tilde{A}^2\Sigma^+ \rightarrow \tilde{X}^2\Pi_{3/2}$ transition of N_2O^+ with a
resolution of 0.004 nm. The top trace was from electron-
impact excitation of a pure N_2O effusive jet while
the bottom trace is of N_2O seeded in a
helium supersonic free jet.

tures of 50 and 100 K. The predicted intensities [20] of the rotational lines of the $2^8_0(\Pi)$ vibronic band at these temperatures are reproduced in the top and bottom traces of figure 8. The middle trace
shows an actual spectrum obtained by electron-impact on a seeded
helium supersonic free jet. Comparison with the calculated intensities yields a rotational temperature between 50 and 100 K, although
an exact temperature cannot be unambiguously determined due to
certain assumptions made in calculating the line strengths [21].
Thus in such cases the spectral distribution obtained from extra-
terrestrial sources can be directly simulated in the laboratory in
the collision free environment of the free jet experiment.

As the final example of this section the emission spectrum of dichloroacetylene cation, $Cl-C\equiv C-Cl^+$, will be considered. This emission system
was recorded some time ago using a thermalised sample and is shown in
the top trace of figure 9 [22]. It was identified as the
$\tilde{A}^2\Pi_{g,\Omega} \rightarrow \tilde{X}^2\Pi_{u,\Omega}$, $\Omega= 3/2, 1/2$, transition. However, no assignment of
the observed bands to specific vibronic transitions could be unambig-
uously made at the time due to the extensive overlap of the two spin-
orbit subsystems. As this is a $^2\Pi_\Omega - ^2\Pi_\Omega$ electronic transition the
$\Omega= 3/2$ and $\Omega= 1/2$ subsystems are separated in energy by an amount
equal to the difference of the spin-orbit splittings in the upper and
lower states. From the photoelectron spectrum [23] only the spin-orbit

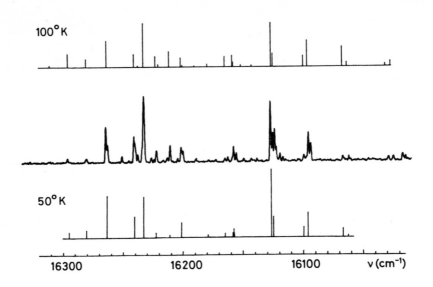

Figure 8
Predicted intensity distribution (top and bottom) within the 2_0^8
vibronic band of the $\tilde{A}^2A_1 \rightarrow \tilde{X}^2B_1$ emission system of H_2O^+.
The middle trace shows an experimental spectrum obtained by
electron-impact excitation of a seeded helium supersonic
free jet with a resolution of 0.045 nm.

splitting of the $\tilde{A}^2\Pi_g$ state could be determined (to within ± 80 cm^{-1});
for the $\tilde{X}^2\Pi_u$ state the resolution of this technique is not adequate.
Also, the emission spectrum shows no predominant pair of bands which
might correspond to the two 0_0^0 transitions, as was observed for the
earlier example of 2,4-hexadiyne cation. Evidently in this case there
is a relatively large change in geometry between the upper and lower
electronic states so that the Franck-Condon factor for the 0_0^0 transition is reduced and long progressions of the vibrational modes
which effect this geometry change are observed.

The bottom trace of figure 9 shows the spectrum obtained with dichloroacetylene seeded in a helium supersonic free jet. Due to the
rotational cooling effect of the free jet expansion the widths of the
individual vibronic transition bands have been greatly reduced,
allowing close-lying bands to be distinguished. Transitions involving
excitation of vibrations which have a significant contribution from
the Cl nuclear motions (e.g. $\nu_2 : \nu_s$(C-Cl)) show the effects of the
natural abundances of the two chlorine isotopes, ^{35}Cl and ^{37}Cl. For
the transitions from the vibrationless level of the upper state, the
splitting of a band corresponds, in a first approximation, to the
vibrational frequency differences due to the isotopes in the ground
state. For dichloroacetylene this splitting is only a few wavenumbers
but is readily discernible in the cooled spectrum. Thus the various
progressions can be identified and the origins of the two spin-orbit
components of the $\tilde{A}^2\Pi_g \rightarrow \tilde{X}^2\Pi_u$ emission system located as shown in
figure 9. Some of the other band assignments are also indicated, the
dominant ones being due to a progression of the ground state ν_2 mode.

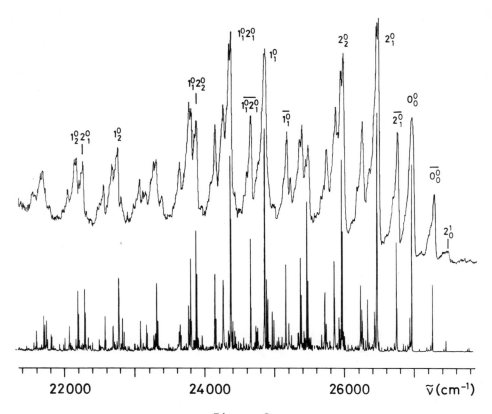

Figure 9

The $\tilde{A}^2\Pi_{\Omega,g} \rightarrow \tilde{X}^2\Pi_{\Omega,u}$, with $\Omega = 3/2$, $1/2$, emission system of di-chloroacetylene cation in the gas phase. The top trace was ob-tained with a thermalised sample at an optical resolution of 0.25 nm. The bottom trace was obtained with a seeded helium supersonic free jet and a resolution of 0.03 nm. Bands of the $\Omega = 1/2$ system are denoted with a horizontal bar.

An alternative means of identifying the origins and the two spin-orbit systems is to compare the emission and laser excitation spectra (vide infra).

5.3. Laser Excitation Spectroscopy

Figure 10 shows laser excitation spectra of the $\tilde{A}^2\Pi_{u,3/2} \leftarrow \tilde{X}^2\Pi_{g,3/2}$ transition of dichlorodiacetylene cation recorded with an optical band-width of 1 cm^{-1}. The top spectrum is observed with thermalised cations [24] whereas the bottom one is the result of cooling the rare gas bath to liquid nitrogen temperature [25]. The most conspicuous changes are the narrowing of the vibronic bands due to the restricted rotational level populations but the effects of vibrational cooling are also clearly apparent from the relative intensity changes of the hot bands (e.g. 3^0_1) and sequence bands (e.g. 3^1_1). In addition to the simplification of some of the congested band structures seen in the

room temperature spectrum, the sharpness of the band maxima in the
bottom spectrum permits one to deduce the vibrational frequencies to
within ±2 cm⁻¹. In higher resolution recordings the isotope splittings
on some of the bands become clearly discernible. The information concerning vibrational frequencies in the excited state and comparison
with the emission spectrum is discussed in section 5.5, where dichlorodiacetylene cation is used to exemplify the complementary nature of
the two techniques.

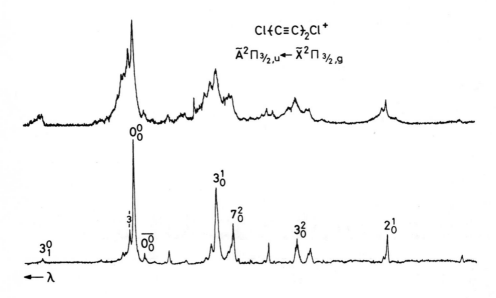

Figure 10
A part of the laser excitation spectrum of the
$\tilde{A}^2\Pi_{3/2,u} \leftarrow \tilde{X}^2\Pi_{3/2,g}$ transition of dichlorodiacetylene
cation at room   (upper trace) and liquid nitrogen (lower trace)
temperature; resolution 0.02 nm.

In the case of dichlorodiacetylene cation the assignment of both the
emission and laser excitation spectra is reasonably straightforward.
The 0_0^0 transitions are dominant since the geometries (assumed linear)
of the cation in the two states are not too different. On the other
hand, in cases where the potential hypersurfaces are displaced in
such a manner that a complex band profile is observed in the emission
spectrum, then the laser excitation spectrum can usually clarify the
situation. The emission spectrum of dichloroacetylene cation is such
a case (see figure 9); an additional complication is that the $\Omega = 3/2$
and $\Omega = 1/2$ components of the $\tilde{A}^2\Pi_g \rightarrow \tilde{X}^2\Pi_u$ transition are intermingled.

A part of the laser excitation spectrum of this transition of dichloroacetylene cation is reproduced in figure 11. Since the initial state
is the cationic ground state, the population is concentrated in the
zeroth level of the $\Omega = 3/2$ component which dominates the spectrum.
The zeroth level of the $\Omega = 1/2$ component lies ≈ 200 cm⁻¹ [22] to
higher energy and consequently at room or liquid nitrogen temperatures
only the $\overline{0_0^0}$ band is perceptible. In the emission spectrum both $\Omega = 3/2$

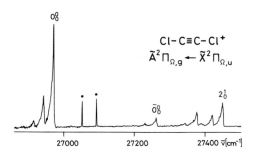

Figure 11
Low energy part of the laser excitation spectrum of the
$\tilde{A}^2\Pi_{\Omega,g} \leftarrow \tilde{X}^2\Pi_{\Omega,u}$ transition of dichloroacetylene cation
at liquid nitrogen temperature; resolution 0.02 nm.

and 1/2 components are about equally populated on ionisation from the
molecular ground state. Comparison with the excitation spectrum, however, enables the origin to be immediately identified.

A similar situation prevails in the $\tilde{A}^2\Pi_\Omega \rightarrow \tilde{X}^2\Pi_\Omega$ emission spectrum of
bromoacetylene cation, which was reported some time ago [13]. Its
complicated vibrational pattern, and the overlap of the two spin-
orbit systems precluded location of the band origins. Rotational
structure could however be partly resolved; the best attainable
resolution was ≈ 0.01 nm, being limited by signal to noise considerations.

Both the origin identification and resolution problems can be overcome by the laser excitation technique. The excitation spectrum shows
clearly the origin band (0^0_0 $\Omega = 3/2$) and by means of the high resolution approach described in section 4.2, this band could be recorded
with an optical resolving power of 400,000 and is shown in figure 12.
The rotational fine structure is very well resolved, even though
there is a complication arising from the presence of the two almost
equally abundant isotopes of bromine. Thus in the case of the few
organic cations for which rotational structure can be resolved (with
resolutions down to 0.001 nm) [6], the laser excitation approach is
the method of choice.

Dieter Klapstein, John P. Maier & Liubomir Misev

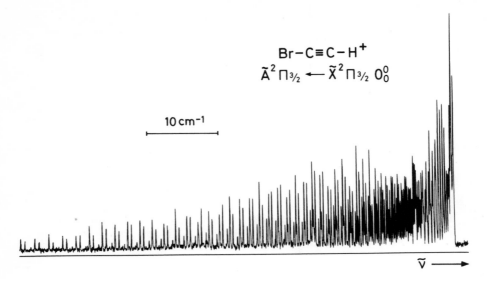

Figure 12
High-resolution laser excitation spectrum of the
$\tilde{A}^2\Pi_{3/2} \leftarrow \tilde{X}^2\Pi_{3/2}$ 0^0_0 transition of
bromoacetylene cation (0.001 nm bandwidth).

5.4. Lifetime Measurements

The emission and laser excitation spectra give primarily structural
information on the ground and first excited electronic states of
radical cations. The apparata described in section 4 can however also
be used to probe the time dependence of the radiative decay process
from the excited state, from which the radiative lifetime of the
state can be determined. This information can then be used in conjunc-
tion with data from other techniques, such as photon-photoelectron
[8] or photon-photoion coincidence [26] spectroscopies, to provide
quantitative information on the radiative and nonradiative channels
by which the excited state is depleted. Two examples are given; one
from each of the techniques.

The well-known $\tilde{A}^2\Sigma^+ \rightarrow \tilde{X}^2\Pi_\Omega$ emission system of N_2O^+ [16] has already
been referred to earlier (cf. fig. 7). Besides the radiative decay,
N_2O^+ in the $\tilde{A}^2\Sigma^+$ excited electronic state may also relax by a fragmentation pathway: the $\tilde{A}$ state is predissociated by a $^4\Sigma^-$ state
yielding $NO^+ + N$ as products. The dynamics of the fragmentation process have been studied by photoelectron-photoion coincidence measurements [27] and photodissociation experiments [28], all of which indicate that only the 0^0 level of the $\tilde{A}^2\Sigma^+$ state is not predissociated.
In response to a theoretical prediction [29] the radiative lifetimes
of selected vibrational levels of the $\tilde{A}$ state were determined from
the electron-impact excited emission bands using a pulsed electron
beam and monitoring the decay by a single photon delayed coincidence
approach [30]. Single vibronic emission lines, corresponding to radiative decay from specific vibrationally excited levels of the $\tilde{A}$ state,
were studied by selecting with a monochromator, with appropriate re-

solution and wavelength settings, suitable emission bands. Since the
radiative decay rate appears constant, $\approx 4.1 \times 10^6$ s^{-1}, over ≈ 2500 cm^{-1}
excess energy above the 0^0 level according to fluorescence quantum
yield and lifetime measurements (but at much lower resolution) [31],
the predissociation rate constants can be inferred for the vibrationally excited levels. The results are shown in figure 13 where the
vertical scale represents the vibrational energy (cm^{-1}) within the
$\tilde{A}^2\Sigma^+$ state. Thus for levels higher than ≈ 1000 cm^{-1} above the 0^0 level
the predissociation rate is faster than the radiative rate, fragmen-
tation being the dominant process depleting the $\tilde{A}$ state levels.

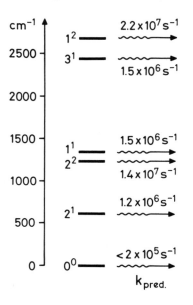

Figure 13
Inferred predissociation rates
for the indicated vibrational
levels in the N_2O^+ $\tilde{A}^2\Sigma^+$ electronic state.

An alternative approach to determine the lifetimes of excited cations
is by pulsed laser excitation. Provided that the laser excitation
spectrum can be obtained, there are advantages in using this method.
These are, on the one hand, the attainable resolution for the measurements (i.e. down to 0.001 nm) without loss of intensity and, on the
other hand, it suffices to produce the cations initially in the ground
state whereas for the emission measurement the excited state has to be
directly populated. The example of C_2^- illustrates both aspects.

The $\tilde{B}^2\Sigma_u^+$ - $\tilde{X}^2\Sigma_g^+$ band system of C_2^- [32] is spectroscopically well
characterized both in the gas phase and in matrices [33]. However
the lifetime of the excited state in the gas phase had not been determined directly partly due to the methods of C_2^- preparation. Since
the laser excitation spectrum of this transition of C_2^- could be detected as a by-product in the Penning ionisation of bromoacetylene
using argon metastables [34], the means to measure the lifetimes of
selected rotational levels became available.

The excitation spectrum of the 0-0 vibronic transition and the semi-
logarithmic plot of the decay curve from the upper-state rotational
level N'= 9 are shown in figure 14. No significant dependence of the
lifetime on the rotational quantum numbers accessible (i.e. up to
N'= 21) was apparent, leading to a mean lifetime for the v'= 0 level

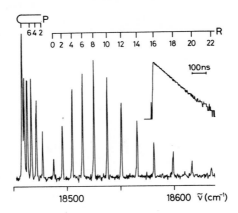

Figure 14
Laser excitation spectrum of the $\tilde{B}^2\Sigma_u^+ \leftarrow \tilde{X}^2\Sigma_g^+$ 0-0 transition of
C_2^- recorded with 0.02 nm resolution. In the inset is shown
the decay curve for the N'= 9 rotational level of the
$\tilde{B}^2\Sigma_u^+$ v'= 0 state plotted on a semi-logarithmic
scale with background subtracted.

of 77±8 ns. The error limit of 10% is given due to the limited range
of pressures (0.1 to 4 mbar) attainable for the measurements for
extrapolation to zero pressure. Of practical interest (e.g. for astro-
physical purposes) is the oscillator strength for the $\tilde{B}^2\Sigma_u^+ - \tilde{X}^2\Sigma_g^+$
transition of C_2^- which can be evaluated from the lifetime value for
the 0-0 origin to be 0.044±0.004 [34]. Recently, the rotational lines
of this transition of C_2^- were searched for in a hydrogen deficient
carbon star [35].

5.5. Complementary Nature of Emission and Laser Excitation Spectros-
     copies: Dichlorodiacetylene Cation

In this section we shall illustrate how the two methods, emission and
laser excitation spectroscopies, complement each other. The example
chosen is the $\tilde{A}^2\Pi_{u,\Omega} - \tilde{X}^2\Pi_{g,\Omega}$, Ω= 3/2, 1/2, transition of dichlorodiacetylene cation (under $D_{\infty h}$ symmetry classification) [36]. Both
the emission and laser excitation spectra were initially recorded

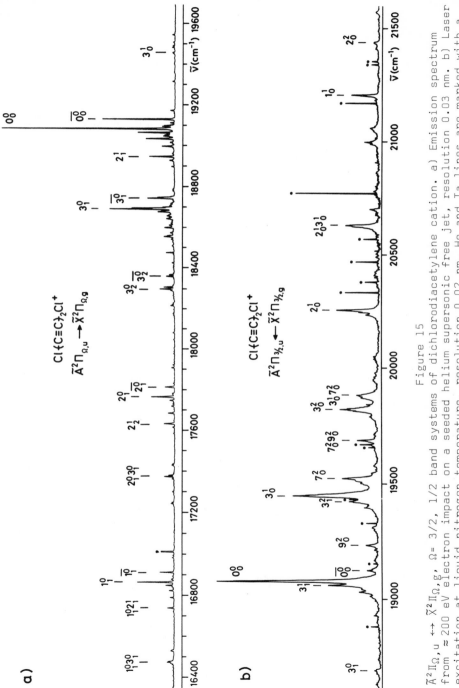

Figure 15

$\tilde{A}^2\Pi_{\Omega,u} \leftrightarrow \tilde{X}^2\Pi_{\Omega,g}$, $\Omega = 3/2$, 1/2 band systems of dichlorodiacetylene cation. a) Emission spectrum from $\approx$ 200 eV electron impact on a seeded helium supersonic free jet, resolution 0.03 nm. b) Laser excitation at liquid nitrogen temperature, resolution 0.02 nm. He and Ta lines are marked with a dot. Transitions of the Ω= 1/2 system are denoted by a horizontal bar.

using a thermalised sample and vibrationally analysed, although the vibrational frequencies had an uncertainty of ±10 cm^{-1} [24]. It was concluded on the basis of the photoelectron and emission spectra that the Ω= 3/2 $\leftrightarrow$ 3/2 and Ω= 1/2 $\leftrightarrow$ 1/2 components were energetically so close that each observed band contained both components. The emission and laser excitation spectra of the rotationally cooled species were subsequently obtained and clearly show the two spin-orbit sub-systems [25].

The top trace of figure 15 shows the $\tilde{A}^2\Pi_{u,\Omega} \rightarrow \tilde{X}^2\Pi_{g,\Omega}$, Ω= 3/2, 1/2, emission spectrum of dichlorodiacetylene cation obtained by electron impact on a seeded helium free jet. With the much reduced widths of the bands due to the restricted rotational level population distribution many more bands are discernible relative to the room temperature spectrum. For each of the bands observed previously there is an accompanying weaker band about 47 cm^{-1} to higher energy which must be due to the Ω= 1/2 $\rightarrow$ 1/2 emission system; these bands appear only as small shoulders on the Ω= 3/2 $\rightarrow$ 3/2 system bands in the room temperature spectrum.

The transitions observed in the emission spectrum correspond mainly to excitation of the three totally symmetric (Σ_g^+) fundamentals [37] of the $\tilde{X}$ state, ν_1 to ν_3. Some of the band assignments are indicated in figure 15. Several weak bands are also observed to higher energy of the 0_0^0 bands which correspond to transitions from excited vibrational levels of the $\tilde{A}$ state. The levels are populated, weakly, in the ionisation process from the vibrationless (due to the cooling) ground state of the molecule. The energies of these emission transitions are confirmed by the laser excitation spectrum. In addition, several sequence bands could be identified in both spectra since the requisite vibrational frequencies were available by considering both sets of data together.

A higher resolution (0.008 nm) recording of the 0_0^0, Ω= 3/2, band is reproduced in figure 16. The sharp, intense band corresponds to the rotational head in the R-branch while the P-branch is more drawn out. The absence of a Q-branch is consistent with a Π-Π transition. That the bands are degraded to the red also implies that, assuming linearity is retained in both states, the rotational constant in the excited $\tilde{A}$ state is less than that in the ground $\tilde{X}$ state.

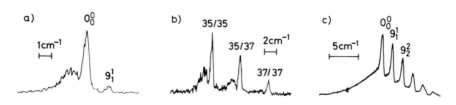

Figure 16
High resolution recordings of some bands of the
$\tilde{A}^2\Pi_{u,\Omega} \leftrightarrow \tilde{X}^2\Pi_{g,\Omega}$ transitions of Cl{C≡C}$_2$Cl$^+$.
a) 0_0^0 and b) 3_1^0 bands by emission from a supersonic
free jet (0.008 nm), c) 0_0^0 laser excitation at
liquid nitrogen temperature (0.005 nm).

The effects on vibrational frequencies due to the chlorine isotopes are also clearly evident in the rotationally cooled spectrum. The natural relative abundances of ^{35}Cl and ^{37}Cl are roughly 3:1 so that for a dichloro substituted molecule the relative abundances of the species $^{35}Cl\{C\equiv C\}_2Cl^{35} : ^{35}Cl\{C\equiv C\}_2Cl^{37} : ^{37}Cl\{C\equiv C\}_2Cl^{37}$ is 9:6:1. The largest isotope effect is observed for the progression in the ν_3 mode (C-Cl stretch); the 3_1^0 band is reproduced in figure 16 and clearly shows the sub-peaks due to the isotope effect. The vibrational frequencies can be inferred to better than ± 1 cm^{-1} when the isotope splittings are considered.

The bottom trace of figure 15 shows the $\tilde{A}^2\Pi_{u,3/2} \leftarrow \tilde{X}^2\Pi_{g,3/2}$ laser excited fluorescence spectrum of $Cl\{C\equiv C\}_2Cl^+$ obtained with the sample cooled to liquid nitrogen temperature. The bandwidths have been sufficiently reduced so that the vibrational frequencies, mainly of the excited $\tilde{A}$ state, can be determined to ± 2 cm^{-1}. Some of the vibrational assignments are indicated in figure 15. The strongest bands correspond to transitions from the zeroth level of the $\tilde{X}^2\Pi_{g,3/2}$ state to the totally symmetric levels of the ν_1 to ν_3 Σ_g^+ fundamentals of the excited $\tilde{A}^2\Pi_{u,3/2}$ state. In addition, the ν_7 and ν_9 degenerate vibrations are each strongly excited in the excited $\tilde{A}$ state by two quanta, which yields a totally symmetric component.

A higher resolution (0.005 nm) excitation spectrum of the 0_0^0 3/2 band given in figure 16 shows a marked sequence structure. A similar band structure, with a slight relative intensity alteration, was observed in the room temperature emission spectrum. This structure has been assigned to sequence transitions of the lowest frequency mode, i.e. 9_n^n, n= 1-4, which has a molecular frequency of 76 cm^{-1} [37]. In the room temperature emission spectrum the $\tilde{A}^2\Pi_{u,\Omega}$ 9^n levels are populated by Franck-Condon allowed transitions from the corresponding levels of the molecular ground state which have a Boltzmann population distribution. Even at around liquid nitrogen temperatures the n= 1-4 levels in the $\tilde{X}^2\Pi_{g,\Omega}$ state still have sufficient population to be directly pumped in the laser excitation experiment. At the lower vibrational temperature of the supersonic free jet experiment only the 9^1 level, of the $\tilde{A}^2\Pi_{u,\Omega}$ state, is still significantly populated. The high resolution spectrum of the 0_0^0 emission band, figure 16, shows only weakly the 9_1^1 sequence band. This allows the unambiguous identification of the 0_0^0 band in the laser excitation spectrum, figure 15.

From the changes in relative intensities of the hot bands and sequence bands compared to the room temperature excitation spectrum the vibrational temperature is estimated to be about 120 K. At this temperature the $\tilde{X}$ 0^0 Ω= 1/2 level has a population only one-tenth that of the Ω= 3/2 level so that only the $\tilde{A}^2\Pi_{u,1/2} \leftarrow \tilde{X}^2\Pi_{g,1/2}$ $\overline{0}_0^0$ transition is weakly observed in the laser excitation spectrum (see figure 15); all other transitions of the Ω= 1/2 system are too weak to be observed.

The relative changes of the vibrational frequencies observed for the $\tilde{A}$ and $\tilde{X}$ states of the dichlorodiacetylene cation compared to the ground molecular state frequencies [37] reflect the molecular orbital description of its electronic structure [38]. Figure 17 shows the electronic configurations of the three electronic states and the observed frequencies for the three Σ_g^+ fundamentals obtained from emission and excitation spectra.

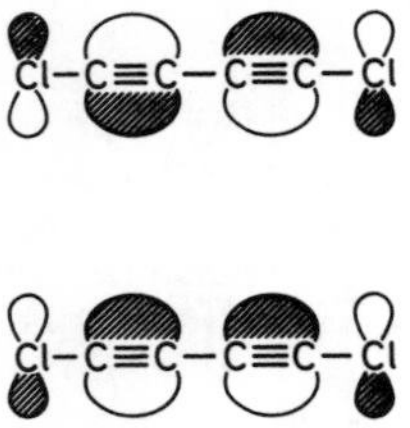

Σ_g^+ – VIBRATIONS

		$\nu_1 : \nu_S(C\equiv C)$	$\nu_2 : \nu_S(C-C)$	$\nu_3 : \nu_S(C-Cl)$
$\cdots \pi_u^4\,\pi_g^4\,\pi_u^4\,\pi_g^4$	$X^1\Sigma_g^+$	2245	1202	330
$\cdots \pi_u^4\,\pi_g^4\,\pi_u^4\,\pi_g^3$	$\tilde{X}^2\Pi_g$	2214	1316	393
$\cdots \pi_u^4\,\pi_g^4\,\pi_u^3\,\pi_g^4$	$\tilde{A}^2\Pi_u$	2125	1179	374

Figure 17

Comparison of the frequencies of the three totally symmetric vibrations of dichlorodiacetylene in the molecular ground state and in
the lowest two cationic states and the main features relating to
the description of the electronic structure (see text).

To the left are drawn pictorial descriptions of the highest occupied
and penultimate molecular orbitals; the shading indicates the relative phases [38]. The highest occupied orbital has bonding characteristics in the C≡C regions and antibonding between the central C-C
and terminal C-Cl nuclei. Upon ionisation of the associated electron,
to give the $\tilde{X}^2\Pi_{g,\Omega}$ state, the frequencies of ν_2(C-C str.) and ν_3(C-Cl
str.) increase whereas that of ν_1(C≡C str.) decreases relative to
the values for the molecular ground state. Thus the $\tilde{X}$ state is
expected to have C-C and C-Cl internuclear distances greater than,
and a C≡C separation smaller than, for the molecule. In contrast, the
penultimate molecular orbital has bonding characteristics between all
the carbon atoms but is antibonding along the C-Cl axis. Thus the ν_1
and ν_2 frequencies are slightly reduced and that of ν_3 increased in
the $\tilde{A}$ cationic state relative to the molecular values. From the observation that the rotational constant of the $\tilde{A}$ state is smaller than
that of the $\tilde{X}$ state, the most apparent change in geometry on passing
from the $\tilde{X}^2\Pi_{g,\Omega}$ to the $\tilde{A}^2\Pi_{u,\Omega}$ cationic state is an increase in the
C-C and C-Cl bond lengths.

6. SUMMARY

The spectroscopic structure of open-shell organic cations in the gas
phase is investigated by emission and laser excitation techniques.
The studies of cations which are rotationally, and vibrationally,
cooled result in emission and excitation spectra which are not only
easier to interpret compared to room temperature spectra but, more
important, the narrowing of the vibronic bands leads to much more
accurate data. The primary information for the organic species are
the vibrational frequencies for the cationic ground and lowest excited
states which can usually be determined to within ±1 cm^{-1}. For the
smaller cations the resolvable rotational structure in the laser excitation measurements provides a further spectral characterisation.
The complementary nature of the methods is evident in the interpreta-

tion of the spectra. These data then provide a means of identifying such cations in inaccessible environments and of probing the physical conditions therein.

Acknowledgement

The studies reviewed in this chapter as a result of the research projects at Basel have been financed throughout the years by the "Schweizerischer Nationalfonds zur Förderung der wissenschaftlichen Forschung".

REFERENCES

[1] G. Herzberg, Quart. Rev. Chem. Soc. 25 (1971) 201.
[2] See for example Kinetics of Ion-Molecule Reactions, ed.
     P. Ausloos (Plenum Press, New York, 1979); Gas-Phase Ion
     Chemistry, ed. M.T. Bowers, Vols. I and II (Academic Press,
     New York, 1979).
[3] J.H. Callomon, Can. J. Phys. 34 (1956) 1046.
[4] D.W. Turner, C. Baker, A.D. Baker and C.R. Brundle, Molecular
     Photoelectron Spectroscopy (Wiley-Interscience, New York,
     1970) and references therein.
[5] J.P. Maier in Kinetics of Ion-Molecule Reactions, ed.
     P. Ausloos (Plenum Press, New York, 1979); J.P. Maier, Ang.
     Chem. Int. Ed. Engl. 20 (1981) 638; J.P. Maier, Acc. Chem. Res.
     15 (1982) 18 discuss these aspects.
[6] J.P. Maier, Chimia 34 (1980) 219; J.P. Maier, O. Marthaler,
     L. Misev and F. Thommen in Molecular Ions, ed. J. Berkowitz
     (Plenum Press, New York, 1982).
[7] T.A. Miller and V.E. Bondybey, J. Chim. Phys. Phys.-Chim. Biol.
     77 (1980) 695 and references therein.
[8] J.P. Maier and F. Thommen, in Ions and Light, Vol. III of
     Gas-Phase Ion Chemistry, ed. M.T. Bowers (Academic Press,
     New York, 1983).
[9] A. Carrington and R.P. Tuckett, Chem. Phys. Letters 74 (1980)
     19; R.P. Tuckett, Chem. Phys. 58 (1981) 151.
[10] T.A. Miller, B.R. Zegarski, T.J. Sears and V.E. Bondybey,
     J. Phys. Chem. 84 (1980) 3154.
[11] D. Klapstein, S. Leutwyler and J.P. Maier, Chem. Phys. Letters
     84 (1981) 534.
[12] D.H. Levy, Ann. Rev. Phys. Chem. 31 (1980) 197.
[13] M. Allan, E. Kloster-Jensen and J.P. Maier, J. Chem. Soc.
     Faraday II 73 (1977) 1406.
[14] J.P. Maier and L. Misev, Chem. Phys. 51 (1980) 311.
[15] M. Allan, J.P. Maier, O. Marthaler and E. Kloster-Jensen,
     Chem. Phys. 29 (1978) 331.
[16] J.H. Callomon and F. Creutzberg, Phil. Trans. Roy. Soc. (London)
     277 (1974) 157.
[17] G. Herzberg, Spectra of Diatomic Molecules (Van Nostrand,
     New York, 1950).
[18] G. Herzberg, Ann. Geophys. 36 (1980) 605.
[19] H. Lew and I. Heiber, J. Chem. Phys. 58 (1973) 1246; H. Lew,
     Can. J. Phys. 54 (1976) 2028.
[20] P.A. Wehinger, S. Wyckoff, G.H. Herbig, G. Herzberg and H. Lew,
     Astrophys. J. 190 (1974) L43
[21] S. Leutwyler, D. Klapstein and J.P. Maier, Chem. Phys., submitted.

[22] M. Allan, E. Kloster-Jensen and J.P. Maier, J. Chem. Soc.
     Faraday II 73 (1977) 1417.
[23] E. Heilbronner, V. Hornung and E. Kloster-Jensen, Helv. Chim.
     Acta 53 (1970) 331.
[24] J.P. Maier, O. Marthaler, L. Misev and F. Thommen, Faraday
     Discuss. 71 (1981) 181.
[25] D. Klapstein, J.P. Maier and L. Misev, J. Chem. Phys. 1983
     in press.
[26] T. Baer in Gas Phase Ion Chemistry, Vol. I, ed. M.T. Bowers
     (Academic Press, New York, 1979).
[27] B. Brehm, R. Frey, A. Küstler and J.H.D. Eland, Intern. J.
     Mass Spectrom. Ion Phys. 13 (1974) 251.
[28] M. Larzilliere, M. Carre, M.L. Gaillard, J. Rostas, M. Horani
     and M. Velghe, J. Chim. Phys. 77 (1980) 689 and references
     therein.
[29] J.A. Beswick and M. Horani, Chem. Phys. Letters 78 (1981) 4.
[30] D. Klapstein and J.P. Maier, Chem. Phys. Letters 83 (1981) 590.
[31] J.P. Maier and F. Thommen, Chem. Phys. 51 (1980) 319.
[32] G. Herzberg and A. Largerqvist, Can. J. Phys. 46 (1968) 2363.
[33] P.L. Jones, R.D. Mead, B.E. Kohler, S.D. Rosner and W.C. Line-
     berger, J. Chem. Phys. 73 (1980) 4419 and references therein.
[34] S. Leutwyler, J.P. Maier and L. Misev, Chem. Phys. Letters,
     1982 in press.
[35] G. Wallerstein, Astron. Astrophys. 105 (1982) 219.
[36] M. Allan, E. Kloster-Jensen, J.P. Maier and O. Marthaler,
     J. Electron Spectry. 14 (1978) 359.
[37] P. Klaboe, E. Kloster-Jensen, E. Bjarnov, D.H. Christensen
     and O.F. Nielsen, Spectrochim. Acta Part A 31 (1975) 931.
[38] E. Heilbronner, V. Hornung, J.P. Maier and E. Kloster-Jensen,
     J. Am. Chem. Soc. 96 (1974) 4252.

Molecular Ions: Spectroscopy, Structure and Chemistry
Terry A. Miller and V.E. Bondybey (editors)
© North-Holland Publishing Company, 1983

The Jahn-Teller Effect in Benzenoid Cations:
Theory and Experiment

by

Terry A. Miller and V. E. Bondybey

Bell Laboratories
Murray Hill, New Jersey 07974

The ground electronic states of the cations of substituted benzenes retaining D_{6h} or D_{3h} symmetry, are doubly degenerate and thus subject to a Jahn-Teller distortion. Recent investigations, particularly laser induced fluorescence studies, have provided a wealth of experimental information about the vibronic structure of the symmetrical ions, $C_6F_6^+$, $C_6H_3F_3^+$, $C_6H_3Cl_3^+$, $C_6Cl_3F_3^+$ and $C_6Br_3F_3^+$. The theory of the Jahn-Teller effect upon a molecule's vibronic structure is reviewed, with particular attention to application to the above ions. The effects of linear and quadratic Jahn-Teller coupling and inter-mode coupling are considered in detail. The experiments and experimental data for these ions are reviewed and tabulated. Finally the theory and experimental data are combined to yield Jahn-Teller stabilization energies. The corresponding distorted geometries are pictorially represented and quantitatively described. It is found that to a first approximation, one obtains a characteristic Jahn-Teller stabilization and distortion, for the removal of one electron from a benzene ring, independent of the substituents on the ring.

I. INTRODUCTION

In 1937 Jahn and Teller wrote [1] what has become a famous paper describing the effect that bears their name. Briefly stated the Jahn-Teller theorem requires that, for any non-linear molecule in a degenerate electronic state, there exists a distortion of the nuclei along at least one non-totally symmetric normal coordinate that results in a splitting of the potential energy function so that the potential minimum is no longer at the symmetrical position. In quantum mechanical terms, this distortion is caused by the existence of non-zero matrix elements of the vibrational Hamiltonian connecting the degenerate electronic states.

While the above description of the Jahn-Teller effect may seems somewhat arcane, the phenomenon has been recognized for many years to play a central role in the structure and spectra of ions of symmetrical molecules. The reason for this is clear. Molecules of high symmetry (C_{3v} or greater) have a number of degenerate molecular orbitals. If an electron is removed from one of these orbitals to form a positive ion, one has a degenerate electronic state and the criteria of the Jahn-Teller theorem are fulfilled. Examples of common molecules whose cations have electronic states subject to Jahn-Teller distortion include ammonia, boron trifluoride, methane, ethane, cyclopropane, cyclobutadiene, cyclobutane, benzene (and substituted benzenes that retain D_{6h} or D_{3h} symmetry), cyclooctatetraene, and coronene. Indeed the ground electronic state of all the above cations is degenerate and thus subject to a Jahn-Teller effect, except for NH_3 and BF_3 where the criteria are fulfilled only in excited electronic states.

In the past, experimental work on the Jahn-Teller effect in ions like the above has employed mainly two techniques, photo-electron spectroscopy [2-6] and electron spin resonance [7-11]. In many instances, anomalous features in the photoelectron or electron spin resonance spectra of symmetrical species have been ascribed to Jahn-Teller interactions. However, it is probably fair to say, that in no instance have these spectra ever given an unambiguous characterization of the Jahn-Teller effect. The reason for this lies in the very nature of the effect and the limitations of the techniques.

The Jahn-Teller effect is a distortion of the potential energy surface so that it no longer has a minimum at the most symmetrical position. The observable quantities that are most sensitive to this distortion are the positions and wavefunctions of the vibrational (vibronic) energy levels. The resolution of even the best photoelectron spectrometers is only about 10 meV ($\sim$80 cm^{-1}). In a molecule of any complexity, the vibrational structure, at this resolution, will be poorly resolved and extremely difficult to interpret.

The resolution of electron spin resonance experiments is, of course, in the MHz ($\leq$.0001 cm^{-1}) regime and would be more than ample to resolve vibrational structure. But electron spin resonance experiments don't measure vibrational structure directly. Rather they center on hyperfine structure, and occasionally fine structure, interactions. The hyperfine or fine structure is only indirectly affected by the Jahn-Teller interaction via changes in its expectation values induced by modifications of the vibronic wavefunctions. On top of this, key details of the hyperfine variations are often masked by averaging effects [9], e.g. tumbling in solution, in the typical electron spin resonance experiment. Thus again one is confronted with the situation that anomalies in the spectra of symmetrical ions are often attributed to Jahn-Teller effects, but inverting the data to fully characterize the underlying Jahn-Teller interaction has thus far proven to be an insurmountable task.

The obvious solution to the above problem would seem to be to obtain direct high resolution information on vibrational structure by infrared or Raman techniques. However the nature of the Jahn-Teller species themselves have generally prevented such an approach. Closed shell species generally do not have degenerate ground electronic states. All ions, and many other free radical species, subject to the Jahn-Teller effect are highly reactive and thus extremely difficult to produce in high enough concentrations to study isolated molecules by infrared or Raman spectroscopy.

Such studies have of course been carried out on Jahn-Teller active species, including ions, in solids. There is in fact an overwhelming literature [12,13] describing such work. While interesting and informative about solids, in general this work does little to characterize Jahn-Teller effects in isolated molecules. The problem is that it is extremely difficult to distinguish between changes in the vibronic potential caused by Jahn-Teller effects and those caused by a lowering of the site symmetry by neighboring species in the solid.

In the last few years, it has become apparent that at least for one very important class of Jahn-Teller active molecules — the benzenoid cations, a combination of techniques based upon optical spectroscopy, provides both the resolution and sensitivity required to obtain the data needed to characterize the Jahn-Teller effect in detail for an isolated molecule. In the remainder of this chapter, we describe first these experiments and their results. We then review the theory necessary for an understanding of the experimentally observed energy levels and wavefunctions. We then combine the theory and experimental results to obtain a detailed characterization of the Jahn-Teller distorted potential surface for the benzenoid cations.

II. EXPERIMENTAL OBSERVATIONS

A. General Considerations

To understand the nature of the optical observations of the Jahn-Teller effect in benzenoid cations, it is useful to briefly review the molecular orbital structure of benzene. Figure 1 shows the three bonding π orbitals, which are completely filled in neutral benzene. The ground state of the benzenoid ions are formed by the removal of one electron from the degenerate pair of orbitals in Fig. 1. In D_{6h} symmetry, these orbitals are of e_{1g} symmetry and the doubly degenerate ground electronic state is denoted as $^2E_{1g}$. If the symmetry of the molecule is lowered to D_{3h}, e.g. by a 1,3,5 substitution, the degeneracy of the highest pair of π orbitals remains. These orbitals transform as e'' representations in D_{3h} symmetry and the removal of one electron from the pair again results in a doubly degenerate electronic state, $^2E''$.

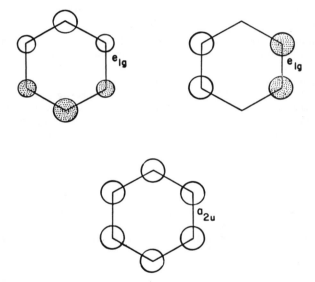

Figure 1.   Pictorial representation of the three highest filled π orbitals in benzene or hexafluorobenzene.

Thus the ground states of benzenoid ions retaining either D_{6h} or D_{3h} symmetry satisfy the criteria for the Jahn-Teller effect. Alternative patterns of substitution that lower the symmetry below D_{3h} break the electronic degeneracy of these orbitals, and no Jahn-Teller effect would be predicted for these ions. These ions have, however, considerable significance for the understanding of the Jahn-Teller effect. When the degeneracy is well split by asymmetrical substitution, the spectra of these ions can serve as valuable "controls" or models of the spectroscopy expected from compounds of this type in the absence of the Jahn-Teller effect. If the electronic splitting is small, then the ions are examples of pseudo-Jahn-Teller effects, i.e. vibronic coupling between near-by electronic states.

The excited state for the ions whose spectra have been observed, is obtained by removing one electron from the non-degenerate penultimately filled π orbital, designated a_{2u} in Fig. 1. This is obviously a π-π transition of a "hole-promotion" type, not found in neutral molecules. Since the upper state is non-degenerate ($^2A_{2u}$ in D_{6h} symmetry and $^2A_2''$ in D_{3h} symmetry), all the Jahn-Teller effects present in the optical spectrum may be ascribed to the ground state. Benzenoid ions, which have observed spectra corresponding to the above described transition include $C_6F_6^+$, 1,3,5 $C_6H_3F_3^+$, 1,3,5 $C_6H_3Cl_3^+$, 1,3,5-2,4,6 $C_6F_3Cl_3^+$, and 1,3,5-2,4,6 $C_6F_3Br_3^+$. The quantum yield for emission from the lowest vibrational levels of the $^2A_{2u}$ or $^2A_2''$ excited state is for all these ions, with the possible exception of the bromo compound, near unity.

It may be surprising that the benzene cation, $C_6H_6^+$, itself is not included in the list of ions with observed spectra. The reason for its absence lies in the fact that the quantum yield for emission from the $^2A_{2u}$ excited state drops to 10^{-5} or below and so its spectrum is undetectable. The explanation for this precipitous drop in quantum yield for $C_6H_6^+$ and many other benzenoid cations resides in the fact that the $^2A_{2u}$ state is no longer the lowest lying excited state. A state derived from a hole in a σ core orbital lies below the $^2A_{2u}$ state. This state apparently promotes extremely efficient radiationless relaxation and the emission spectra of species having this orbital ordering generally are not observed.

Without going into the theoretical detail of Section III, it is well to anticipate some of the experimental manifestations of the Jahn-Teller effect in optical spectra. In the absence of Jahn-Teller coupling, the only vibrational modes expected to form progressions for an electronically allowed transition are the totally symmetric ones [14]. For molecules with D_{6h} (D_{3h}) symmetry there are 2(4) a_{1g} (a_1) such totally symmetric modes. Clearly if there were only 2 or 4 progression forming modes (assuming weak overtones of other symmetries as is the case) the vibrational structure of the π-π transition of the benzenoid cations would be rather simple.

However the existence of a Jahn-Teller effect allows the forming of progressions by modes of other symmetries. For D_{6h} (D_{3h}) symmetry, it is easily shown [14] that e_{2g} (e') modes will also form progressions. An easy way to visualize this result is to realize that in the reduced symmetry of the Jahn-Teller distortions, these modes also become totally symmetric. The net result of the Jahn-Teller distortion is therefore to increase the complexity of the spectrum by introducing progressions in up to 4(7) e_{2g} (e') modes. Moreover, as can be clearly seen by the results in Section III, the spacing between successive members in the Jahn-Teller progressions will not be equal. Instead they will follow a complicated pattern determined by the details of the Jahn-Teller coupling. Thus one can see that the spectrum of the Jahn-Teller distorted molecular ions can become very complicated. Extremely high quality spectral data are required for untangling such a spectrum.

B. Experimental Results

Apparently the first realization that any of the benzenoid cations' excited states emitted with reasonable quantum yield occurred about ten years ago. The group of Turner noted [15] a blue emission associated with $C_6F_6^+$ when it was produced by photo-ionization. Since that initial observation, there have been a succession of experiments of ever increasing detail and sophistication devoted to the elucidation of the spectrum of $C_6F_6^+$ and the other symmetrical, emitting cations. These experiments have culminated in an unprecedentedly complete characterization of the Jahn-Teller effect for these ions.

Maier and co-workers [16-20] were the first to follow-up the initial observations of $C_6F_6^+$ fluorescence with studies of spectroscopic significance for a number of halobenzene cations. In these studies samples of neat halobenzene at a pressure of 10^{-3}-10^{-5} torr were subject to bombardment by electrons of controlled energy (20-40 eV). The emission was dispersed by a scanning monochromator of moderate resolution, i.e., 10 to a few tens of cm^{-1}. While pinpointing the optical transitions involved, this experiment permitted only poor resolution of vibrational structure. For all the ions studied only a handful of vibrational frequencies were reported. Indeed, apparently no evidence for, or even mention of, the Jahn-Teller effect is present in these early papers. Nonetheless, they constitute a critical experimental step on the road to understanding.

The work in Maier's group was expanded [21-24] upon by the Orsay group, headed by Leach. They photographed the emission of several halobenzenes using a discharge source. The instrumental precision in measuring line positions was quoted to be 0.1 cm^{-1} in these experiments. However in several cases, observed bands approached 10 cm^{-1} in width due to intrinsic spectral congestion because of the density of emitting vibrational and rotational levels in the excited electronic states of the relatively hot, large ions. Nonetheless these spectra have lead to a much more complete and elaborate vibrational analysis than found in Maier's original papers.

During the same time period, different optical experiments were being carried out at Bell Laboratories by Miller, Bondybey, and co-workers [25-45]. These experiments were built around laser excitation and laser excited, wavelength resolved emission spectra of the halobenzene cations. Both the Orsay and Bell Labs groups recognized, in their first publications, evidence for the Jahn-Teller effect in the vibrational structure of their spectra.

However, as noted above, the vibrational structure of these spectra are very complicated, precluding an immediate, complete analysis. Both groups made substantial progress in analysis in the 2-3 year period after their initial publications. However by far the most complete interpretation both in terms of experimental data and spectral analysis — has been put forth by the Bell group in several publications describing the different ions — $C_6F_6^+$ (ref. 45) $C_6H_3F_3^+$ (refs. 38 and 44), $C_6H_3Cl_3^+$ (refs. 38 and 44), $C_6F_3Cl_3^+$ (ref. 38), and $C_6F_3Br_3^+$ (ref. 40).

The Orsay group did produce Jahn-Teller analyses of considerable detail for two ions, $C_6H_3F_3^+$ (ref. 23) and $C_6H_3Cl_3^+$ (ref. 24) based on their discharge spectra. Their analysis for $C_6H_3Cl_3^+$ agrees in every substantial way with the independent Bell analysis. For a considerable length of time, there was a controversy [46,47] between the Jahn-Teller interpretations of the Bell and Orsay groups for $C_6H_3F_3^+$. Several experiments were performed [32-34] by the Bell group that supported their interpretation. Very recently the Orsay group has published [48] an analysis of the $C_6D_3F_3^+$ spectrum. They find that the $C_6D_3F_3^+$ data strongly support the Bell interpretation, and that their $C_6H_3F_3^+$ spectrum is indeed consistent with the Bell interpretation. Happily then, there are no longer any significant discrepancies in the analyses of the spectra of the symmetrical benzenoid cations and there is no longer any controversy surrounding the interpretation of the Jahn-Teller effect.

While the original references should be consulted for details, it is instructive to see roughly how the vibronic structure in these spectra was assigned and analyzed. Consider for example the most symmetrical of the ions, $C_6F_6^+$. Because of congestion, little experimental information characterizing the Jahn-Teller effect, was obtained from the Orsay discharge spectra, so we will concentrate on the laser experiments of the Bell group.

The first published report [27] of a laser excitation spectrum of gaseous $C_6F_6^+$ by the Bell group was in early 1979. The ion was produced by Penning ionization in a flow system. This spectrum was relatively badly congested and overlapped because of the population at room temperature of many low lying vibrational and rotational levels (see Fig. 2). However the

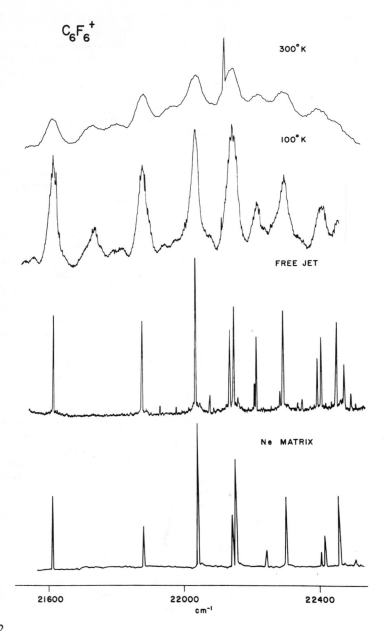

Figure 2.

Laser excitation spectra of $C_6F_6^+$ taken under a variety of conditions. Preceeding upward from the bottom (i) $C_6F_6^+$ in a Ne matrix at 4°K, (ii) $C_6F_6^+$ produced via 2-photon ionization by an ArF laser in a free jet expansion, (iii) $C_6F_6^+$ produced by Penning ionization with the flow system at liquid N_2 temperature, (iv) same as (iii) except flow system at room temperature. The Ne matrix spectrum has been shifted in absolute frequency so that its origin coincides with those of the gas phase spectra.

spectrum did show unambiguous progressions in the two lowest frequency e_{2g} modes, ν_{17} and ν_{18} — the nominal C-C-F bend and the C-C-C bend motions. This observation is only consistent with a lowering of the symmetry in either the upper or lower state involved in the electronic transition. Presumably this lower symmetry results from a Jahn-Teller distortion of the ground state.

It was realized at that time that further progress in the characterization of the Jahn-Teller effect would require better spectra, i.e., higher resolution, better S/N, etc. An obvious way to simplify the spectrum and obtain better resolution is to lower the ion's temperature. The first successful approach in this direction involved producing the ions in a solid inert gas matrix, first Ar [26] and then Ne [36], at 4K. In the Ne matrix, very high resolution laser excitation (see Fig. 2), and laser excited, wavelength resolved emission, spectra were obtained for $C_6F_6^+$. These spectra served as the basis for the first realistic attempts to interpret the Jahn-Teller effect in this ion.

If the matrix spectra could be established as the true low temperature limit of the gas-phase spectra, then they could serve as an ideal blueprint for unraveling the Jahn-Teller effect. However, as we have said before it is often impossible to distinguish between true Jahn-Teller effects and deviations in site symmetry in a solid. While several indirect tests suggested that, especially in Ne, the vibronic structure of ions was unperturbed, the only way to be certain was to obtain the same low temperature spectra in the gas phase.

The first important step in this direction was reported in 1981 by Sears, et al., [35] for $C_6F_6^+$. They observed both laser excitation (see Fig. 2), and laser excited, wavelength resolved emission spectra for $C_6F_6^+$ cooled to ~100K. This was accomplished by cooling with liquid N_2 the Penning ionization apparatus used in the initial studies. The resolution in this experiment, while still not as good as in the matrix experiment, was greatly improved compared to the earlier Penning ionization experiments [27]. These experiments showed that to a high degree the vibronic structure observed in the Ne matrix was unperturbed. (There is, however, an overall electronic shift in the Ne matrix of ~60 cm^{-1} to the red.)

The laser excited, wavelength resolved emission experiments on $C_6F_6^+$, reported in the same paper, gave key evidence for a similar lack of perturbation in the Ne matrix emission spectra. At the same time, this work provided important results not available from the earlier matrix experiments. In a solid matrix, vibrational relaxation is usually much faster than electronic emission. Thus regardless of which vibrational level of the $\tilde{B}\,^2A_{2u}$ state of $C_6F_6^+$ is excited, most of the emission occurs from the vibrationless level. Unfortunately selection rules prohibit the combination of certain key, Jahn-Teller shifted levels in the $\tilde{X}\,^2E_{1g}$ state with the vibrationless level of the $\tilde{B}\,^2A_{2u}$ state. Therefore the early matrix spectra provided no information on these levels.

In the gas phase the bulk of the emission occurs from the initially excited level. Thus by pumping vibrationally excited levels of the $\tilde{B}\,^2A_{2u}$ state, transitions involving these key vibronic levels of the $\tilde{X}\,^2E_{1g}$ state could be directly observed. The positions of these levels were subsequently independently established by monitoring, in a time resolved fashion, extremely weak unrelaxed emission in the matrix [33,41].

The gas phase low temperature limit appears to have recently been reached with the obtainment of spectra of $C_6F_6^+$ produced in a supersonic expansion [37,42,49]. In the first reports [37,49], a greatly simplified emission spectrum from the $\tilde{B}\,^2A_{2u}$ state was produced by electron impact excitation in the jet. More recently laser excitation spectra [42,43] have been obtained for $C_6F_6^+$ cooled to near 0K in a supersonic expansion (see Fig. 2). In this case the ion was produced by 2-photon photoionization of the neutral with an ArF laser.

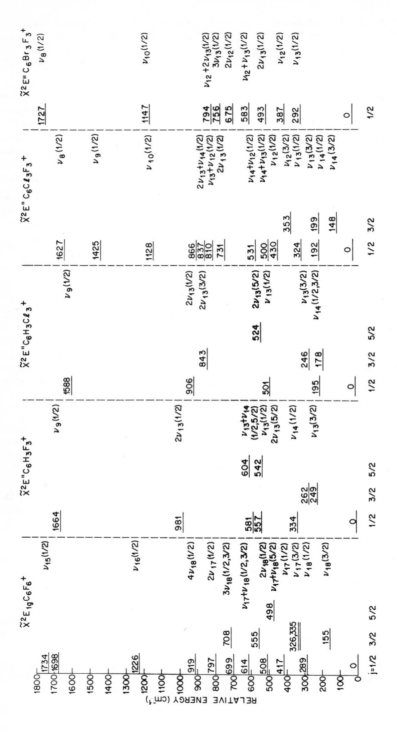

Figure 3. Experimentally determined positions of the Jahn-Teller active, vibronic energy levels for the ground states of $C_6F_6^+$, $C_6H_3F_3^+$, $C_6H_3Cl_3^+$, $C_6F_3Cl_3^+$, and $C_6F_3Br_3^+$.

The obtainment of the spectra of the other symmetrical benzenoid ions have followed a similar pattern to that for $C_6F_6^+$. In Fig. 3, we reproduce the experimentally determined positions for the Jahn-Teller active vibrational energy levels, i.e. the e_{2g} levels of $C_6F_6^+$ and the e'' levels of $C_6H_3F_3^+$, $C_6H_3Cl_3^+$, $C_6F_3Cl_3^+$, and $C_6F_3Br_3^+$. Clearly the data is relatively incomplete for $C_6F_3Br_3^+$, but for the other ions, a fairly large number of energy levels have been well located experimentally.

III. THEORETICAL BACKGROUND

A. Jahn-Teller Hamiltonian

We need to develop a theoretical framework for the understanding of the experimentally observed Jahn-Teller shifted levels in the symmetrical (D_{3h} or higher) halobenzene cations as shown in Fig. 3. Our procedure for doing this is to construct a Hamiltonian describing the nuclear motion in a degenerate electronic state. Our results will be specially tailored for the halobenzene cations of Fig. 3 but carry considerably wider applicability. Little of what we present here is new; it is based primarily on the work of Longuet-Higgins and co-workers [50-52] and to a lesser extent on the somewhat parallel treatments of Liehr [53-57]. These early works have since been considerably modified and refined [14,44,45,58-62].

We consider a doubly electronic state, e.g. $^2E_{1g}$ in D_{6h} or $^2E''$ in D_{3h} symmetry, with wavefunctions ψ_+ and ψ_-. The Hamiltonian for the nuclear motion within this state can be written

$$\mathcal{H} = \mathcal{H}_T + \hat{V} \tag{1}$$

$\mathcal{H}_T$ represents the kinetic energy of the nuclei and has its usual form. For the moment the potential, $\hat{V}$, in which the nuclei move is the most interesting. In vibrational problems, it is traditional to express the potential $\hat{V}$ in some sort of Taylor expansion of the nuclear coordinates.

In general for a large molecule such an expansion can be rather complicated owing to the large number, 3N-6, of degrees of freedom for the vibrational motion. This expansion can be put into convenient form if we make the following definitions and assumptions (which will not limit the usefulness of our approach for the molecules that have been studied experimentally). Let's assume that there are p doubly degenerate modes of the symmetry group that is Jahn-Teller active. Further let's assume that the remaining 3N-6-2p motions can be adequately represented by harmonic restoring forces. In this case, we have, up to quadratic power in Q,

$$\hat{V} = V_e^o + \frac{1}{2} \sum_{i=1}^{3N-6-2p} \lambda_i |Q_i|^2 + \sum_{i=1}^{p} \sum_{r=+,-} \left\{ \frac{1}{2} \hat{\lambda}_i |Q_r^i|^2 + \hat{k}_i Q_r^i \right.$$

$$\left. + \hat{g}_i^r (Q_r^i)^2 + \sum_{s>i} \hat{g}_{is} Q_r^i Q_r^s \right\} \tag{2}$$

Clearly even this expression has considerable complexity.

However, Eq. 2 is written in such a form that each term in the expansion has a fairly well defined physical significance. The first term represents the electronic potential energy at the symmetrical nuclear configuration. The next term represents the harmonic restoring potential for each of the non Jahn-Teller active modes. Given the vibrational quantum numbers of these modes, the first two terms simply constitute a fixed energy and need not concern us further. (This statement is strictly true only if one is not concerned with sequence transitions of a non-linearly active Jahn-Teller mode. Scharf, et al., [63] show that these neglected terms can lead to anomalous sequence structure for such transitions. However, as such sequence transitions have not directly entered into the analysis of the spectra in the benzenoid cations, we will omit consideration of these interactions.)

The remaining four terms in the expansion represent contributions from the p Jahn-Teller active modes. The first of these terms is the harmonic potential for these modes. There is fundamentally no difference between this term and the corresponding ones for the non Jahn-Teller active modes. However, it is worthwhile looking at this familiar term in slightly more detail, because it introduces some notation that we will use extensively for the Jahn-Teller active modes. We define the complex conjugate forms of the normal coordinates by

$$Q^i_\pm = Q^i_1 \pm \sqrt{-1}\, Q^i_2 \qquad (3)$$

where (Q^i_1, Q^i_2) represents the real, orthogonal normal coordinates of a doubly degenerate Jahn-Teller active mode. (For the molecular ions studied all the Jahn-Teller active modes are doubly degenerate.)

The Hamiltonian in Eq. (1) was noted to be appropriate for the subspace consisting of the pair of degenerate electronic wavefunctions and the vibrational wavefunctions corresponding thereto. The normal coordinates Q are obviously operators with respect to the vibrational wavefunctions, but not with respect to the electronic ones. The "hatted" parameters, e.g. $\hat{\lambda}_i$, preceding the nuclear coordinates are operators with respect to the electronic wavefunctions but not with respect to the vibrational wavefunctions. For example the explicit form for $\hat{\lambda}_i$ is

$$\frac{1}{2}\hat{\lambda}_i = \left(\frac{\partial^2 \mathcal{H}_e}{\partial Q^i_+ \partial Q^i_-} \right)_0 \qquad (4)$$

where $\mathcal{H}_e$ is the operator giving the electronic eigenvalues for the state of interest. The derivatives are to be evaluated with the nuclear configuration at the symmetrical configuration, as indicated by the subscript zero.

The fourth term in the expansion of Eq. 2 represents the *linear* Jahn-Teller interaction. The fifth term represents the *quadratic* Jahn-Teller interaction. The final term represents *quadratic* coupling among the Jahn-Teller active modes. This term was introduced by Cossart-Magos and Leach [61] in their treatment of the Jahn-Teller effect. They were not able to find any evidence for the observability of such a term. Most importantly, a detailed analysis of the matrix elements of this term shows that it will have only off-diagonal matrix elements coupling states quite different in energy. Unless quadratic coupling is large, this term will have no observable effect on the molecular spectrum. For this reason we shall neglect further consideration of this complicated term. The interested reader is referred to the original work of Cossart-Magos and Leach [61]. It is extremely important to note that by neglecting this term we have not neglected mode mixing effects [58,60], only their quadratic form. Mode mixing is implicitly contained (as we shall see in more detail later) in the fourth and fifth terms of V.

It shall be to our advantage to have a potential operator that only operates explicitly in the space of the vibrational wavefunctions. Thus we pre and post multiply the terms of $\hat{V}$ by the electronic eigenfunctions, ψ_+ and ψ_-, of $\mathcal{H}_e$. This reduces the electronic operators multiplying the nuclear coordinates in $\hat{V}$ to a simple product of an electronic constant times a projection operator specifying the electronic states that give a non-zero matrix element. Therefore neglecting the first two constant terms, and the negligible last term of $\hat{V}$, we obtain the portion of $\hat{V}$, $\hat{V}_{JT}$, appropriate to the Jahn-Teller vibrational problem,

$$\hat{V}_{JT} = \sum_{i=1}^{P} \hat{V}_{JT}^i \tag{5}$$

where

$$\hat{V}_{JT}^i = \frac{\lambda_i}{2} Q_+^i Q_-^i + k_i^{-+} Q_+^i + k_i^{+-} Q_-^i$$

$$+ \frac{g_i^{-+}}{2} Q_-^2 + \frac{g_i^{+-}}{2} Q_+^2 \tag{6}$$

In the above the parameters are defined as

$$\frac{1}{2}\lambda_i = \langle\psi_+|(\partial^2\mathcal{H}_e/\partial Q_+^i\partial Q_-^i)_0|\psi_+\rangle = \langle\psi_-|(\partial^2\mathcal{H}_e/\partial Q_+^i\partial Q_-^i)_0|\psi_-\rangle \tag{7a}$$

$$k_i^{+-} = \langle\psi_+|(\partial\mathcal{H}_e/\partial Q_+^i)_0|\psi_-\rangle = \langle\psi_-|(\partial\mathcal{H}_e/\partial Q_-^i)_0|\psi_+\rangle \tag{7b}$$

$$g_i^{+-} = \langle\psi_+|(\partial^2\mathcal{H}_e/\partial Q_-^i)_0|\psi_-\rangle = \langle\psi_-|(\partial^2\mathcal{H}_e/\partial Q_+^i)_0|\psi_+\rangle \tag{7c}$$

The superscript $+-$ denotes that the values of k_i and g_i are non-zero only between the ψ_+ and ψ_- states while the absence of any superscript on λ_i indicates it has only equal, non-zero, expectation values for both the ψ_+ and ψ_- states. The subscript zeros indicate the derivatives are to be evaluated at the symmetrical position.

B. The Effective Potential

The $\hat{V}_{JT}^i$ represents the effective, Jahn-Teller distorted, potential in which the nuclear vibrational motion occurs. Before proceeding to the actual solution of the dynamical problem, it is worthwhile looking at the potential surface that $\hat{V}_{JT}^P$ represents. We know that $\hat{V}_{JT}^i$ has both diagonal and off-diagonal matrix elements in the $\psi_\pm$ basis set, i.e. the eigenfunctions of V_e^o. Thus if we consider a particular mode i, we will obtain a matrix of the form

$$\begin{bmatrix} \frac{\lambda}{2}\rho^2 & \rho k e^{\sqrt{-1}\phi} + \rho^2 g e^{-2\sqrt{-1}\phi} \\ \rho k e^{-\sqrt{-1}\phi} + \rho^2 g e^{2\sqrt{-1}\phi} & \frac{\lambda}{2}\rho^2 \end{bmatrix} \tag{8}$$

where in the above, we have dropped the $+-$ superscripts on the k and g and switched to cylindrical coordinates,

$$Q_\pm = \rho e^{\pm\sqrt{-1}\phi}$$

If we denote the eigenvalues of Eq. 8 by U' and U'', then

$$U'^{('')} = \frac{1}{2}\lambda\rho^2 \pm \rho k \left[1 + \frac{2g\rho}{k}\cos 3\phi + \frac{g^2\rho^2}{k^2} \right]^{\frac{1}{2}}$$

$$\approx \frac{1}{2}\lambda\rho^2 \pm (k\rho + g\rho^2\cos 3\phi) \tag{9}$$

where in the expansion of the radical only terms up to second order in ρ have been retained.

   In the discussion of the form of this potential surface we will assume that the upper sign corresponds to U'' and the lower to U' and that k and g are both positive. The classic form of this potential is shown in Fig. 4. The simplest form of the potential is obtained for the quadratic coupling parameter g=0. In this case, the lower energy surface U' has a maximum at the symmetrical position ($\rho=0$) and its minimum corresponds to an equal-potential moat. The value of the radius, ρ_{1m}, of the moat and its depth, ϵ^1, are found by setting the differential of U' equal to 0, thus (for each mode i)

$$\rho_{1m} = k/\lambda \tag{10}$$

$$\epsilon^1 = k^2/2\lambda \tag{11}$$

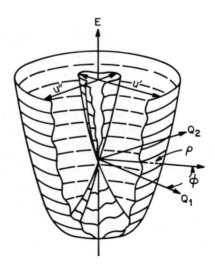

Figure 4.   Schematic representation of the Jahn-Teller distorted potential surfaces, U' and U''.

   If we introduce a non-zero quadratic coupling parameter g we see that the form of the potential in Fig. 4 becomes more complicated. U' will now have local maxima (saddle points) in the moat at $\phi = \pi/3$, π, and $5\,\pi/3$, while minima will occur at $\phi=0$, $2\,\pi/3$, and $4\,\pi/3$. We can again calculate the values of ρ at the saddle points, ρ_{sp}, and the quadratic minima, ρ_{qm},

$$\rho_{qm} = \pm \frac{k}{(\lambda-g)} = \rho_{lm}\,(1-K)^{-1} \tag{12}$$

and

$$\rho_{sp} = \mp \frac{k}{(\lambda+g)} = \rho_{lm}\,(1+K)^{-1} \tag{13}$$

where we have introduced the dimensionless quadratic coupling parameter K

$$K = g/\lambda \tag{14}$$

which is just the ratio of the coupling constant, g, to the harmonic force constant, λ.

If we substitute ρ_{qm} and ρ_{sp} into the expressions for U', we find the energy at the minima or saddle points below the symmetrical configuration

$$\epsilon^Q = \frac{k^2}{2\lambda}\,(1\mp K)^{-1} = \epsilon^l(1\mp K)^{-1} \approx \epsilon^l(1\pm K) \tag{15}$$

where in the last, approximate equality K is assumed small. It is clear that for small K, the separation in energy between quadratic minima and saddle points is $2K\epsilon^l$.

C. The Dynamical Problem

To this point we have considered the potential in which the nuclei move in some detail. To solve the spectroscopic problem we must consider simultaneously all the terms of the Jahn-Teller Hamiltonian, $\mathcal{H}_{JT}$. We now rewrite Eq. 1 in the following manner

$$\mathcal{H}_{JT} = \mathcal{H}_T + \hat{V} = \sum_i^p \left\{ \mathcal{H}_T^i + \hat{V}_{JT}^i \right\}$$

$$= \sum_i^p \left\{ \mathcal{H}_i^h + \mathcal{H}_i^l + \mathcal{H}_i^Q \right\} \tag{16}$$

where

$$\mathcal{H}_i^h = \mathcal{H}_T^i + \frac{\lambda_i}{2}\,Q_+^i Q_-^i \tag{17a}$$

$$\mathcal{H}_i^l = k_i^{+-} Q_+^i + k_i^{-+} Q_-^i \tag{17b}$$

$$\mathcal{H}_i^Q = \tfrac{1}{2}\left\{ g_i^{-+}\,(Q_-^i)^2 + g_i^{+-}\,(Q_+^i)^2 \right\} \tag{17c}$$

Written in this form $\mathcal{H}_i^h$ represent the usual harmonic oscillator Hamiltonian, $\mathcal{H}_i^l$ represents the linear Jahn-Teller coupling, and $\mathcal{H}_i^Q$ represents the quadratic Jahn-Teller coupling for each of the p Jahn-Teller active modes.

What we now wish to do is to construct the Hamiltonian matrix in a suitable basis set. For a doubly degenerate vibration the operator $\mathcal{H}_i^h$ is diagonal in the basis set of the 2-dimensional isotopic harmonic oscillator wavefunctions [52]. We'll denote a member of the basis set, η_{v_i,l_i} where v_i is the usual vibrational quantum number and l_i its associated angular momentum.

The eigenvalues of $\mathcal{H}_i^h$ in this basis are $h\omega_i(v+1)$ where the oscillator frequency ω_i is defined by

$$\omega_i = (2\pi)^{-1}(\lambda_i/M_i)^{1/2} \tag{18}$$

Thus we define our overall basis set as

$$|\pm,v_1,l_1,v_2,l_2...v_p,l_p> \; = \; |\psi_\pm > \; \prod_i |\eta_{v_i,l_i}> \tag{19}$$

The matrix elements of $\mathcal{H}_{JT}$ can be simply calculated by reference to those for powers of the coordinates for a 2-dimensional isotopic harmonic oscillator [44,61,59,64]. These matrix elements are given in Table I, in terms of the previously defined parameters, K_i, ω_i, and D_i which is given by

$$D_i = \epsilon_i/h\omega_i = \frac{k_i^2}{2\hbar}\left[\frac{M_i}{\lambda_i^3}\right]^{1/2} \tag{20}$$

D_i represents a dimensionless linear Jahn-Teller distortion parameter analogous to K_i the quadratic one. As Eq. 20 shows, D has the physical significance of being the ratio of the moat depth to the harmonic frequency. It is often said that for $D<1$, one has a dynamic Jahn-Teller effect while for $D>1$, one has a static Jahn-Teller effect.

The matrix elements in Table I upon examination reveal several important physical aspects of the Jahn-Teller interaction. Let us define a quantum number Λ which has values of $+1$ or -1 for ψ_+ or ψ_- respectively. We notice that as long as we consider only harmonic and linear Jahn-Teller terms, $\mathcal{H}^h + \mathcal{H}^l$, we can define [44,52] a new, conserved quantum number j for the matrix where

$$j = \frac{1}{2}\,\Lambda + \sum_i l_i \tag{21}$$

It has been shown [65,66] that the operator, whose eigenvalue is j and that commutes with $\mathcal{H}^h + \mathcal{H}^l$, corresponds to the total angular momentum, both electronic and nuclear, about the three (D_{3h}) or six (D_{6h}) fold symmetry axis. A glance at Fig. 4 shows the necessity of having a conserved projection of the total angular momentum. In the absence of quadratic coupling the potential clearly has cylindrical symmetry. The existence of quadratic coupling creates saddle points and minima around the moat, destroying the cylindrical symmetry. The quadratic matrix elements in Table I similarly clearly do not conserve j.

However, in those cases where the quadratic coupling is small or negligible, the blocking of the matrix by the quantum number j can lead to considerable simplification. If the elements of $\mathcal{H}^Q$ cannot be neglected, one can still see from the matrix elements of Table I

TABLE I

Matrix Elements of the Jahn-Teller Hamiltonian, Eq. 16.

Harmonic Oscillator

$$< \pm, \Pi_j(v_j, l_j) \, |\mathcal{H}^h| \pm, \Pi_j(v_j', l_j') > \; = \sum_i \left\{ \hbar\omega_i (v_i+1) \right.$$

$$\times \left. \left[\delta_{v_i, v_i'} \, \delta_{l_i, l_i'} \left(\prod_{j\neq i} \delta_{v_j', v_j} \, \delta_{l_j, l_j'} \right) \right] \right\}$$

Linear Jahn-Teller

$$< \mp, \Pi_j(v_j, l_j) \, |\mathcal{H}^l| \pm, \Pi_j(v_j', l_j') > \; = \sum_i \hbar\omega_i \left\{ \left[D_i(v_i \mp l_i + 2) \right]^{\frac{1}{2}} \right.$$

$$\times \left. \left[\delta_{v_i+1, v_i'} \, \delta_{l_i \mp 1, l_i'} \left(\prod_{j\neq i} \delta_{v_j', v_j} \, \delta_{l_j, l_j'} \right) \right] \right\}$$

$$< \pm, \Pi_j(v_j, l_j) \, |\mathcal{H}^l| \mp, \Pi_j(v_j', l_j') > \; = \sum_i \hbar\omega_i \left\{ \left[D_i(v_i \pm l_i)^{\frac{1}{2}} \right] \right.$$

$$\times \left. \left[\delta_{v_i-1, v_i'} \, \delta_{l_i \pm 1, l_i'} \left(\prod_{j\neq i} \delta_{v_j', v_j} \, \delta_{l_j, l_j'} \right) \right] \right\}$$

Quadratic Jahn-Teller

$$< \pm, \Pi_j(v_j, l_j) \, |\mathcal{H}^Q| \mp, \Pi_j(v_j', l_j') > \; = \sum_i \hbar\omega_i \left\{ \frac{K_i}{4} \left[(v_i \pm l_i)(v_i \pm l - 2) \right]^{\frac{1}{2}} \right.$$

$$\times \left. \left[\delta_{v_i-2, v_i'} \, \delta_{l_i \mp 2, l_i} \left(\prod_{j\neq i} \delta_{v_j, v_j'} \, \delta_{l_j, l_j'} \right) \right] \right\}$$

$$< \pm, \Pi_j(v_j, l_j) \, |\mathcal{H}^Q| \mp, \Pi_j(v_j', l_j') > \; = \sum_i \hbar\omega_i \left\{ \frac{K_i}{2} \left[(v_i \mp l_i + 2)(v_i \pm l_i) \right]^{\frac{1}{2}} \right.$$

$$\times \left. \left[\delta_{v_i, v_i'} \, \delta_{l_i \mp 2, l_i} \left(\prod_{j\neq i} \delta_{v_j, v_j'} \, \delta_{l_j, l_j'} \right) \right] \right\}$$

$$< \pm, \Pi_j(v_j, l_j) \, |\mathcal{H}^Q| \mp, \Pi_j(v_j', l_j') > \; = \sum_i \hbar\omega_i \left\{ \frac{K_i}{4} \left[(v_i \mp l_i + 4)(v_i \mp l_i + 2) \right]^{\frac{1}{2}} \right.$$

$$\times \left. \left[\delta_{v_i+2, v_i'} \, \delta_{l_i \mp 2, l_i} \left(\prod_{j\neq i} \delta_{v_j, v_j'} \, \delta_{l_j, l_j'} \right) \right] \right\}$$

that only states of j differing by |3| are coupled. So that even in this worst case some symmetry blocking according to j is still possible.

Another fundamental property of the Jahn-Teller interaction, illustrated by the matrix elements of Table I, is the property of mode mixing among states of Jahn-Teller active symmetry [58]. The Jahn-Teller coupling clearly mixes different modes even though there is no explicit term in the Hamiltonian, or associated molecular parameter, coupling different modes.

To see how this comes about note only the terms of the linear Jahn-Teller Hamiltonian, $\mathcal{H}^l$. Consider the following simple example involving three levels with $j = \frac{1}{2}$, a) the vibrationless level, b) the level with ν_1 excited one quantum, all other levels unexcited, and c) the level with ν_2 excited one quantum, all other levels unexcited. Consulting Table I, we see that there exists an element $h\omega_1(2D_1)^{\frac{1}{2}}$ connecting levels a and b while levels a and c are coupled by $h\omega_2(2D_2)^{\frac{1}{2}}$. Clearly unless D_2 is very small, it is impossible to correctly predict the positions of levels a and b while neglecting the effects of D_2. Even though most early Jahn-Teller calculations involved only one mode, Table I shows that if values of the D_i for other modes are non-negligible, this is clearly an incorrect approach.

IV. CONCLUSIONS

A. Spectra Analysis

In Section II, especially Fig. 3, we have summarized the experimental observations for the symmetrically substituted benzene cations. Table I gives the matrix elements of the Hamiltonian in terms of the adjustable parameters, ω_i, D_i, K_i for each mode. Thus in principle the problem reduces to diagonalizing, numerically on a computer, the Hamiltonian matrix and adjusting the parameters until the observed and calculated energy levels agree. Of course, the vibronic basis set is infinite but this is often the case, and one simply truncates it at a point where the accuracy of the calculated eigenvalues is undisturbed to the desired precision. The problem is that the size of the truncated matrix required to obtain the experimental precision is quite large for the Jahn-Teller problem in these ions.

To be specific consider the case of $C_6F_6^+$. As Fig. 3 shows there have been experimentally determined (with an accuracy of order of a few cm^{-1}) energy levels up to about 2000 cm^{-1} above the vibrationless levels. Intuitively (and it can be verified by explicit calculation) one would hope to have basis functions in the matrix several vibrational levels above the highest observed one.

However, 2000 cm^{-1} corresponds to approximately the 7th and 4th overtone respectively of the two lowest frequency modes. Suppose we were to take a basis set including all vectors with $0 \leqslant v_i \leqslant 6$ for $i = 1$-4. Such a basis set would have $\sim 10^7$ members leading to a matrix containing $\sim 10^{14}$ elements! Clearly diagonalizing such a matrix is impossible even with today's computers.

One thus has great motivation for trying to reduce the size of the required matrices. It was remarked earlier that if terms of the quadratic coupling Hamiltonian, $\mathcal{H}^Q$, could be neglected, then the matrix could be blocked according to the then good quantum number, j.

For $C_6F_6^+$, there are 4 Jahn-Teller active e_{1g} modes, labelled 15 through 18 according to the Herzberg-Teller convention. 15 corresponds to the highest frequency mode and 18 the lowest. In practice the overtones of the lower frequency (17 and 18) modes are not followed experimentally all the way to 2000 cm^{-1} above the origin. Indeed it was found [45] that an accuracy of better than 10 cm^{-1} could be obtained for all the eigenvalues, corresponding to observed transitions, for a basis set consisting of all vectors with $0 \leqslant v_{18} \leqslant 5$, $0 \leqslant v_{17} \leqslant 7$,

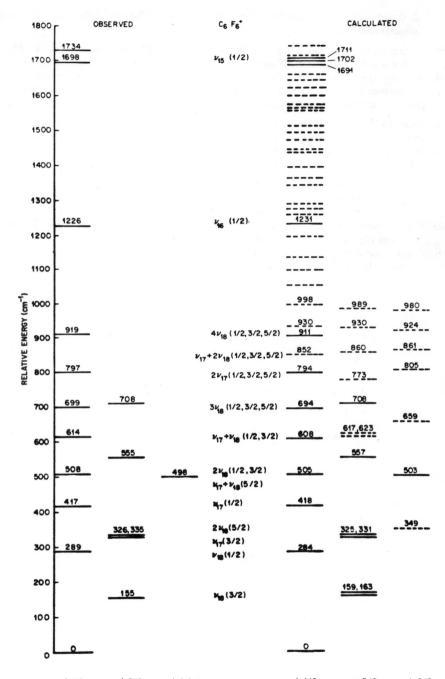

Figure 5.   Comparison of observed and calculated vibronic energy levels for $C_6F_6^+$. The dashed levels are calculated to have been involved in no transitions observable in the previous experiments. See ref. 45 for more details.

$0 \leqslant v_{16} \leqslant 2$, $0 \leqslant v_{15} \leqslant 3$. For the j=½ block this matrix is 7236 × 7236. For j=3/2 and 5/2, it is of comparable size.

Even this eigenvalue problem is not trivial. The lowest ~40 eigenvalues of each j block are required of a matrix with some 50 million elements. However, fortunately many of these elements are zero and there have recently been developed very high speed sparse matrix eigenvalue techniques [67]. These techniques executed [45] on a state-of-the-art computer, a CRAY-1, provide the desired eigenvalues using ~150 sec of CRAY-1 time for each j block.

It was found [45] impossible to include quadratic coupling into the 4 mode problem, but it could be explicitly included when only three modes were considered. Thus a single, most weakly Jahn-Teller active, mode was dropped and the effect of quadratic coupling directly calculated. The error in this approach is negligible.

In Fig. 5 we show a comparison of the computed and observed eigenvalue results for $C_6F_6^+$. As Fig. 5 shows there is excellent agreement between the computed and observed eigenvalues. Further proof of the adequacy of the calculation is given in Table II. The observed relative intensities of a number of transitions in $C_6F_6^+$ are listed in Table II. Also given are the calculated relative intensities predicted by the eigenvectors of the same calculation that gave the correct frequency predictions shown in Fig. 5. One can again see excellent agreement between the calculated and observed results.

TABLE II

Comparison of Observed and Calculated Transition
Probabilities for $C_6F_6^+$ (ref. 45)

Transition	Observed	Calculated
$v_0' - v_0''$	100	100
$v_{18}' - v_0''$	26	28
$v_{17}' - v_0''$	52	55
$2v_{17}' - v_0''$	16	14
$3v_{17}' - v_0''$	6	3
$v_{15}' - v_0''$	20	26
$v_0' - v_{18}''$	22	14
$v_0' - v_{17}''$	18	19
$v_0' - 2v_{18}''$	23	27
$v_0' - 3v_{18}''$	6	6
$v_0' - 2v_{17}''$	7	8
$v_0' - v_{15}''$	11	15

The spectral analysis for the other ions followed much the same pattern as discussed above for $C_6F_6^+$. We list in Table III the resulting "best" values for the molecular

TABLE III

Summary of Jahn-Teller Parameters for Benzenoid Cations (refs. 38, 40, 44, 45).
The units are as follows: i, D_i, K_i - dimensionless; ϵ_i, ω_i - cm^{-1}; ρ'_{lm} - a.m.u.$^{\frac{1}{2}}$ Å.
The column headings for the modes are only qualitatively descriptive.
The total stabilization energy $\sum_i \epsilon_i$ is given beside each ion.

	C-C Stretch	C-F (Cl) Stretch	C-C-C Bend	C-F Bend	C-Cl Bend
$C_6F_6^+$ ($\sum_i \epsilon_i = 821$)					
i	15	16	17	18	...
ω	1610	1215	425	265	...
D	0.23	0.05	0.68	0.38	...
K	...	...	0.006	...	...
ϵ	370	61	289	101	...
ρ'_{lm}	0.098	0.053	0.328	0.310	...
$C_6H_3F_3^+$ ($\sum_i \epsilon_i = 922$)					
i	9	12	13	14	...
ω	1570	960	480	335	...
D	0.35	0.01	0.73	0.03	...
K	...	...	0.007	...	...
ϵ	550	12	350	10	...
ρ'_{lm}	0.122	0.029	0.319	0.078	...
$C_6H_3Cl_3^+$ ($\sum_i \epsilon_i = 563$)					
i	9	11	13	...	14
ω	1540	1060	420	...	190
D	0.18	0.02	0.62	...	0.03
K	...	...	$\leqslant$0.005	...	...
ϵ	277	21	260	...	5
ρ'_{lm}	0.089	0.036	0.315	...	0.094
$C_6Cl_3F_3^+$ ($\sum_i \epsilon_i = 575$)					
i	8	...	12	13	14
ω	1550	...	390	310	185
D	0.15	...	0.60	0.32	0.05
K	...	...	...	...	...
ϵ	233	...	234	99	9
ρ'_{lm}	0.081	...	0.321	0.263	0.135
$C_6Br_3F_3^+$ ($\sum_i \epsilon_i = 660$)					
i	8	10	12	13	...
ω	1560	1100	385	260	...
D	0.15	0.10	0.33	0.75	...
K	...	...	...	...	...
ϵ	234	110	127	195	...
ρ'_{lm}	0.080	0.078	0.240	0.440	...

parameters, ω_i, D_i, ϵ_i, ρ_{lm}^i, and K_i obtained from the calculation. Besides the results for $C_6F_6^+$, we also give the results for the other symmetrical ions, $C_6F_3H_3^+$, $C_6Cl_3H_3^+$, $C_6Cl_3F_3^+$, and $C_6Br_3F_3^+$. These parameters were obtained from entirely analogous 4-mode calculations and the fits between the calculated and the experimentally observed results were comparable to the case of $C_6F_6^+$. For the case of $C_6Br_3F_3^+$, the available experimental data were not as extensive and thus the accuracy of the quoted results are considerably lower.

For the ions other than $C_6F_6^+$, the nominal symmetry is lowered to D_{3h}. There are thus 7 possible e', Jahn-Teller active, modes. For all these observed spectra there are from 4-6 e' modes identified to have some activity. In all cases at least 2 or 3 of these modes are only weakly active. Thus the neglect in the Hamiltonian matrix of modes beyond the four most active should have only a small effect on the determination of the parameters for the four most active modes.

B. Stabilization Energy

The "best" molecular parameters, D_i, ω_i, etc. for the five studied cations are given in Table III. A few general comments concerning Table III are in order. The columns are labeled by the symmetry mode designations C-C stretch, C-C-C bend, etc. However, all the numerical results are for the actual normal modes (e' in D_{3h} symmetry, numbered 8-14, and e_{2g} in D_{6h} symmetry, numbered 15-18). The symmetry mode designations are usually reasonably accurate, but certainly not exact, descriptions of the normal modes. Generally speaking the description is probably least accurate for the heavier species, e.g., in $C_6F_3Br_3^+$, modes 12 and 13 are strong mixtures of C-C-C bend and C-F bend.

Inspection of Table III shows that the total Jahn-Teller distortion energy is similar for all the cations studied. It falls into the range 500-1000 cm^{-1}, 1.4-2.8 Kcal/mole. It can also be seen that for all the ions most ($\geq 80\%$) of the stabilization energy can be ascribed to two modes, the nominal C-C stretch and C-C-C bend, both motions characteristic of the ring itself. One then reaches the conclusion that there is a characteristic Jahn-Teller stabilization energy of $\sim$500-1000 cm^{-1} whenever one removes an electron from the highest filled π molecular orbital. This idea is reinforced by a recent ab initio calculation [68] of the Jahn-Teller stabilization energies in $C_6F_6^+$ and $C_6H_6^+$. It was found that the calculated Jahn-Teller stabilization energy for $C_6H_6^+$ was less than 10% different from that of $C_6F_6^+$.

One should take care before drawing too detailed a conclusion because of the $\sim$10-20% uncertainty (depending on the species) in the total stabilization energy for the ions. However if one looks closely, it appears that there is somewhat less stabilization for the ions with the heavier halogen substituents. This relatively small substituent effect is indeed consistent with the notion of a characteristic Jahn-Teller stabilization energy if one removes an electron from the highest filled π orbital. In the heavier species, there is the possibility of some delocalization of the π electron onto the halogen. Thus in some sense, when the π electron is removed from the heavier molecules to form the ion, the net loss to the ring is somewhat less than one electron.

C. Distorted Geometry

The final and in some ways most interesting result from the Jahn-Teller experiments is the determination of the geometric distortion of the ions at the minimum of the potential surface. The direct experimental measures of this distortion are the values of ρ_{lm}' (we here neglect the small quadratic effects) for each normal mode i, listed in Table III. These are calculated via Eqs. 10 and 20 from the experimental values of D_i and ω_i, i.e.

$$\rho_{lm}^i = k_i/\lambda = \left[\frac{\hbar D_i}{\pi M_i \omega_i} \right]^{1/2} \tag{22a}$$

and

$$\left(\rho'_{lm}\right)^i = M_i^{1/2}\,\rho^i_{lm} = \left[\frac{\hbar D_i}{\pi \omega_i}\right]^{1/2} \tag{22b}$$

Again one sees, in general, the largest values of ρ'_{lm} for the two ring modes that contributed the bulk of the stabilization energy. Furthermore one finds a remarkable constancy of the size of the distortion. If one neglects $C_6F_3Br_3^+$ where mode mixing is likely strong, then one finds that the mode designated as the C-C-C bend has a value of $\rho'_{lm} = 0.32\ AMU^{1/2}Å$ with a variation of only $\pm 0.02\ AMU^{1/2}Å$. In a similar way within $\pm 0.02\ AMU^{1/2}Å$ the value of ρ'_{lm} for the C-C stretch mode is $0.10\ AMU^{1/2}Å$ for all the ions experimentally studied. This extreme similarity of the ρ'_{lm} values is yet another example of the characteristic Jahn-Teller distortion for all the benzenoid ions.

To get a more physical description of the Jahn-Teller distortion of the ions, it is necessary to inquire into the nature of the normal coordinates, ρ_i. For this discussion we shall again consider explicitly the most symmetrical ion, $C_6F_6^+$.

A good analytical approximation to the four normal coordinates, ρ_i, for $C_6F_6^+$ can be obtained from the symmetry coordinates [69], S_a^i, S_b^i constructed from the C-C stretch, t, C-C-C bend, α, C-F stretch, s, and C-C-F bend, β. In Table IV, we list these coordinates. These coordinates have two advantages. They have simple analytical representations and they are reasonably good approximations to the actual normal coordinates.

TABLE IV

Analytical representation of the symmetry coordinates
for $C_6F_6^+$. $\hat{t}_i$, $\hat{s}_i$, $\hat{\alpha}_i$, and $\hat{\beta}_i$ represent unit vectors
respectively for the C-C stretch, C-F stretch, C-C-C bend, and $\hat{\beta}_i = (\hat{\theta}_i - \hat{\theta}_{i+1})/2$
where θ is the C-C-F bend angle. The length of the
vector is determined by the respective multipling fractors,
Δt, Δs, $\Delta \alpha$, and $\Delta \beta$.
For the last two entries, the changes in bond angles,
$\Delta \alpha$, $\Delta \beta$, are multiplied by R, the bond
length to give common units to all four vectors.

	S_a	S_b
C-C Stretch	$-\Delta t\,(\hat{t}_1 - 2\hat{t}_2 + \hat{t}_3 + \hat{t}_4 - 2\hat{t}_5 + \hat{t}_6)/\sqrt{12}$	$-\Delta t\,(\hat{t}_1 - \hat{t}_3 + \hat{t}_4 - \hat{t}_6)/2$
C-F Stretch	$\Delta s\,(2\hat{s}_1 - \hat{s}_2 - \hat{s}_3 + 2\hat{s}_4 - \hat{s}_5 - \hat{s}_6)/\sqrt{12}$	$\Delta s\,(\hat{s}_2 - \hat{s}_3 + \hat{s}_5 - \hat{s}_6)/2$
C-C-C Bend	$-R\Delta\alpha(2\hat{\alpha}_1 - \hat{\alpha}_2 - \hat{\alpha}_3 + 2\hat{\alpha}_4 - \hat{\alpha}_5 - \hat{\alpha}_6)/\sqrt{12}$	$-R\Delta\alpha(\hat{\alpha}_2 - \hat{\alpha}_3 + \hat{\alpha}_5 - \hat{\alpha}_6)/2$
C-C-F Bend	$R\Delta\beta(\hat{\beta}_2 - \hat{\beta}_3 + \hat{\beta}_5 - \hat{\beta}_6)/2$	$R\Delta\beta(2\hat{\beta}_1 - \hat{\beta}_2 - \hat{\beta}_3 + 2\hat{\beta}_4 - \hat{\beta}_5 - \hat{\beta}_6)/\sqrt{12}$

The correspondence between the Jahn-Teller distorted potential of Fig. 4 in terms of ρ and

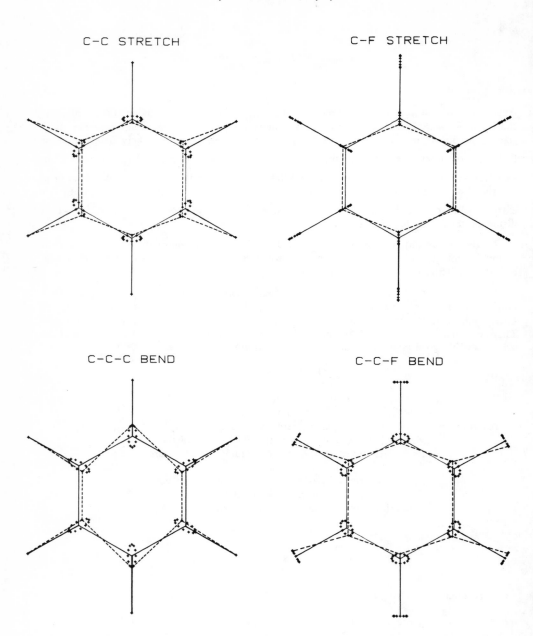

Figure 6.   Schematic representation of the motions of the atoms in $C_6F_6^+$ for the potential energy surface in Fig. 4. The solid line structure shows the D_{6h} configuration. The dashed line connects the distorted configuration at the minimum of the potential energy surface, i.e. for $\rho = \rho_{lm}$ and with $\phi = 0$. In this case the normal coordinates Q_1 and Q_2 are approximated by the symmetry coordinates of Table IV.

MODE 15　　　　　　　　　　　MODE 16

MODE 17　　　　　　　　　　　MODE 18

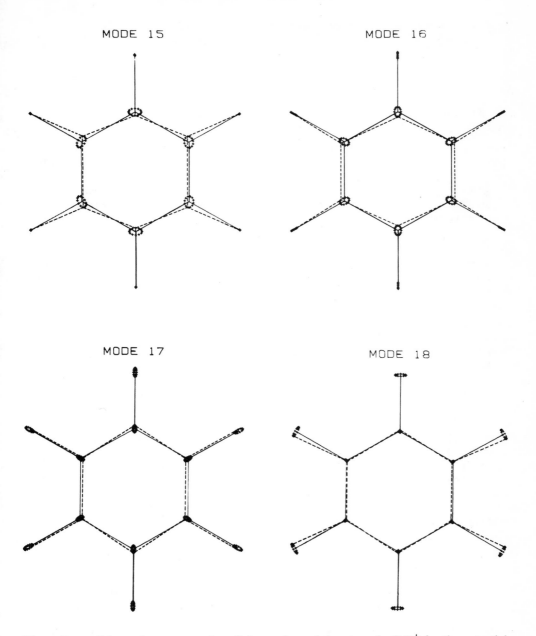

Figure 7.　Schematic representation of the motions of the atoms in $C_6F_6^+$ for the potential energy surface in Fig. 4. The solid line structure shows the D_{6h} configuration. The dashed line connects the distorted configuration at the minimum of the potential energy surface, i.e. for $\rho = \rho_{lm}$ and with $\phi = 0$. In this case the normal coordinates Q_1 and Q_2 are determined by a force field calculation for $C_6F_6^+$ (see text for details).

ϕ and the motions of the atoms is made by noting that

$$S_a^i \approx Q_1^i = \rho \cos \phi \qquad\qquad (23a)$$

$$S_b^i \approx Q_2^i = \rho \sin \phi \qquad\qquad (23b)$$

where Q_1^i and Q_2^i are respectively the real and imaginary parts of the normal coordinates vectors $Q_\pm^i$, introduced in Section II, Eq. 3.

It is probably worthwhile commenting on a few aspects of the symmetry coordinate depiction of the atomic motion that seem to have been discussed only with brevity previously. An e_{2g} distortion of $C_6F_6^+$ lowers its symmetry from D_{6h} to D_{2h}. The distortions described by the S_a^i functions in Table IV all conserve D_{2h} symmetry. The S_b^i distortions however breach D_{2h} symmetry, thus allowing the molecule to move from a configuration corresponding to one irreducible representation, e.g., B_{3g}, of D_{2h} symmetry to another, e.g., B_{2g}.

Figure 6 shows the path of the atomic motion as a function of the phase angle ϕ of the Jahn-Teller potential, Fig. 4, using the symmetry coordinates of Table IV as basis vectors. Consistent with the experimental observations, the quadratic coupling is taken to be negligible in Fig. 6. This results in the atomic motions following regular ellipses (which occasionally degenerate into straight lines). If quadratic coupling were not negligible then the actual motion of the atoms would meander about these ellipses.

Some readers may be perplexed by the fact that the ellipses are not circles, for the atomic motions of Jahn-Teller molecules have previously been depicted [12,55,57] as purely rotatory. The only way one obtains purely circulatory motion for a benzene ion is to use a special set of symmetry coordinates, *constructed* to that end [56]. For the realistic normal coordinates depicted in Fig. 7 and the symmetry coordinates of Fig. 6, which are reasonable approximations to them, the motion is elliptical.

As an aid to the reader, we show in Fig. 8, the twelve configurations of $C_6F_6^+$ obtained by connecting the appropriate points of the atomic loci given in Fig. 6 for the C-C stretch mode. Figure 8 exhibits clearly the pseudo-rotatory nature of the bond length alternation inherent in this mode. One can also see that there are only six unique angles where the molecular shape corresponds to D_{2h} symmetry, the quadratic minima at $0°$, $120°$, and $240°$ and the corresponding saddle points at $60°$, $180°$, and $300°$. It is also obvious from Fig. 8 that the three angles indicated for each set of extrema must have equal energies. The molecular distortions for these angles are identical, being obtained from one another by simple rotation of the entire molecule.

If one, for the moment, assumes that the $S_{a,b}^i$ are true normal coordinates, it is easy to see the relationship between the change in ρ_i and the corresponding change in the Δt, Δs, etc. Correlate mode 15 with the C-C stretch in $C_6F_6^+$. Then from Eqs. 22 and 23 and Table IV with $\phi=0$ we see that $\Delta t = (\rho_{lm})_{15} = (\rho'_{lm})_{15} M^{-\frac{1}{2}}$ since except for Δt, S_a is simply a unit vector. We note here that this value of Δt is appropriate to characterize the bond length distortion (indicated by dashed lines) shown in Fig. 6. It is important to note that the distorted structure of Fig. 6 corresponds to Fig. 4 with $\phi=0$, so that this value of ϕ is most appropriate for determining Δt. We have [44,45], however, in the past, arbitrarily chosen $\phi=45°$ to divide equally the distortion between S_a and S_b. Hence those results were reduced from the present ones by a factor of $\sqrt{2}$.

While it is instructive, and gives one considerable physical feel, to consider the distortions when the normal coordinates are approximated by the above symmetry coordinates, such a

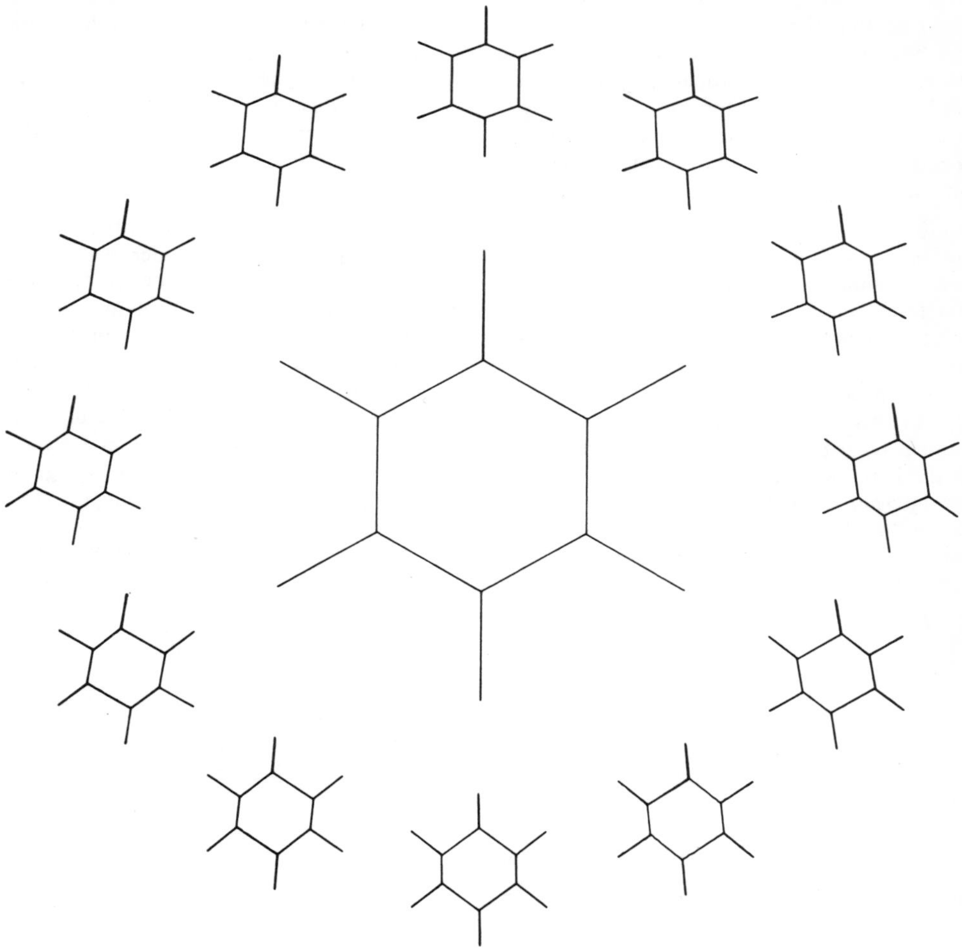

Figure 8.   Pictorial representation of the distorted $C_6F_6^+$ molecule for 12 values of the angle ϕ in Fig. 4. The values chosen are 0, 30, 60, 90. . .330°, with 0° at the top of the diagram and preceeding in a clockwise fashion. The distortions are depicted for the C-C stretch mode of Fig. 6. The distorted (dashed line) structure of Fig. 6 is equivalent to the distorted $C_6F_6^+$ shown at the top of the figure, i.e. for $\phi = 0$.

result is certainly only qualitatively correct. The actual e_{2g} normal coordinates are indeed linear combinations of the four symmetry coordinates in Table IV. A force field calculation has been performed [45] to obtain the true normal coordinate representations for $C_6F_6^+$. While force field calculations of this type are always subject to some uncertainly, it certainly produces a reasonable approximation to the true normal coordinates. We show in Fig. 7 the positions of the atoms for a fixed ρ_{lm} and variable ϕ for this, our best, representation of the normal coordinates. Comparison of Figs. 6 and 7 shows that the atomic motions are indeed similar but not precisely the same.

A simple way of thinking about the comparison between the symmetry coordinates and our best approximation to the true normal coordinates is as follows. Consider both to be linear combinations of the symmetry coordinate functions in Table IV. However for the simple symmetry coordinates all the values of Δt, Δs, $\Delta\alpha$, $\Delta\beta$ are zero, except for the one value uniquely correlated with that symmetry coordinate. The normal coordinate can then be viewed as a linear combination of all the symmetry coordinates S^i with the coefficients in the expansion being the values of Δt, Δs, $\Delta\alpha$, $\Delta\beta$, which in general will all be non-zero for each Q_i. In this way the values of Δt, $\Delta\alpha$, etc. will be a direct measure of the amount of bond stretch, bond bend, etc., distortion present in that mode. Table V gives these values for each of the four calculated normal modes. (Here values were determined by noting the change in the internal coordinates, Δt, Δs, $\Delta\alpha$, $\Delta\beta$, for a unit change in the normal coordinate. The values listed in Table V then correspond to the same percentage change in these internal coordinates as ρ_{lm} represents to the unit change.) As noted before, we now choose our phase convention at $\phi=0$ rather than $45°$ so the values listed are $\sqrt{2}$ times those given in refs. 44 and 45.

TABLE V

Values of the changes (ref. 45) for the internal coordinates corresponding to the mimimum of the Jahn-Teller distorted potential at the bottom of the moat (Fig. 4) with $\phi = 0$ for $C_6F_6^+$.

Mode	$|\Delta t|$ Å	$|\Delta s|$ Å	$|\Delta\alpha|$ deg	$|\Delta\beta|$ deg
15	.013	0.006	0.7	0.8
16	.001	.006	0.4	0.0
17	.001	.004	1.3	0.7
18	0.0	0.0	0.4	1.4

The results in Table V are explicitly only for $C_6F_6^+$. The quality of the available force fields for the benzene cations with heavier halogen substituents is not sufficiently high to justify a similar analysis for these species. However, from the similarity of the ρ_{lm} for all the cations in Table IV, it is clear that qualitatively the distortions for all the cations are indeed closely related. Thus, just as there appears, to a first approximation, to be a characteristic energy stabilization for the removal of an electron from the benzene ring, there is a corresponding characteristic distortion of the ring.

In conclusion, one can note that the symmetrical benzenoid cations represent the premier example of the Jahn-Teller effect in an isolated molecule. An unprecented amount of experimental information has been amassed which characterizes the interaction. A detailed analysis has been performed for five different cations, yielding a quantitatively accurate description of the Jahn-Teller distorted potential energy surface. It is found that all the substituted benzenoid cations suffer similar distortions.

Acknowledgement: The authors acknowledge the extensive contributions, both theoretical and experimental, of Dr. Trevor Sears to the subject matter of this article. We also acknowledge substantial contributions to this work by Dr. Michael Heaven and B. R. Zegarski.

REFERENCES

1. H. A. Jahn and E. Teller, Proc. Roy. Soc. (Lond) Ser. A **161**, 220 (1937).

2. J. W. Rabalais, T. Bergmark, L. O. Werme, L. Karlsson and K. Siegbahn, Phys. Scripta **3**, 13 (1971).

3. L. Asbrink, E. Lindholm, and O. Edquist, Chem. Phys. Lett. **5**, 609 (1970).

4. P. C. Engelking and W. C. Lineberger, J. Chem. Phys. **67**, 1412 (1977).

5. J. W. Rabalais, "Principles of Ultraviolet Photoelectron Spectroscopy," Wiley-Intersceince, New York, 1977, and references therein.

6. J. Dyke, N. Jonathan, E. Lee, A. Morris, J. Chem. Soc. Faraday II 1386 (1976).

7. S. I. Weissman, T. R. Tuttle, and E. de Boer, J. Phys. Chem. **41**, 26 (1957).

8. S. I. Weissman and E. de Boer, J. Am. Chem. Soc. **80**, 4549 (1958).

9. W. D. Hobey and A. D. McLachlan, J. Chem. Phys. **33**, 1695 (1960).

10. J. H. van der Waals, A. M. P. Berghuir, and M. S. de Groot, Mol. Phys. **13**, 301 (1967) and **21**, 497 (1971) and references therein.

11. B. Sharf and J. Jortner, Chem. Phys. Lett. **2**, 68 (1968) and references therein.

12. R. Englman, "The Jahn-Teller Effect in Molecules and Crystals," Wiley-Interscience New York, 1972.

13. M. D. Sturge, *Solid State Physics* **20**, 91 (1967).

14. G. Herzberg, "Molecular Spectra and Molecular Structure III," Van Nostrand Reinhold Co., New York, 1966.

15. J. Daintith, R. Dinsdale, J. P. Maier, D. A. Sweigart, and D. W. Turner, Molecular Spectroscopy 1971 (Institute of Petroleum, 1972) p. 16.

16. M. Allan and J. P. Maier, Chem. Phys. Lett. **34**, 442 (1975).

17. M. Allan, J. P. Maier, and O. Marthaler, Chem. Phys. **26**, 131 (1977).

18. J. P. Maier and O. Marthaler, Chem. Phys. Lett. **32**, 419 (1978).

19. J. P. Maier, O. Marthaler, M. Mohraz, and R. H. Shiley, Chem. Phys. **47**, 295 (1980).

20. J. P. Maier, O. Marthaler, and M. Mohraz, Chem. Phys. **47**, 307 (1980).

21. C. Cossart-Magos, D. Cossart, and S. Leach, Mol. Phys. **37**, 793 (1979).

22. C. Cossart-Magos, D. Cossart, and S. Leach, J. Chem. Phys. **69**, 4313 (1978).

23. C. Cossart-Magos, D. Cossart, and S. Leach, Chem. Phys. **41**, 345 (1979).

24. C. Cossart-Magos, D. Cossart, and S. Leach, Chem. Phys. **41**, 363 (1979).

25. T. A. Miller and V. E. Bondybey, Chem. Phys. Lett. **58**, 454 (1978).

26. V. E. Bondybey, J. H. English, and T. A. Miller, J. Am. Chem. Soc. **100**, 5251 (1978).

27. V. E. Bondybey and T. A. Miller, J. Chem. Phys. **70**, 138 (1979).

28. V. E. Bondybey, T. A. Miller, and J. H. English, J. Am. Chem. Soc. **101**, 1248 (1979).

29. T. A. Miller, V. E. Bondybey, J. H. English, J. Chem. Phys. **70**, 2919 (1979).

30. V. E. Bondybey, T. A. Miller, and J. H. English, J. Chem. Phys. **71**, 1088 (1979).

31. T. J. Sears, T. A. Miller, and V. E. Bondybey, J. Am. Chem. Soc. **102**, 4864 (1980).

32. T. J. Sears, T. A. Miller, and V. E. Bondybey, J. Chem. Phys. **72**, 6749 (1980).

33. V. E. Bondybey, T. A. Miller, and J. H. English, Phys. Rev. Lett. **44**, 1344 (1980).

34. V. E. Bondybey, T. J. Sears, J. H. English, and T. A. Miller, J. Chem. Phys. **73**, 2063 (1980).

35. T. J. Sears, T. A. Miller, and V. E. Bondybey, J. Am. Chem. Soc. **103**, 326 (1981).

36. V. E. Bondybey and T. A. Miller, J. Chem. Phys. **73**, 3053 (1980).

37. T. A. Miller, B. R. Zegarski, T. J. Sears, and V. E. Bondybey, J. Phys. Chem. **84**, 3154 (1980).

38. T. J. Sears, T. A. Miller, and V. E. Bondybey, Discuss. Faraday Soc. **71**, 175 (1981).

39. V. E. Bondybey, J. Chem. Phys. **71**, 3586 (1979).

40. V. E. Bondybey, T. J. Sears, T. A. Miller, C. Vaughn, J. H. English, and R. H. Shiley, Chem. Phys. **61**, 9 (1981).

41. V. E. Bondybey, J. H. English, and T. A. Miller, J. Phys. Chem., in press.

42. M. Heaven, T. A. Miller, and V. E. Bondybey, J. Chem. Phys. **76**, 3832 (1982).

43. T. A. Miller and V. E. Bondybey, Phil. Trans. Roy. Soc. (Lond) A, **307**, 617 (1982).

44. T. J. Sears, T. A. Miller, and V. E. Bondybey, J. Chem. Phys. **72**, 6070 (1980).

45. T. J. Sears, T. A. Miller, and V. E. Bondybey, J. Chem. Phys. **74**, 3240 (1981).

46. S. Leach and C. Cossart-Magos, Discuss. Faraday Soc. **71**, 336 (1981).

47. T. A. Miller, Discuss. Faraday Soc. **71**, 341 (1981).

48. C. Cossart-Magos, D. Cossart, and S. Leach, J. P. Maier, and L. Misev, J. Chem. Phys., **78**, 3673 (1983).

49. R. Tuckett, Chem. Phys. **58**, 151 (1981).

50. H. C. Longuet-Higgins, U. Opik, M. H. L. Pryce, and R. A. Sack, Proc. Roy. Soc. (Lond) Ser. A **244**, 1 (1958).

51. M. S. Child and H. C. Longuet-Higgins, Proc. Roy. Soc. (Lond) Ser. A **245**, 33 (1961).

52. H. C. Longuet-Higgins, Adv. Spectros **2**, 429 (1961).

53. W. Moffitt and A. D. Liehr, Phys. Rev. **106**, 1195 (1956).

54. A. D. Liehr, Zeit. Phys. Chem. New Fol. **9**, 338 (1956).

55. A. D. Liehr, Rev. Mod. Phys. **32**, 436 (1960).

56. A. D. Liehr, J. Phys. Chem. **67**, 389 and 471 (1963).

57. A. D. Liehr, Ann. Rev. Phys. Chem. **13**, 41 (1962).

58. C. S. Sloane and R. Silbey, J. Chem. Phys. **56**, 6031 (1972).

59. E. R. Bernstein and J. D. Webb, Mol. Phys. **36**, 1113 (1978).

60. D. Purins and H. F. Feeley, J. Mol. Struct. **22**, 11 (1974) and refs. therein.

61. C. Cossart-Magos and S. Leach, Chem. Phys. **48**, 329 (1980).

62. C. Cossart-Magos and S. Leach, Chem. Phys. **48**, 349 (1980).

63. B. Scharf, R. Vitenberg, B. Katz, and Y. Band, J. Chem. Phys. **77**, 4903 (1982).

64. C. Di Lauro, J. Molec. Spectros. **41**, 598 (1972).

65. W. Moffitt and W. Thorson, *Calcul. der Functions D'Onde Moleculaire*, Edition du Centre National de la Recherche Scientifique **82**, Paris, 1958.

66. C. J. Ballhausen, "Molecular Electronic Structures of Transition Metal Complexes" McGraw-Hill, New York, 1972.

67. B. N. Parlett and D. S. Scott, Math Comput **33**, 217 (1979).

68. K. Raghavachari, R. C. Haddon, T. A. Miller, and V. E. Bondybey, J. Chem. Phys., in press.

69. E. B. Wilson, Jr., J. C. Decus, and P. C. Cross, "Molecular Vibrations," McGraw-Hill, New York, 1955.

Molecular Ions: Spectroscopy, Structure and Chemistry
Terry A. Miller and V.E. Bondybey (editors)
© North-Holland Publishing Company, 1983

PHOTOFRAGMENTATION OF MOLECULAR IONS

Robert C. Dunbar

Chemistry Department
Case Western Reserve University
Cleveland, Ohio
U.S.A.

Ion photodissociation is discussed from the two
points of view of obtaining spectroscopic information and of understanding the processes of
photofragmentation. Photofragmentation yields
information about the identity of fragmentation
products, rate of fragmentation, kinetic energy
and angular dependence. Measuring rates of
photodissociation as a function of wavelength
indicates spectroscopic absorptions of ions, and
structures and rearrangements can be inferred.

INTRODUCTION

Photofragmentation can mean various things, and we will take a
rather broad view. Most narrowly, the word has been used to
designate the experiment, well established for neutral molecules, in
which isolated molecules are photodecomposed, and the products are
analyzed for identity, kinetic energy, and angular distribution; we
will describe numerous such experiments below. These experiments
are indeed much easier with ions than with neutrals, and are being
successfully carried out in both ion-beam and ion-trap configurations. More loosely, we can take it to denote another widespread
experiment in which photodissociation is followed as a function of
wavelength to give information of spectroscopic character. This
leads to the elegant fast-ion-beam experiments realizing some of the
highest spectroscopic resolutions ever achieved, and it leads also
to low-resolution experiments which have displayed the spectral
characteristics of a great variety of previously uncharacterized
ionic molecules. Further afield, we will include multiphoton ion
chemistry in which dissociation is only one of several kinetic
processes occurring. In fact, we will feel free to gather under
this title all the various chemistry and physics which has developed
from irradiating gas-phase ions and observing their dissociation,
(but excluding electron photodetachment of negative ions, and giving
only relatively little attention to diatomic ions, which are quite
extensively discussed in other chapters.)

The intent is not to provide a comprehensive review, since reviews
of various aspects of the topic have been frequent [1-7]. Rather, a
variety of ideas, possibilities and perspectives on the information
available from studying ion photofragmentation will be explored and
illustrated. After some examination of the experimental approaches,
the discussion will be divided into two broad topics: first, the
nature, rates and other details of the various photophysical
processes occurring after photon absorption by an ion; and second,
the spectroscopic information obtained from ion photodissociation,

its relation to ion structures, and the characterization of ion structures and isomerism that it makes possible.

EXPERIMENTAL METHODS

Ion photodissociation can be studied in any instrument capable of forming and observing ions, and many varieties of mass spectrometer have been used. The instruments most widely used have been ion beams, the drift-tube mass spectrometer, the tandem quadrupole mass spectrometer, and the ion cyclotron resonance spectrometer; the rf quadrupole ion trap also has promise, but has been little used for this purpose. In Table I are listed some of the laboratories using each of these techniques, along with the approximate year of inception and an early literature reference. Some characteristics, virtues and drawbacks of these different approaches may be noted briefly:

Ion beams. Ion beam experiments are simple in principle: A beam of mass-selected ions is formed and passed through an interaction region in which light intersects the ion path, either along or perpendicular to the ion flight path. Fragment ions are sorted by mass and detected. Fragment detection at different points along the path permits a straightforward measurement of fragmentation rate of light-induced metastable ions, while analysis of peak broadening arising from fragment velocity spread gives a determination of kinetic energy release. Polarized light also gives angular dependence information when the light beam is perpendicular to the ion beam, if there is sufficient kinetic energy release to give a noticeable angular asymmetry in the fragment ion velocity distribution.

TABLE I

SOME PHOTODISSOCIATION LABORATORIES

Group	Year	Instrument Type	Reference
Washington	1962	rf quadrupole trap	8
Boulder	1964	beam	9
Stanford	1971	ICR ion trap	10,11
Paris	1972	fast beam	12
CWRU	1973	ICR ion trap	13
Amsterdam	1974	fast beam	14
Houston[a]	1974	tandem quadrupole	15
Cal Tech	1974	ICR ion trap	16
NBS	1974	ICR ion trap	17
SRI	1975	drift-tube mass spectrometer	18
Southampton	1976	fast beam	19
Aberdeen	1976	drift-tube mass spectrometer	20
SRI	1977	fast beam	21
Hanscomb AFB	1977	beam	22
Purdue	1978	ICR ion trap	23
La Trobe	1978	tandem quadrupole	24
Florida	1979	ICR ion trap	25
Leyden	1979	ICR ion trap	26
Lyon	1980	fast beam	27
Rochester	1981	beam	28
Swansea	1981	fast beam	29

[a]Moved to Houston from Utah.

Most results have been obtained using homebuilt ion beam instruments. However, since commercial mass sectrometers have most of the needed capabilities, it is attractive to avoid the formidable difficulty of construction of a high quality beam instrument, as at least one laboratory has done [29].

A great deal of interest in ion beam photodissociation has centered around the extraordinary optical resolution possible in Doppler-tuned fast beam studies (reviewed by Moseley and Durup [3,4]). In an ion beam accelerated to kilovolt energies, the phenomenon of kinematic compression reduces Doppler broadening of the optical absorptions by a large factor, and virtually Doppler-free spectroscopy at optical resolutions of tens of MHz has been carried out in this way on a number of small ions. The corresponding kinematic expansion which takes place on decelerating the fast beam to low energies also means that kinetic energy release occurring before deceleration is greatly magnified: This effect, which has long been used to study metastable ion decompositions [30], is readily applied to measuring photodissociation energy releases in the meV range.

Drift-Tube Mass Spectrometers. In these instruments, ions generated by electron impact drift through several cm of neutral gas at relatively high pressure (of the order of 10^{-1} Torr). The strength of the electric field driving this drift is low enough so that the drift velocity is much less than thermal velocity, minimizing ion heating or collisional excitation. Just before the exit aperture from the drift tube the ions are intersected by the laser beam, following which they are extracted and mass analyzed by a conventional quadrupole mass analyzer. Two great virtues of this technique are, first that the ions are very thoroughly thermalized by a large number of collisions so that there is no possibility of excess internal energy; and second that the long traversal of high-pressure neutrals makes it easy to generate a large variety of interesting cluster ions and other species formed in termolecular or very slow bimolecular ion-molecule reactions.

The conditions in such a drift tube can provide a very convincing simulation of regions of the earth's ionosphere. The SRI group in particular has taken advantage of this in a very extensive survey of the spectroscopic and photodissociation characteristics of ions of atmospheric interest, such as for instance small ions clustered to water molecules.

Tandem Quadrupoles. Placing three quadrupole mass filters in tandem gives an instrument of moderate cost and complexity with very high sensitivity to photodissociation. The first quadrupole in the sequence serves to mass select the target ions; the second quadrupole, operated in a mode giving low ion velocity and little mass selectivity, serves to contain the ions and transport them through the light-interaction region; and the third quadrupole serves to mass analyze the product-ion distribution. The very high transmission factor of the quadrupole mass filters contributes to high sensitivity, and allows the use of inefficient ion sources which can be chosen to give thermalized ions or to produce low-abundance species. Also contributing to the high sensitivity is the low ion velocity in the interaction region, which gives each ion a long exposure to light in comparison with most beam experiments.

Ion Cyclotron Resonance Ion Traps. Photodissociation in the ICR ion trap exploits the ability of the trap to contain ions for tens of

seconds or minutes, and then mass analyze the residual ion population. In a standard mode of experiment, ions are produced by an
electron beam pulse through the neutral gas at 10^{-8} or 10^{-7} torr.
Following trapping and optical irradiation for a time which might be
about 3 s, the photodissappearance of the parent ion is ascertained
by an ICR detect pulse or by a Fourier-transform excitation-
detection sequence, and then the cell is purged for a new cycle.
Fig. 1 shows an example of a photodissociation spectrum obtained by
repeating this process at a series of wavelengths: The ion
illustrated here, methylcyclopentadiene parent ion, is in fact
interesting, because the photodissociation spectrum is identical
with that of 1,3-cyclohexadiene parent ion, and no doubt this
reflects complete rearrangement of the former ion to the latter,
which is far more stable [31].

Using the ICR double resonance capabilities to eject selectively ion
species of chosen mass, a variety of more elaborate experiments can
be arranged to sort out details of fragmentation products and their
subsequent ion-molecule reactions and photochemistry.

The spatial anisotropy of the ICR cell can also be used to determine
kinetic energy releases and angular dependences of photodissociation processes [32]. This is based on the fact that ion motion
along the lines of magnetic field is constrained only by a small
electrostatic potential barrier which can be surmounted by ions with
tenths of an eV of kinetic energy, while ion motion in the
directions cutting lines of magnetic field is highly constrained.
Anisotropic photodissociation in polarized light thus leads to
varying fragment-ion loss from the cell, depending on the amount of
kinetic energy release, on the orientation of the light polarization, and on the angular characteristics of the fragmentation.

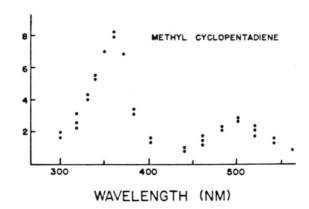

Fig. 1. ICR photodissociation spectrum showing the disappearance of
the parent ion of methylcyclopentadiene as a function of wavelength
[31].

DYNAMICS AND KINETICS OF PHOTOFRAGMENTATION

Fragmentation Products

At the outset of thinking about ion photofragmentation chemistry, it is natural to ask which product ions are formed, and what factors go into determining this. In the absence of other ideas, one looks for the least endothermic product to be favored. For very small ions, this simple idea does not have very convincing predictive power. Thus, although the lowest-energy S^+ product ion is indeed predominant in the photodissociation of CS_2^+ and COS^+ [33], on the other hand the predominant O^+ product of O_3^+ photodissociation [34,35] is about 1.5 eV less favorable than the minor O_2^+ product. Direct, and exceedingly difficult, dynamical calculations on the potential surface seem mandatory for predictive theoretical consideration of such small ions.

Not unexpectedly, larger ions present a more regular picture, and a substantial degree of understanding of the fragmentation patterns of such ions has been the fruit of the long efforts of mass spectrometrists. So it may be surprising that fragmentation patterns in photodissociation have played a minor role in the interests of investigators in this field. Partly this is a question of opportunities still to be pursued: Fragmentation patterns of ions with the well controlled internal energy given by photon excitation, will eventually be a fruitful source of new information about the details of ion fragmentation mechanisms. Partly, also, this reflects a characteristic of photon excitation, that in a typical experiment the photon energy gives the ion only one or two eV of internal energy above the fragmentation threshold, and as a general rule, the lowest-energy fragmentation channel is predominant at these low excitation levels.

Numerous cases are already recorded of photodissociation giving more than one product ion. Butene ion loses $H^{\cdot}$ and $CH_3^{\cdot}$ with nearly equal activation energy [36]. A wavelength-dependent study of this competition showed the applicability of statistical (quasiequilibrium theory) concepts to the photodissociation chemistry, and gave an early impetus to exploiting the precise energy control of photon excitation for controlled ion-fragmentation studies. Among other ions that have shown comparable yields of two or more different product ions at ordinary photon energies, trifluorotoluene and para-fluoro-trifluorotoluene both give an array of different product ions with near-UV photons [37], and protonated mesitylene gives three different products [38].

Some intriguing prospects of multiple-fragment ion dissociations are raised by cases where the product distribution is wavelength dependent because of dissociation from excited electronic states. Morgenthaler and Eyler [39] found that dissociation of $C_2H_5Cl^{+\cdot}$ yields the two products corresponding to loss of $Cl^{\cdot}$ and HCl from the parent (Eq. 1). The branching ratio was strongly wavelength depen-

$$C_2H_5Cl^{+\cdot} \xrightarrow{h\nu} \begin{cases} C_2H_5^+ + Cl^{\cdot} \\ \\ C_2H_5^{+\cdot} + HCl \end{cases}$$

(1)

dent, and was strikingly at odds with the predictions of RRKM theory, giving strong evidence that energy equilibration within the ion does not precede dissociation.   One or more excited electronic states of the parent ion are certainly involved directly in the dissociation.

Brauman's group [40,41] have reported an exciting set of results on photodissociation of butyrophenone ion and related ions: Two products are observed in different wavelength regions, as displayed in Eq. 2, and the two fragmentation paths can be rationalized in a very

$$(2)$$

convincing way in terms of ground and excited electronic state dissociation paths. Although there is a degree of uncertainty about the experimental interpretation [42], these experiments are convincing in their evidence for two products, and this work implies the attractive prospect that photochemical understanding on the basis of molecular electronic state behavior may evolve for ions as it has already for many neutral organic molecules.

Recent photodissociation studies of n-butylbenzene ion make an interesting contrast between beam and trap methods.   Beynon's group [43-47] have reported a series of studies of this dissociation in an ion beam.   The dissociation products are $C_7H_7^+$ (the direct-cleavage product) and $C_7H_8^+$ (a rearrangement product), according to the scheme

$$(3)$$

The $C_{10}H_{16}^+$ ions emerging from the ion source have a spread of internal energies from 0 to 1.4 eV, and many of the more excited ions have suffficient energy to yield the more energetic m/e 91 product. Accordingly, a mixture of photo-produced m/e 91 and m/e 92 is observed, with the relative amount of m/e 91 increasing at higher photon energy.   The ratio of products is in accord with the RRKM calculations of the group.

In our Fourier transform ICR spectrometer, on the other hand, no primary photoproduct m/e 91 is observed at visible wavelengths [48], although the m/e 91 ion is formed as a secondary product according to the scheme

$$C_{10}H_{14}{}^{+\cdot} \xrightarrow[-C_3H_6]{h\nu \text{ (Visible)}} C_7H_8{}^{+\cdot} \xrightarrow[-H\cdot]{h\nu} C_7H_7{}^{+}$$

(134)   (92)   (91)

(4)

(UV)

The relative extent of formation of m/e 92 and m/e 91 as a function of light intensity is in quantitative accord with this scheme. The exclusive formation of m/e 92 as the primary product is just as expected for thermal parent ions. With UV irradiation near 330 nm, however, the m/e 91 product predominates as a direct photoproduct, again as expected. The contrast with the beam results illustrates the important point that the ions in the ICR trap are highly thermalized (by collisional and radiative processes) relative to the ions in typical beam experiments, and the photodissociation processes are typically those of ions with little excess internal energy. Interpretation of beam results must take into account the unthermalized internal energy of the parent ions, as Beynon's group have done, to the extent that it can be estimated from the characteristics of the ion source.

Discussion of ion fragmentation patterns often resolves into the question of whether the statistical theory (quasiequilibrium, or RRKM, theory) [49] provides a useful description. Central to this theoretical description is the expectation that fragmentation will only depend on the internal energy of the activated ion, not on its method of preparation (except that the angular momentum has some, perhaps a major, effect). From this point of view, a set of results [50] comparing fragmentation patterns for photodissociation with patterns for collision-induced dissociation is of considerable interest, although the implications are probably not yet understood. Collision-induced dissociation [51] involves preparation of activated ions with a few eV of internal energy by grazing impact with neutrals, and the resulting fragmentation could be expected to be qualitatively similar to photodissociation of the same ion. Table II reproduces data from Ref. [50] showing the most intense (followed by the second most intense) product ion using these two techniques on some protonated aromatic molecules. The interesting and surprising feature is that loss of H$^\cdot$ is a prominent feature in

TABLE II

PHOTODISSOCIATION AND COLLISION-INDUCED DISSOCIATION PRODUCTS

Protonated Molecule	Photodissociation Product	Collision-Induced Dissociation Product
Benzene	$-H_2$	$-H_2,H$
Toluene	$-CH_4$	$-H_2,H$
Mesitylene	$-H_2,2H_2,CH_4$	$-H,CH_4$
Benzaldehyde	$-CO$	$-H,H_2$
Pyridine	$-H_2$	$-H,H_2CN$
Cyanobenzene	$-HCN$	$-H,HCN$

the CID spectra of all these ions, but is not observed in photodissociation; and there is only poor correspondence between the other product channels as well.   No fragmentation channel yielding radical products is observed in photodissociation, while these are common in CID.   It is possible that these results can be reconciled with statistical theory by invoking extreme angular momentum effects, along the lines suggested by Franchetti et al. [50].   It is also possible, even likely, that some state-specific mechanisms are involved in one or both of the processes for these ions.   The eventual understanding of these effects will be a significant step in understanding fragmentation patterns.

<u>Kinetic Energy Release</u>

We can make a useful division into two types of fragmentation as it affects kinetic energy release, (as well as angular anisotropy). Small ions, (and at least some larger ones as well,) usually dissociate directly from an excited electronic state, in what can be called <u>direct</u> fragmentation.   (Often the excited state is predissociated by intervention of another, repulsive electronic state, as in the predissociation of bound b-state levels of O_2^+ via the f state [3-5].   This will not affect the observed kinetic energy release, although it may conceivably reduce angular anisotropy if the predissociation is slow enough for rotational averaging to occur.) One expects a large fraction of the available energy to be released as kinetic energy in this case, as the fragments recoil along the repulsive electronic potential surface, so that kinetic energies of tenths of an eV are normal.   Angular anisotropy is also normal, since dissociation is fast compared with rotational averaging.   The other case we might call <u>metastable</u> fragmentation, in which the initial excess energy deposited in the ion is degraded by

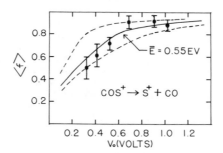

Fig. 2.   The fraction of S^+ fragments which remain trapped in the ICR cell after production by photofragmentation of COS^+ ions by isotropically polarized light near 300 nm, plotted as a function of trapping well depth.

internal conversion to vibrational excitation of the ground electronic state. The hot ground-state ion then dissociates by vibrational predissociation. In the usual situation where the excess energy is distributed among a number of vibrational modes of a polyatomic ion, the dissociation process (often described by the ideas of quasiequilibrium theory [49]) is slow, usually much slower than rotation, and the fraction of available energy partitioned into kinetic energy of the fragments is small, usually less than 10%. So metastable photofragmentation is characterized by kinetic energy release in the meV range, long fragmentation times, and rotationally averaged angular distributions.

The large kinetic energy releases of direct dissociations are readily measured in either ion-trap or ion-beam instruments. As an example of the first, Fig. 2 shows the ion-trap data [52] for COS^+ photodissociation at about 4 eV photon energy, where the total available excess energy in the products is about 1.7 eV. The fraction of product ions retained in the trap increases with increasing electrostatic trapping voltage in a predictable way [52,53]: The solid theoretical curve shows the prediction assuming .55 eV of kinetic energy in the S^+ product ion, while the dashed curves are plotted assuming .30 and .80 eV to suggest the precision of the determination. The conclusion that S^+ is formed with $.55 \pm .2$ eV of kinetic energy rules out its formation in the excited doublet state, and indicates a spin-orbit-coupled crossover to the quartet manifold during dissociation.

Ion beams, and particularly kilovolt energy beams, benefit from the fortunate advantage of kinematic velocity expansion, allowing kinetic energy releases of as little as 10^{-5} eV in the center-of-mass frame to be easily observed in the laboratory frame. They are thus preeminently suited to studying metastable fragmentation processes, as well as direct fragmentations. An illustration of a highly structured photofragment energy spectrum is shown in Fig. 3 [54]. The bilateral symmetry around W=0 arises from the symmetric effect of fragment ion velocity released along the forward or back-

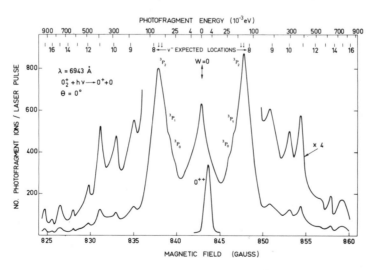

Fig. 3. Laser photofragment spectrum of O_2^+ at 694.3 nm [54].

ward beam direction, and the great expansion of the energy scale
near $W=0$ is evident (W being the total kinetic energy of the
photofragments in the center-of-mass frame). For the O_2^+ case
illustrated, the upper spectroscopic state is the repulsive $f^4\Pi_g$
state, and the progression of vibrational peaks in the kinetic
energy spectrum corresponds to Franck-Condon transitions to this
repulsive curve from the series of v'' vibrational levels of the
lower spectroscopic state, $a^4\Pi_u$. The position and spacing of these
Franck-Condon peaks lead to plots of the upper- and lower-state
potential curves such as have been made for several diatomic ions,
including Ar_2^+ and Kr_2^+ in addition to O_2^+.

1,3-Butadiene ion would be a natural candidate as a prime example of
a metastable photodissociation system, both because the optically-
allowed Π excited state seems almost certain to be a bound state,
and also because the kinetic energy released in dissociative
photoionization is in accord with RRKM predictions, having the low
values expected for metastable fragmentation [55]. Thus the recent
ion-beam photofragmentation results of Preuninger and Farrar [28]
are a considerable surprise. The most likely kinetic energy release
for the process

$$C_4H_6^{+\cdot} \xrightarrow{\ h\nu\ } C_3H_3^+ + CH_3\cdot$$

$$(5)$$

in 1,3-butadiene ion with light at 514.5 nm (.12 eV excess energy
for ground-state ions) was found to be over 50 meV, with substantial
depletion of low-kinetic-energy fragments. The only uncertainty in
interpreting this result is in the assignment of the internal energy
of the ions emerging from the ion source, which must be estimated
from the photoelectron spectrum, a plausible but not well tested
procedure. Aside from this one residual uncertainty, these results
show unambiguously that too much kinetic energy is released in this
photodissociation to be consistent with metastable-type fragmentation, and the authors suggest that the excited electronic ^2A_u state
couples to a repulsive potential surface in a direct fragmentation
process. Similar, and also surprising, results were observed for
three of the other $C_4H_6^+$ isomers.

More in line with one's preconceived expectations are the observations of Beynon's group on the kinetic energy releases in photodissociation of a number of aromatic molecular ions [56,57]. In
most cases the kinetic energy release is linear in the photon
energy, as expected for RRKM behavior. The ions emerge from the ion
source with substantial excitation, as indicated by the appearance
of metastable fragmentations (in the absence of light) releasing
25-100 meV of kinetic energy. Taking the large implied initial
internal energy into account, however, the kinetic energy releases
from photodissociation have reasonable magnitude for metastable-type
dissociation.

In the ion beam photodissociation experiments of Beynon's group, and
perhaps of Preuninger and Farrar as well, the kinetic energy
releases confirm that the photon serves to add an increment of
energy to an ion already containing a substantial amount of excess
internal energy. In interesting work on the SRI fast beam [58,59],
it was found that even the small energy of an IR photon from a CO_2
laser is sufficient to produce photodissociation in highly excited
ions, and these one-photon IR photodissociation processes give

TABLE III

OBSERVED AND RRKM QUANTITIES FOR IR ONE-PHOTON FRAGMENTATION

	METASTABLE FRAGMENTATION		IR-INDUCED FRAGMENTATION	
	ENERGY RELEASE (MEV)	AVERAGE LIFETIME (μSEC)	ENERGY RELEASE (MEV)	AVERAGE LIFETIME (μSEC)
CF_3I^+				
OBS.	—	—	4.4	<<1
CALC.	—	—	24	<<1
$C_2F_5I^+$				
OBS.	2.8	26	10	2.2
CALC.	2.5	25	12	2.8
$C_3F_7I^+$				
OBS.	10	>100	14	5.5
CALC.	10	170	21	5.5
$C_6H_5I^+$				
OBS.	29	>100	28	N.A.
CALC.	39	250	50	30

insight into the characteristics of the more widely studied multi-photon IR processes. In the experiments at SRI, ions produced by electron impact are dissociated by infrared radiation collinear with the ion beam, with an ion-laser interaction time so short that absorption of at most one IR photon is possible. Thus only those ions are dissociated which already have internal energy nearly equal to the dissociation energy. It may be said that this photodissociation corresponds to the <u>last</u> photon in a multiphoton IR photodissociation experiment.

The ion-beam experiment yields a wealth of information about the photofragmentation process, including kinetic energy release and fragmentation rates. Table III shows such data for a few polyatomic ions which have been carefully analyzed. Corresponding metastable fragmentations (laser-independent) are also tabulated. The calculated values are from RRKM calculations using the assumptions of statistical theory, which imply complete randomization of internal energy prior to dissociation. For the larger ions, the RRKM model evidently gives a very satisfactory description of the fragmentation process. However, for CF_3I^+ the kinetic energy release is much too small, and it is concluded that this dissociation occurs by a non-statistical mechanism. The slightly counterintuitive observation that $C_3F_7I^+$ releases more kinetic energy than $C_2F_5I^+$ is an interesting feature of the beam experiment, arising because many of the $C_3F_7I^+$ ions arrive at the photon interaction region with substantially more internal energy than the dissociation threshold, while such above-threshold ions are more nearly depleted with $C_2F_5I^+$. The lifetime values given in the table are obtained by comparing the amount of photoproduct in the beam at two different points downstream from the laser-interaction region. In practice, because of the spread of ion internal energies, the dissociation is not characterized by a single "lifetime," but by a range of different dissociation rates, of which the reported value represents an average. The "calculated" values are based on an extensive simulation of the conditions of the beam, using RRKM rates. The excellent accord of both energy release and dissociation rate observations with RRKM predictions makes a good case for metastable-type photofragmentation behavior in these cases.

Angular Dependence

Some of the most specific and useful detailed information about photofragmentation dynamics comes from the angular distribution of fragment trajectories; this angular dependence is observed relative to the reference direction established by the polarization direction of a polarized light source. In the simplest case of a direct fragmentation of a diatomic, the optical transition moment must be either parallel or perpendicular to the molecular axis, while the fragments always separate (neglecting corrections for rotational effects) along the axis. So the fragmentation is always characterized by a preference for fragment paths either along or perpendicular to the polarization direction, and from this the symmetry of the optical transition follows at once. For polyatomic

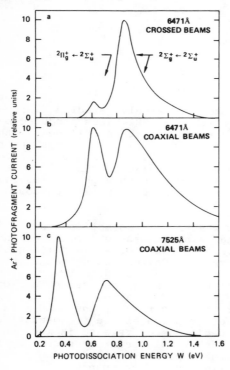

ions, the situation is of course more complex, but similar in principle, and in fact ions like the methyl halides can be considered as pseudo-diatomics. While any photodissociation will have an angular distribution in principle, measurements have so far been limited to cases where the fragment kinetic energy is tenths of an eV, so that the oriented part of the fragment motion stands out easily from random thermal contributions.

One excellent way to observe these effects is in an ion beam with very high angular beam resolution, crossed by a perpendicular laser. Low angular acceptance for fragment ions will mean that only fragments ejected approximately parallel to the beam path will be collected, those flying off perpendicular to the beam being lost from the accepted beam trajectory. The SRI fast beam, with its excellent angular resolution, exemplifies this capability, as shown in Fig. 4 [60]. The dissociation of Ar_2^+ is observed to proceed with comparable probablitiy through two repulsive excited-state potential curves. Dissociation through the $^2\Sigma_u^+$ state has parallel polarization of the transition moment, and gives greatest collected product abundance with light polarized along the beam, as in Fig. 4a. The $^2\Pi_g^+$

Fig. 4. Photofragment kinetic energy release spectra for Ar_2^+ with light polarized along (a) and perpendicular to (b,c) the direction of the ion beam [60].

state, on the other hand, has perpendicular optical polarization, and gives most product with light polarized perpendicular to the beam, as in the coaxial-beams geometry of Fig. 4b and 4c. As expected, both photofragment peaks shift to lower energy at the longer wavelength of Fig. 4c, as there is less excess energy deposited in the ion. Using this angular information to identify the excited states, along with the kinetic-energy release information also contained in spectra like those of Fig. 4, potential energy diagrams for Ar_2^+ and Kr_2^+ have been refined [60,61].

The ion whose angular properties have been most thoroughly studied by ICR ion trap methods is methyl chloride ion [62]. The kinetic energy release for 360 nm light is more than half an eV, making excellent results possible with the anisotropic ion trap technique. Fig 5. shows some results. On the left is plotted the anisotropy index I as a function of angle of orientation of the light polarization. I is defined by I = [(perpendicular dissociation)-(dissociation at angle ϕ)]/(perpendicular dissociation). On the right is a plot of I versus the electrostatic trapping well depth. In both plots, the solid curve is the theoretical expectation based on kinetic energy release of .58 eV and parallel orientation of the optical transition moment. As can be seen, the fits of data to theory are satisfactory. The large kinetic energy release indicates that photodissociation is of the direct type: as noted above, Morgenthaler and Eyler's results on ethyl chloride ion also indicate dissociation from an excited electronic state, which may thus by a general feature of photodissociation of the $C-Cl^+$ bond. The orientation of the transition moment found in methyl chloride indicates participation of the $^2E \rightarrow \, ^2E$ transition, even though there is evidence from other techniques to suggest involvement of the excited ^2A state in the dissociation. It seems most likely that optical excitation to ^2E is followed by predissociation through ^2A.

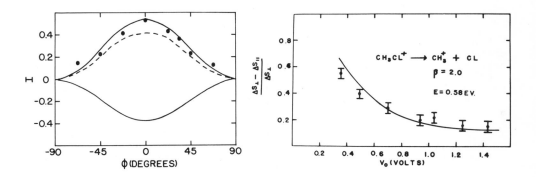

Fig. 5. (Left side.) Photofragmentation anisotropy index I as a function of polarization angle for methyl chloride ion irradiated at 360 nm. The upper solid curve is the calculated curve assuming perfect parallel orientation of the transition moment. (Right side.) Anisotropy as a function of trapping well depth for light polarized parallel to the magnetic field. The calculated fit is for kinetic energy release of .58 eV and parallel orientation of the transition moment [62].

Fe(CO)$_4^-$ makes a particularly interesting application of the ICR
angular dependence technique [63].   It is somewhat surprising that
the ion dissociates under irradiation, as electron photodetachment
has been the more common dominant outcome observed in anions.   But
most interesting, the dissociation process is observed to occur with
substantial angular anisotropy.   This indicates a direct type of
dissociation process with substantial kinetic energy release; and it
proves that the ion is not tetrahedrally symmetric, as was indeed
predicted on the basis of Jahn-Teller distortion of a potentially
tetrahedral species.

Relaxation of Excited Ions

Questions of lively recent interest concern the mechanisms and rates
by which ions with vibrational excitation dissipate their energy.
The ICR ion trap is preeminently well suited to such studies,
because of its long trapping time scale, its control of collision
rates from essentially zero up to high rates, and the wide
flexibility it offers.   Relaxation experiments imply three distinct
steps: first, creation of the excited ions; second, the relaxation
period; and third, a probe of remaining excitation.   Various ideas
have appeared for implementing these three steps.   The excitation
may be supplied by initial ionization, or it may be added by
visible or IR irradiation.   The relaxation may be by collisionsl, or
by infrared fluorescence (the whole area of relaxation of electronic
excitation by visible-UV fluorescence will not be touched on in this
chapter.)   And the final probe may utilize hot-band changes in the
ion spectrum, or it may depend on internal energy enhancement of
photodissociation.

In one recent notable experiment, Morgenthaler and Eyler [64]
followed the collisional relaxation of excited 1,3,5-hexatriene
ions.   The excited ions were formed by the electron-impact ionization, and were allowed to relax in a collisional bath gas for a
period of tens of ms.   Then the ions were probed by a dissoociating
pulse from a pulsed dye laser.   The probe made use of the v"=1 to
v'=Ø hot band, which is resolved in the hexatriene ion spectrum, and
has a peak near 646 nm.   As the hot ions cool, depleting the v"=1
vibrational state, the intensity of the hot-band peak in the
photodissociation spectrum decreases, while the intensity of the Ø-Ø
transition at 619 nm remains relatively constant.   This method gave
nice decay curves for the energy relaxation process, and indicated
that collisions of the ion with its parent neutral gave excitation
quenching on nearly every collision, while about 5Ø collisions with
argon were required for quenching.

A little more elaborate are two-photon experiments in which both the
excitation and probe steps involve photon absorption.   This was
among the many new possibilities for ion photochemistry which have
followed Freiser and Beauchamp's discovery in 1975 [65] that
continuous-wave light of modest power could drive two-photon ion
dissociation according to the sequential kinetic scheme

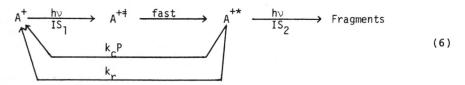

(6)

where A^+ is an electronically excited ion; A^{+*} is a vibrationallly excited ground-state ion; I is the light intensity; S_1 and S_2 are optical absorption cross sections; P is the neutral pressure; k_c is the bimolecular collisional quenching rate constant; and k_r is the infrared radiative relaxation rate constant. The photon absorption cross sections S_1 and S_2 need not be equal, but it has often been assumed that since they both refer to the same electronic excitation they will not differ by a large factor; this greatly simplifies interpreting experiments, but one must be cautious not to draw conclusions which would be sensitive to a very possible difference of a factor of two or three. van Velzen and van der Hart [66] have shown that given a sufficient amount of high quality kinetic data, an analysis separating S_1 and S_2 is possible, but this has not been attempted for many ions. Our interest in this two-photon chemistry has centered on the access it gives us to the two relaxation processes, represented by k_c, the collisional relaxation rate constant, and k_r, the radiative relaxation rate constant.

If we neglect radiative relaxation for the moment, the extent of dissociation after time t is easily written from the solution of the kinetic scheme (6):

$$\frac{1}{D} = \frac{IS_1 + IS_2 + k_c\,P}{I^2 S_1 S_2 t} \tag{7}$$

Since 1/D appears as a linear function of pressure, we can find the slope and intercept of a plot of 1/D vs. P, giving two equations for the unknowns k_c and IS (assuming $S_1=S_2=S$). The collisional relaxation rate for photo-excited bromobenzene ions was measured in this way using a number of different gases as collisional quenching partner [67]. More recent experiments on iodobenzene ion [68] incorporate an improvement: by using ICR line broadening techniques to measure the ion-neutral collision rate, and two-photon relaxation techniques to measure the collisional relaxation rate in the same instrument, the relaxation rate can be compared with the collision rate free of uncertainties in pressure measurements and in collision-rate calculations from theory. Table IV shows the values obtained.

The most efficient quenching is provided by the parent neutral, which takes two or three collisions for iodobenzene ion; this is less efficient than the quenching measured for bromobenzene ion in its parent neutral, where quenching was found to occur as fast as the orbiting collision rate. Assuming that energy transfer is efficient only in orbiting ion-molecule collisions, the extreme limit of efficient quenching will occur when the internal energies of the ion and neeutral are fully equilibrated within each orbiting collision complex. Making the simple calculation based on assuming that the iodobenzene ion and the neutral reach a common vibrational temperature at each collision, we may find the number of collisions needed to bring the iodobenzene ion from its initial photo-excited energy of 2 eV down to below 0.4 eV, which is the energy at which one-photon dissociation is just possible. Calculating this equilibration limit, iodobenzene bath gas requires two to three collisions, cyclohexane about two collisions, and methane about 15 collisions. From Table IV it is seen that parent neutral as bath gas matches this full-equilibration limit. This is not surprising,

TABLE IV

Bath Gas	Measured Laser Quenching Rate $(x10^{-10}\ cm^3/molec-sec)$	Collisions Needed for Quenching
C_6H_5I	7	2.5
CH_4	.08	50
cyclo-C_6H_{12}	0.8	7

since the match of vibrational levels, along with the possibility of efficient symmetric charge exchange, make efficient energy transfer very likely within the collision complex. On the other hand, quenching by either cyclohexane or methane bath gases is about a quarter as fast as the full-equilibration prediction: these collision complexes are far from reaching statistical sharing of internal energy between ion and neutral.

These quenching results are in line with the findings of Brauman's group, looking at collisional quenching in the multiphoton infrared dissociation of anions. These studies are similar in approach to the work of Morgenthaler and Eyler described above, in that ions are formed initially with excess vibrational energy, are allowed to relax, and are then probed by photodissociation. However, the probe step in this case is a pulse of infrared radiation from a CO_2 TEA laser which induces multiphoton IR dissociation of the ions. Since this multiphoton process is affected by ion internal energy, it provides an effective energy probe. In these studies the trapping time was 1 s, and pressures ranged down to about $5x10^{-8}$ Torr, so that radiative relaxation processes were accessible as well as collisional processes.

Collisional relaxation of CF_3O^- requires between 7 collisions (CF_2O neutral) and 30 collisions (N_2 neutral) [69]; Collisional quenching of CH_3OHF^- in $HCOOCH_3$ requires tens of collisions [70]. Radiative quenching of CF_3O^-, on the other hand, was quite fast, at 20 s^{-1} [69]: We will see below that the larger ions we have studied radiate their energy considerably more slowly.

If the pressure of neutral gas is in the low 10^{-8} Torr range, coollisions become rare even on a time scale of seconds, and even slow infrared radiative relaxation becomes the dominant process. The ICR ion trap has proven to be uniquely effective in working on the long time scales involved, and two-photon kinetic analysis via Eq. (6) again provides convenient access to the relaxation rate constant. van Velzen and van der Hart [66] first showed the possibility of doing this in their pioneering study of bromobenzene ion two photon dissociation at extreme low pressure. However, easier analyses and more reliable results can be obtained using pulsed light sources to drive the two-photon process.

When the light source is not continuous, but delivers light in spaced pulses, an interesting interplay develops between the occasional photoexcitation of the ions and their steady relaxation by collisions and by infrared fluorescence. Our group [71] explored the effect of changing from continuous to repetitively pulsed radiation in the two-photon dissociation of iodobenzene ion. The effect is simply understood in a qualitative way: If the laser pulse repetition rate is fast compared with all the ion relaxation rates, then the light source excites ions just as if it were actually continuous, and the time-average light intensity is the important variable. But if the pulse rate becomes slow compared with the fastest relaxation, then the ions can relax between each pulse, and the experiment is equivalent to a series of relaxation-free single pulse experiments for which the pulse energy and number of pulses, but not the pulse shape or repetition rate, are the important variables.

Recently results of higher precision have been obtained for bromo-benzene ion, using a cw laser chopped at rates varying from 1 to 12 s^{-1}, and the radiative and collisional parts of the relaxation have been clearly separated [72]. Using a chopper, the average light intensity is conveniently kept constant, while the repetition rate (and pulse length) are varied, so that the dissociation should be pulse-rate-independent at high pulse rates and low relaxation rates. Fig. 6 shows the strong dependence of dissociation on pulse rate at rather high pressure, compared with the theoretical curve calculated using the collisional relaxation rate already determined from the quenching studies. At high pulse rates (4 s^{-1} and above) the dissociation is strongly pressure-dependent in the $2-30 \times 10^{-8}$ torr region. However, at 1 s^{-1} there is little pressure dependence, and it is clear that at this low repetition rate, relaxation by infrared radiation is largely completed between each pulse. At 2 s^{-1} and 3 s^{-1} the pressure dependence is intermediate, indicating that at these pulse rates radiative relaxation is partial, allowing some collisional relaxation in addition to the radiative relaxation. Using a radiative relaxation rate of around 2 s^{-1} the basic kinetic scheme Eq. (6) gives a good account of the pressure and pulse rate

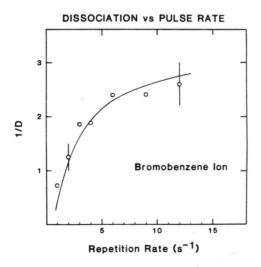

DISSOCIATION vs PULSE RATE

Bromobenzene Ion

1/D

Repetition Rate (s^{-1})

Fig. 6. Photodissociation of bromobenzene ion as a function of pulse rate at 2×10^{-7} torr. The solid line is the theoretical curve assuming a radiative relaxation rate of 1.5 s^{-1}.

dependences in this experiment, and firmly establishes the radiative
rate k_r in the range 1-3 s^{-1}.   The various experiments described in
this section are noteworthy, in demonstrating the remarkable new
capability one has of directly observing the energy relaxation of
collision-free isolated molecular species on a time scale of
seconds.

Multiphoton Photochemistry With Infrared Photons

An interesting variation on the visible two-photon dissociation
idea gives us access to some new aspects of the infrared spectroscopy and photochemistry of ionic molecules.   An infrared laser can
be used to deliver the energy represented by the S_1 step in Eq.   6,
following the new scheme

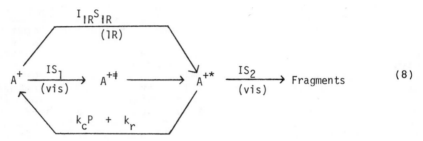

This effect is clearly observed as a strong increase in dissociation
(typically up to a factor of three increase) when a CO_2 laser of
several Watts power is added to the visible irradiation [73].

We know that the IR radiation serves to to give initial excitation
to the ions, rather than serving to push excited ions over the brink
of dissociation, from recent sequential-excitation experiments [74]:
If the ions (iodobenzene or bromobenzene ions were studied) are
first irradiated with the IR laser and then immediately with the
visible laser, the IR enhancement is practically undiminished; while
if the visible-laser irradiation precedes the IR irradiation, no
enhancement is seen, nor does IR irradiation alone have any effect.
(Wight and Beauchamp [75] have reported a very small IR enhancement
effect in cyanobenzene ion which seems to require simultaneous
irradiation by both lasers, and very likely is a different sort of
effect than the large effect discussed here).

For iodobenzene ion, which has been studied the most extensively,
photodissociation data were taken as a function of pressure and of
the intensities of both lasers. Given the five independently
adjustable rate constants in the kinetic scheme (8) (with the added
complication that these constants may be dependent on the internal
energy level of the ion), it is not surprising that agreement can be
forced between theory and experiment, but in fact we found only a
narrow choice of these parameters which were both reasonable and
gave a satisfactory fit of the kinetic equations to all of the data
[76].   Based on extensive exploration of the parameter values,
several features of the iodobenzene kinetics were concluded to be
essential features of a successful model:
      Most or all of the ions are susceptible to both infrared and
   visible up-pumping processes at the wavelengths used.
      The infrared photoexcitation has a bottleneck somewhere below
   4000 cm^{-1}, and is also too slow to compete with relaxation
processes near the dissociation threshold at 20,000 cm^{-1}.

Infrared radiative relaxation is significant at pressures of a few times 10^{-8} torr, but collisional relaxation dominates above about 10^{-7} torr. The radiative relaxation time constant is around 500 ms.

The visible photoexcitation cross section at 610 nm decreases by a factor of 2 to 5 with increasing internal energy from zero to 20,000 cm^{-1}.

In view of the high light intensities used to effect infrared multiphoton photodissociation of neutral molecules, the observation of Beauchamp's group that a modest power from a cw infrared laser could easily dissociate some fairly weakly bound ionic species in the ion trap [7] was widely unanticipated. However, this chemistry is not actually surprising on reflection, because under the unique almost collision-free conditions of the ion trap experiment, energy accumulation by absorption of successive IR photons into the ion is counterbalanced chiefly by the slow process of infrared radiative relaxation [77]. If the laser has sufficient power to pump the ion up to the dissociation threshold within perhaps a few hundred ms, dissociation will be observed. As Jasinski et al. [78] showed for two representative ions, $C_3F_6^+$ and diethyl ether proton-bound dimer ion, pulsed-laser and continuous-laser IR dissociations are qualitatively similar, reflecting the same basic kinetic steps on their different time scales.

PHOTODISSOCIATION SPECTROSCOPY OF IONS

Photodissociation spectroscopy observes the optical absorption peaks by measuring photodissociation as a function of wavelength, with the simple idea that dissociation can only occur at wavelengths where light is absorbed. This straightforward plan has led to a revolutionary development of knowledge of gas-phase ion spectroscopy: a decade ago spectroscopic properties of virtually no ions were known, while by now hundreds of ion spectra have been observed, some at high resolution and in great detail, others in a more limited way. In concert with the parallel development of emission-spectroscopic methods, this has amounted to the rapid emergence of broad, deep and systematic understanding of the spectroscpy of gas-phase ions. We will single out for attention ion-beam spectroscopy at extreme resolution; lower-resolution systematic spectroscopy of organic ions; infrared spectroscopy; and the identification of isomeric ion structures and the rearrangements among them.

Ion Spectra at Extreme High Resolution

The dramatic new spectroscopic capabilities of Doppler-tuned ion beam methods has attracted wide attention. The phenomenon of kinematic compression in a fast beam reduces the thermal velocities of the ions to total insignificance in the coordinate frame moving with the ions, so that a laser beam travelling coaxially with the ion beam sees essentially stationary ions. This makes it not just possible, but easy to observe spectroscopic transitions free of thermal Doppler broadening. Furthermore, with a stabilized, fixed-wavelength, narrow band laser, the spectrum is readily scanned over an appreciable wavelength range by Doppler tuning through varying the velocity of the ion beam. Fig. 7 shows an example of the remarkable data obtained principally at the fast beam instruments at SRI, Southampton and Lyon during the last half-dozen years. In this

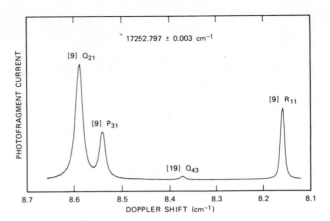

Fig.7.   Doppler-tuned ion beam spectrum of the predissociating
transition O_2^+ $(a^4\Pi_u, v''=4)$ $\longrightarrow$ $(b^4\Sigma_g^-, v'=4)$.   Nominal   ion beam
kinetic energy was 3600 eV, with laser wavelength 17252.797 cm^{-1}
[79].

particular   case,   the   linewidths   of   200-300 MHz are real natural
linewidths, the instrumental resolution being of the   order   of   100
MHz (.003 cm^{-1}, or two parts in 10^7). O_2^+ has received the bulk of
the attention, with hundreds   or   thousands   of   peaks   having   been
observed, giving extremely high accuracy for the resulting potential
energy curves. CH^+ has also been successfully studied; and Ref. [5]
gives further access to the literature on this technique.

Spectroscopy of Organic Ions

Many photodissociation spectra have been reported at low   resolution
(about   10   nm),   of   which a few will be discussed below.   Spectra
taken at   laser   resolution   (better   than   1   nm)   are   especially
interesting because of the resolution of vibrational fine structure,
but   these   have   been rarer, partly because of the difficulty of the
experiments,   and partly   because   the   vibrational   structure   often
appears   to   be   smeared by a combination of lifetime broadening and
spectral congestion.   Among ions whose laser   spectra   have   shown
notable vibrational structure we might mention the well known cases
of CH_3I^+ [80], hexatriene ion [81], and CO_3^- [82],   and   the   more
recently reported $C_3H_5Cl^+$ [83] and $Fe(CO)_3$ [84].

Some   recent experiments [85] have suggested the possibility of what
might   be called "proton-labelling" spectroscopy, in which   a   proton
is   attached   to a gas-phase molecule, to obtain a photodissociation
spectrum of the protonated species.   Assuming that the spectroscopic
properties of the protonated molecule are   interpretable   in   a   way
similar   to   neutral   molecules, the useful capabilities of optical-
spectroscopic identification of   samples   developed   long   ago   for
neutral   molecules   might   be applied to the extremely small samples
which can be handled in mass spectrometers.   These experiments   are
based   on   the   observation   of Freiser and Beauchamp [86] that in a
variety   of   unsaturated   organic   molecules   the   photodissociation
spectrum   of   the   protonated molecule closely resembles the optical
spectrum of the neutral, but with   a   shift   to   longer   wavelength.

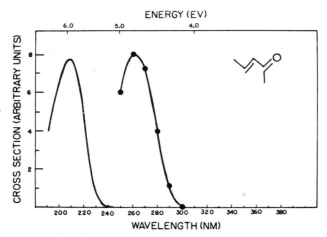

Fig. 8. Photodissociation spectrum of the protonated ion of 3-penten-2-one (——•——) compared with the neutral absorption spectrum (——), showing a red shift of 1.17 eV [85].

They related this observation to the difference in ground- and excited-state proton affinities via a Forster-cycle argument.

The effects of substituents on the peak positions in the neutral spectra of acrolein derivatives are very well characterized and understood. In Fig. 8 is shown an illustration from our results on the protonated acroleins, exhibiting the typical red shift upon protonation. A series of protonated acrolein derivatives showed substituent effects on the peak positions closely similar to the neutral-molecule patterns. The extent of red shift depends on the nature of the pi system, and is about 1.15 eV for simple acroleins, 1.35 eV for triene analogs, and 1.6 eV for aromatic derivaties. Within each of these classes of compound, the peak positions varied with substituent in the same way as with the neutrals. It seems reasonable to hope that the well worked out spectroscopy of many neutral chromophoric groups will be directly applicable to interpreting photodissociation spectra of proton-labelled molecules in this way.

The task of relating ion structure and spectroscopy, through a systematic connection between an ion's optical absorptions (reflected by its photodissociation spectrum) and its electronic structure, orbitals and chromophores, is challenging and interesting. These questions will be touched on to some extent below; more of our own reflections on them can be found in the references (for instance [881,87,88-91]).

Infrared Spectroscopy

From the point of view of the ion chemist, an infrared spectrum of an ion would be of even more value than a visible spectrum for identification and structure characterization. Unfortunately, this is a very demanding request. Part of the problem is the lack of intense, broadly tunable light sources; and part, the small energy contained in an IR photon, which makes IR photodissociation a difficult proposition. However, some ideas and some experiments have appeared, and progress can be charted toward the goal of chemically useful IR spectroscopy of ions. Following our arbitrary restriction to dissociation experiments, we will not discuss, but should note, the promising IR spectral information also being obtained by IR multiphoton electron detachment of anions, first characterized by Brauman's group [92] and recently pursued as an ion structural characterization tool by Beauchamp's group [93].

The IR-visible two laser experiment discussed above offers one route to IR spectroscopy, and is attractive in that most of the dissociation energy is supplied by the visible photons, so that absorption of one or a few IR photons by the ion may produce an observable photodissociation enhancement. The IR enhancement effect depends on the absorption of IR photons by near-thermal ions, as we have seen, so the wavelength dependence of the effect gives an indirect route to finding the IR absorption peaks of thermal gas-phase ions.

Fig. 9 shows an example of the spectra we have been able to obtain [94]: the spectrum does indeed show wavelengths of high and low IR enhancement, and the main obstacle to measuring a complete IR spectrum is the sparse wavelength coverage of the CO_2 laser. Such IR-enhancement spectra of several halobenzene and halotoluene ions all show sharply wavelength-dependent effects. In most cases there is strong IR absorption in the ion at wavelengths longer than any of the ring vibrations in the coresponding neutral molecules. This seems to reflect some red shift of the ring vibrational frequencies upon ionization, not surprising since it is a bonding ring pi electron which is removed.

IR multiphoton dissociation directly by IR irradiation is a straightforward approach to the problem. $C_3F_6^+$ is an outstanding example of this possibility. One of the first ions studied by Beauchamp's group, it gives an excellent spectral peak in the CO_2 laser tuning range, at 1043 cm^{-1} [7]. This is very close to the C-F stretch at 1037 cm^{-1} in the neutral molecule. So this ion is evi-

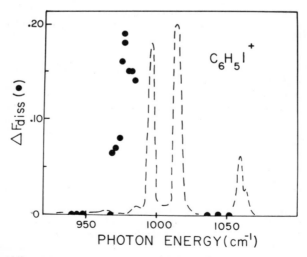

Fig. 9. IR enhancement spectrum of iodobenzene ion. The visible wavelength was 610 nm.

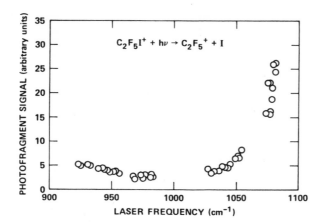

Fig 10. Fast-ion-beam one-photon IR spectrum of highly vibrationally excited $C_2F_5I^+$ [59].

dently a favorable case in which the IR multiphoton dissociation spectrum gives a strong, sensible and potentially useful spectrum.

A most interesting study of Bomse and Beauchamp [95] compared the IRMPD spectra of three isotopic analogs of the dimethyl chloronium ion $(CH_3)_2Cl^+$. Only the $(CD_3)_2Cl^+$ ion gave dissociation, showing a broad maximum near 980 cm^{-1}. It is a matter of speculation why the $(CH_3)_2Cl^+$ and $(CH_3)(CD_3)Cl^+$ ions failed to dissociate. It is possible, but not very believable, that they do not absorb light in the laser tuning range; and Bomse and Beauchamp argue instead that the higher vibrational state density of $(CD_3)_2Cl^+$ gives easier intramolecular energy flow and faster photodissociation.

Infrared photodissociation or enhanced photodissociation experiments like those described above yield information about the lower part of the IR excitation ladder, since the important IR absorption events are normally those involving ions far below the dissociation threshold. The one-photon IR photodissociation experiment in the SRI ion beam described above [58,59] is complementary, in the sense that the single IR photon is absorbed by an ion already containing the maximum possible internal excitation. Thus a spectrum of this process as a function of wavelength gives a spectroscopic characterization of very hot ions.

As is illustrated in Fig. 10 for a typical polyatomic ion, the spectra typically show well defined peaks in the spectra, and it is certainly true that the presence of one eV or more of internal energy in the ion does not destroy the IR spectral peaks of the cool ions. This is easily understood in a nearly-harmonic picture of polyatomics, because in such large molecules the individual vibrational modes are not highly excited. The v=0 to v=1 transitions of the IR active modes are still dominant, although possibly broadened and shifted by anharmonic perturbations.

One molecule for which a comparison has been made between the one-photon hot-ion spectra of the present experiments and a cool-ion multiphoton IR dissociation spectrum is $C_3F_6^+$. The multiphoton IR spectrum shows its peak at 1043 cm^{-1}, as noted above, while the one-photon beam experiments show the peak red-shifted to 1030 cm^{-1}, no doubt due to anharmonic coupling between the IR active mode and the other excited modes of the hot ion.

ION STRUCTURES AND REARRANGEMENTS

Among the areas of widest interest and application of the capabilities of photodissociation spectroscopy has been the linking of spectra with ion structures. A frequent point of view takes the spectra as reliable fingerprints of the structures of different isomeric ions; more demanding and ambitious efforts attempt to assign specific structures to ions, based on systematic understanding of their spectral features.

A key point about spectroscopic ion structure determination has been made often, but leads to such frequent confusion that it can be emphasized yet again. The parent-ion photodissociation spectrum characterizes the structures of the intact parent ions at the moment of irradiation. Ions which rearrange and fragment in the initial ionization process are not relevant--since it is exactly these latter which are characterized by the common techniques of mass spectral fragmentation studies, it is important to remember that it is the non-decomposing ions whose structures are probed by photodissociation. A further point to bear in mind is that rearrangements which are induced by the absorbed photon are also not relevant in interpreting the photodissociaton spectrum of the parent ion: once the ion has absorbed a photon of sufficient energy it is going to dissociate regardless of how it may rearrange during the process. (This last statement may not always be true, but few exceptions are yet known. Of course, also, ion rearrangements following photon absorption can be and have been followed by isotope-labelling techniques.)

These ideas were nicely illustrated by one of the very first ion-structure studies [96] on $C_7H_8^+$ ions. Mass spectral studies indicate that for a large variety of C_7H_8 neutral precursors the mass spectral cracking pattern is governed by an initial rearrangement of the parent ion through a common intermediate structure. However, probing the non-decomposing parent ions from toluene, cycloheptatriene and norbornadiene by photo- dissociation showed that the structures do not rearrange. One way to express this result is to say that the activation energy for rearrangement on the $C_7H_8^+$ potential surface is similar to or larger than the activation energy for dissociation.

Some more recent studies will illustrate various aspects of the relation of ion structures and spectra, and of ion rearrangement chemistry:

Diene Radical Ions. The conjugated diene ion chromophore is one of the most intense and easily recognized [88]. These ions universally show a pair of peaks: one in the visible around 500 nm with cross section of the order of 5×10^{-18} cm^2; and one in the UV around 350 nm with cross section typically an order of magnitude larger. Some recent work has expanded our understanding of rearrangements of some of these ions and the relation of the spectra to structural details.

The spectra of isomeric hexadiene ions [97] provide a good example
of the mapping of ion structure/rearrangement chemistry by photodissociation. The conjugated isomers 1,3- and 2,4-hexadiene give
ions whose spectra clearly identify them as conjugated ions
according to the pattern noted above. (These two isomers do give
distinguishable spectra, however, and do not interconvert.) The 1,4
isomer gives an ion spectrum identical with that of 2,4-hexadiene
ion, and clearly the 1,4 to 2,4 rearrangement following electron
impact ionization is rapid compared with the 1-second time scale of
the experiment. Such a 1-3 hydrogen shift to bring a double bond
into conjugation seems to be universal: Among many other instances
known, 1,4-octadiene and 1,4-pentadiene ions similarly rearrange
into conjugation [98]. On the other hand, the 1,5-hexadiene
isomer, and similarly the 1,5-, 2,6- and 1,7-octadiene isomers,
yield spectra clearly characteristic of the isolated double bond
chromophore in radical ions [88], with strong dissociation in the
far UV, and certainly do not rearrange into conjugation to an
appreciable extent.

More subtle is the question of cis-trans isomerization in these
ions. The three cis-trans isomers of 2,4-hexadiene ion are just
barely distinguishable from their low-resolution spectra, but laser
photodissociation at about 1 nm resolution resolves the three
structures comfortably (Fig. 11). Observing clearly distinct peaks
for these three isomers rules out their interconversion by rotation
about the double bonds (which is far easier than in the neutral
diene). Molecular orbital calculations prove to be useful in
predicting the peak shifts due to cis-trans isomerism in these ions:
The ordering of the photodissociation peaks for the three cis-trans
isomers and the spacing between them of about 7 nm, is reproduced
exactly by a Koopmans' Theorem calculation using MINDO/3.

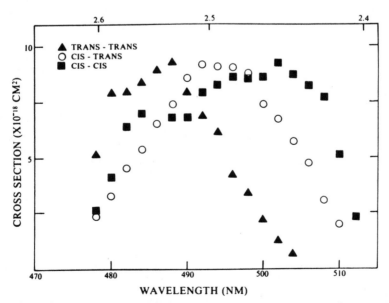

Fig. 11. Laser photodissociation spectra of the three cis-trans
isomers of 2,4-hexadiene parent ion [97].

Separate cis-trans isomers of 1,3-hexadiene have not been studied, but the visible-light spectrum of the ions from a mixed-isomer sample shows two peaks, which may well corrrespond to the two non-interconverting isomers. The visible-region spectrum of the 1,4-hexadiene sample is also interesting, in showing three peaks which suggest the formation of the trans-trans, cis-trans and cis-cis isomers of 2,4-hexadiene ions in a ratio of 3:2:1 when the initially-formed 1,4-hexadiene ions rearrange by bond migration to the conjugated structure.

$C_6H_6O^+$ A careful study [99] has untangled much structural chemistry in the $C_6H_6O^+$ system. Two ion structures, the phenol structure 1 and the dienone structure 2 (see Eq. 9) were clearly identified

$$(9)$$

by their photodissociation spectra, structure 1 giving characteristically strong far-UV dissociation, and structure 2 strong near-UV and weaker visible peaks. The m/e 94 ion was generated from a number of neutral precursors, and the extents of production of the two ion isomers were found to be as shown schematically in Eq. 9. The evidence was against slow tautomerism of 1 and 2, indicating that upon ionization a given molecule decides rapidly between possible ion isomers.

Benzyl Chloride. The structural study of the ion generated from benzyl chloride has been a long-developing story. The first observations [100] used the original time-resolved photodissociation approach, in which a population of ions was built up in the ion trap whose dissociation was then followed as a function of time until either all the ions were dissociated, or those remaining were photo-inert. Benzyl chloride was one of the first cases found in which part of the ion population did not photodissociate, and it was

concluded that electron impact ionization forms a mixture of
structures, 30% of which (probably unrearranged parent ions) dissociated rapidly at 600 nm, and 70% of which (perhaps having a
chlorotoluene structure) were photo-inert at this wavelength.

This sensible picture was overturned by Morgenthaler and Eyler's
observation [101] that in their pulsed-laser instrument, the ions
behaved as a homogeneous population which dissociated cleanly at 600
nm. The contradiction between these two apparently valid experiments was resolved with a further experiment [102] using pulsed ion
production followed by dissociation with gated lasers at varying
times after ion formation. It was found that the two ion structures
are indeed present, but are not both formed by the initial electron
impact. The chemistry is as shown in Eq. 10:

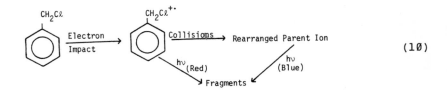

$$(10)$$

Electron impact forms the red-absorbing ions observed in both
previous studies. At high pressures and long times, a collisional
process converts these into ions of different structure which
dissociate in the blue but not in the red, requiring about 50
collisions for the conversion to happen. Fig. 12 shows clearly the
increase in blue photodissociation and decrease in red
photodissocia- tion with time, as the isomerization proceeds.
Morgenthaler and Eyler, working at short times and modest pressures,
naturally saw no collisionally-converted ions.

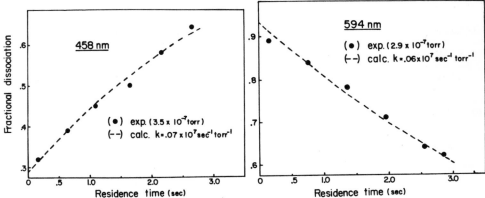

Fig. 4. Comparison of $C_7H_7Cl^+$ photodissociation at red and blue
wavelengths as a function of ion residence time. Following ion
formation, ions are stored and undergo collisions for the indicated
residence time, and are then sampled by a brief (300 ms) pulse of
laser irradiation. The increase in blue dissociation and decrease
in red dissociation match the expectation for a simple collisinal
rearrangement process (dashed lines).

<u>Cyclic Dienes</u>.   The photodissociation peaks of the typical open-
chain diene ion are at 345 nm and 490 nm. Cyclizing seems to give a
significant red shift to the UV peak, while the effect on the
visible peak is unpredictable [103].   Thus 1,3-cyclohexadiene has
its peaks at 360 nm and 489 nm, compared with about 325 nm and 495
nm for an average of hexadiene ion isomers.   1,3-cyclooctadiene
shows peaks at 360 nm and 560 nm, compared with 330 nm and 500 nm
for the straight-chain octadienes. There is no possibility that the
cyclohexadiene ring opens to the triene structure, since the spectra
of these two ions are completely orthogonal in the visible region.

These highly characteristic spectra make isomer distinctions very
easy in the cyclic diene ions.   Both 1,4-cyclohexadiene and
methylcyclopentadiene give parent ions having spectra identical in
peak positions and intensities with the 1,3-cyclohexadiene spectrum,
and there is no doubt that in both cases the ions are rearranging to
the most stable 1,3-cyclohexadiene structure: This rearrangement is
downhill by about 40 kcal/mole in the methylcyclopentadiene case.
1,4-cyclooctadiene ion similarly rearranges completely to a
1,3-cyclooctadiene structure (Eq. 11):

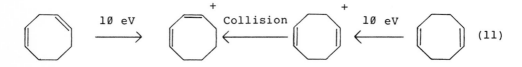

$$\text{(11)}$$

1,5-cyclooctadiene, on the other hand, yields a spectrum that looks
like a mixture of 1,5- and 1,3- isomers; Recent work with a gated
light source has shown that the absorption due to the 1,3- isomeric
component of the mixture decreases with lower pressure and shorter
reaction time.   Apparently this is a case similar to that of benzyl
chloride, in which the parent ion is formed by electron impact with
unrearranged parent structure, and then undergoes a collision-
induced rearrangement to the more stable structure.

<u>Cis-Trans Isomerism: Chloropropene and Stilbene Ions</u>.   The question
of cis-trans isomerism in radical cations was discussed above
briefly for the dienes.   Another case for which there is some
information is the chloropropene ions. The cis and trans isomers of
1-chloropropene ion are readily distinguished in their laser photodissociation spectra by their well resolved and distinctive vibrational structure [104,105], and it is clear that rotation about the
double bond does not occur.

A classic test system for problems of this sort is the cis and trans
stilbenes, 3 and 4, and recently Gooden and Brauman [106] have

(3)   (4)

succeeded in obtaining excellent photodissociation spectra of the parent ions from both isomers. The spectra of the two were identical, both showing a strong, sharp peak at 460 nm and a broad, weak absorption in the red. For these ions high quality matrix spectra are also available, and comparison shows that the gas-phase ions have the trans structure: the peak in the matrix spectrum of 4 is similar to the photodissociation spectra (with a matrix shift of 10 nm to the red), while the matrix spectrum of 3 is quite different. Clearly the cis ion rearranges to trans in the gas phase.

Dissociation With Rearrangement. The characteristic kinetic behavior of two-photon dissociation gives an unexpected means of viewing some rearrangement-dissociation chemistry. For example, the dissociation of p-iodotoluene, 5, yields $C_7H_7^+$. The simple bond-cleavage pro-

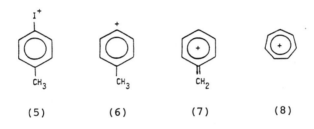

(5)   (6)   (7)   (8)

duct, tolyl ion 6, has its one-photon threshold near 530 nm (by analogy with the accurately known iodobenzene thermochemistry). The

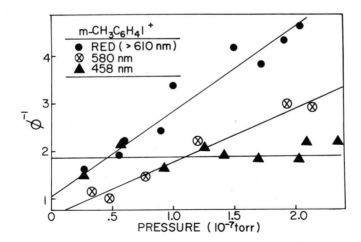

Fig. 13. Pressure dependence of the iodotoluene dissociation at wavelengths of 458 and 580 nm.

rearrangement dissociation to yield benzyl (7) or tropylium (8)
product ions is much easier, and is accessible by one-photon
dissociation at all visible wavelengths. As shown in Fig. 13,
photodissociation at 458 nm is a one-photon process, as expected
(signalled by the lack of pressure dependence). But the strong
pressure dependence at 580 nm indicates a clear two-photon dissociation at this wavelength, and rules out the possibility of the
rearrangement dissociation to yield 7 or 8 as a one-photon product.

A striking contrast is seen with bromotoluene, whose thermochemistry
is similar to that of iodotoluene ion, except that the threshold for
one-photon dissociation to tolyl ion 2 is around 440 nm. For this
ion, pressure independent one-photon dissociation is seen not only
in the blue, but also at wavelengths longer than 600 nm. So
dissociation at long wavelengths must proceed through a rearrangement path to yield more stable products, presumably 3 or 4. It
is not known, and will not be easy to find out, whether the intact
parent ion rearranges after absorption of the second photon and then
dissociates, or whether there is a concerted rearrangement/dis-
sociation process.

ACKNOWLEDGEMENTS

Gratitude is extended to all those collaborators who contributed to
the work of our laboratory described here. The author expresses
appreciation for the support of the National Science Foundation and
of the donors of the Petroleum Research Fund, administered by the
American Chemical Society.

REFERENCES

1. R.C. Dunbar, In Physical Methods of Modern Chemical Analysis (T.
Kuwana, Ed., Academic Press, 1980, Vol. 2).

2. R.C. Dunbar, In Gas Phase Ion Chemistry (M.T. Bowers, Ed.,
Academic Press, 1979, Vol. 2).

3. J.T. Moseley and J. Durup, Ann. Rev. Phys. Chem. 32 (1981) 53.

4. J.T. Moseley and J. Durup, J. Chim. Phys. 77 (1980) 673.

5. J.T. Moseley, In Applied Atomic Collision Physics (H.S.W. Massey,
B. Bederson and E.W. McDaniel, Eds., Academic Press, 1982, Vol. 5).

6. R.J. Saykally and R.C. Woods, Ann. Rev. Phys. Chem. 32 (1981)
403.

7. R.L. Woodin, D.S. Bomse and J.L. Beauchamp, In Chemical and
Biochemical Applications of Lasers (C.B. Moore, Ed., Vol. 4).

8. H.G. Dehmelt and K.B. Jefferts, Phys. Rev. 125 (1962) 1318.

9. G.H. Dunn, In Atomic Collision Processes (M.R.C. McDowell, Ed.,
North Holland Publ., 1964).

10. R.C. Dunbar, J. Am. Chem. Soc. 93 (1971) 4354.

11. J.H. Richardson, L.M. Stephenson and J.I. Brauman, J. Am. Chem.
Soc. 96 (1974) 3671.

12. J.-B. Ozenne, D. Pham and J. Durup, Chem. Phys. Lett. 17 (1972) 422.

13. R.C. Dunbar, J. Am. Chem. Soc. 95 (1973) 472.

14. N.P.F.B. van Asselt, J.G. Maas and J. Los, Chem. Phys. Lett. 24 (1974) 555.

15. M.L. Vestal and J.H. Futrell, Chem. Phys. Lett. 28 (1974) 559.

16. B.S. Freiser and J.L. Beauchamp, J. Am. Chem. Soc. 96 (1974) 6260.

17. J.R. Eyler and G.H. Atkinson, Chem. Phys. Lett. 28 (1974) 217.

18. P.C. Cosby, R.A. Bennett, J.R. Peterson and J.T. Moseley, J. Chem. Phys. 63 (1975) 1612.

19. A. Carrington, D.J. Milverton and P.J. Sarre, Mol. Phys. 32 (1976) 297.

20. R.A. Beyer and J.A Vanderhoff, J. Chem. Phys. 65 (1976) 2313,

21. B.A. Huber, T.M. Miller, P.C. Cosby, H.D. Zeman, R.L. Leon, J.T. Moseley and J.R. Peterson, Rev. Sci. Instrum. 48 (1977) 1306.

22. T.F. Thomas, F. Dale and J.P. Paulson, J. Chem. Phys. 67 (1977) 793.

23. D.A. McCrery and B.S. Freiser, J. Am. Chem. Soc. 100 (1978) 2902.

24. D.C. McGilvery and J.D. Morrison, Int. J. Mass Spectrom. Ion Phys. 28 (1978) 81.

25. R.J. Dugan, L.N. Morgenthaler, R.O. Daubach and J.R. Eyler, Rev. Sci. Instrum. 50 (1979) 691.

26. P.N.T. van Velzen and W.J. van der Hart, Chem. Phys. Lett. 62 (1979) 135.

27. M. Larzilliere, M. Carre, M.L. Gaillard, J. Rostas, M. Horani and M. Velghe, J. Chim. Phys. Phys. Chim. Biol. 77 (1980) 689.

28. F.N. Preuninger and J.M. Farrar, J. Chem. Phys. 74 (1981) 5330.

29. E.S. Mukhtar, I.W. Griffiths, F.M. Harris and J.H. Beynon, Int. J. Mass Spectrom. Ion Phys. 37 (1981) 159.

30. R.G. Cooks, J.H. Beynon, R.M. Caprioli and G.R. Lester, Metastable Ions (Elsevier, 1973).

31. J.D. Hays, Ph.D. Thesis, Case Western Reserve Univ. (1980).

32. R.C. Dunbar and J.M. Kramer, J. Chem. Phys. 58 (1973) 1266.

33. R.G. Orth and R.C. Dunbar, Chem. Phys. 45 (1980) 195.

34. J.T. Moseley, J.-B. Ozenne and P.C. Cosby, J. Chem. Phys. 74 (1981) 337.

35. J.F. Hiller and M.L. Vestal, J. Chem. Phys. 77 (1982) 1248.

36. M.T. Riggin, R.G. Orth and R.C. Dunbar, J. Chem. Phys. 65 (1976) 3365.

37. R.B. Cody and B.S. Freiser, Anal. Chem. 51 (1979) 547.

38. B.S. Freiser and J.L. Beauchamp, J. Am. Chem. Soc. 98 (1976) 3136.

39. L.N. Morgenthaler and J.R. Eyler, J. Chem. Phys. 71 (1979) 1486.

40. R. Gooden and J. I. Brauman, J. Am. Chem. Soc. 99 (1977) 1977.

41. R. Gooden and J.I. Brauman, J. Am. Chem. Soc. 104 (1982) 1483.

42. M.S. Kim, R.C. Dunbar and F.W. McLafferty, J. Am. Chem. Soc. 100 (1978) 4600.

43. I.W. Griffiths, E.S. Mukhtar, F.M. Harris and J.H. Beynon, Int. J. Mass Spectrom. Ion Phys. 38 (1981) 333.

44. I.W. Griffiths, E.S. Mukhtar, F.M. Harris and J.H. Beynon, Int. J. Mass Spectrom. Ion Phys. 39 (1981) 125.

45. I.W. Griffiths, F.M. Harris, E.S. Mukhtar and J.H. Beynon, Int. J. Mass Spectrom. Ion Phys. 41 (1981) 83.

46. E.S. Mukhtar, I.W. Griffiths, F.M. Harris and J.H. Beynon, Int. J. Mass Spectrom. Ion Phys. 37 (1981) 159.

47. E.S. Mukhtar, I.W. Griffiths, I.W. March, F.M. Harris and J.H. Beynon, Int. J. Mass Spectrom. Ion Phys. 41 (1981) 61.

48. J.H. Chen and R.C. Dunbar, To be published.

49. P.J. Robinson and K.A. Holbrook, Unimolecular Reactions (Wiley, New York, 1972).

50. V. Franchetti, B.S. Freiser and R.G. Cooks, Org. Mass Spectrom. 13 (1978) 106.

51. C. Koppel and F.W. McLafferty, J. Am. Chem. Soc. 98 (1976) 8293.

52. R.G. Orth and R.C. Dunbar, Chem. Phys. 45 (1980) 195.

53. R.G. Orth, R.C. Dunbar and M.T. Riggin, Chem. Phys. 19 (1977) 279.

54. A. Tabche-Fouhaille, J. Durup, J.T. Moseley, J.-B. Ozenne, C. Pernot and M. Tadjeddine, Chem. Phys. 17 (1976) 81.

55. C.E. Klots, D. Mintz and T. Baer, J. Chem. Phys. 66 (1977) 5100.

56. I.W. Griffiths, F.M. Harris, E.S. Mukhtar and J.H. Beynon, Int. J. Mass Spectrom. Ion Phys. 38 (1981) 127.

57. E.S. Mukhtar,I.W. Griffiths, I.W. March, F.M. Harris and J.H. Beynon, Int. J. Mass Spectrom. Ion Phys. 41 (1981) 61.

58. M.J. Coggiola, P.C. Cosby and J.R. Peterson, J. Chem. Phys. 72 (1980) 6507.

59. M.J. Coggiola, P.C. Cosby, H. Helm, J.R. Peterson and R.C. Dunbar, J. Chem. Phys. (Submitted for publication, 1982).

60. J.T. Moseley, R.P. Saxon, B.A. Huber, P.C. Cosby, R. Abouaf and M. Tadjeddine, J. Chem. Phys. 67 (1977) 1659.

61. R. Abouaf, B.A. Huber, P.C. Cosby, R.P. Saxon and J.T. Moseley, J. Chem. Phys. 68 (1978) 2406.

62. R.G. Orth and R.C. Dunbar, J. Chem. Phys. 68 (1978) 3254.

63. J.H. Richardson, L.M. Stephenson and J.I. Brauman, J. Am. Chem. Soc. 96 (1974) 3671.

64. L.N. Morgenthaler and J.R. Eyler, J. Chem. Phys. 74 (1981) 4256.

65. B.S. Freiser and J.L. Beauchamp, Chem. Phys. Lett. 35 (1975) 35.

66. P.N.T. van Velzen, Ph.D. Thesis Univ. of Leyden (1981).

67. M.S. Kim and R.C. Dunbar, Chem. Phys. Lett. 60 (1979) 247.

68. N.B. Lev and R.C. Dunbar, J. Phys. Chem., In press (1983).

69. J.M. Jasinski and J.I. Brauman, J. Chem. Phys. 73 (1980) 6191.

70. R.M. Rosenfeld, J.M. Jasinski and J.I. Brauman, J. Am. Chem. Soc. 101 (1979) 3999.

71. N.B. Lev and R.C. Dunbar, Chem. Phys. Lett. 84 (1981) 483.

72. R.C. Dunbar, J. Phys. Chem. In press (1983).

73. R.C. Dunbar, J.D. Hays, J.P. Honovich and N.B. Lev, J. Am. Chem. Soc. 102 (1980) 3950.

74. J.P. Honovich and R.C. Dunbar, J. Am. Chem. Soc., 104 (1982) 6220.

75. C.A. Wight and J.L. Beauchamp, Chem. Phys. Lett. 77 (1981) 30.

76. N.B. Lev and R.C. Dunbar, To be published.

77. R.C. Dunbar, Spectrochim. Acta 31A (1975) 797.

78. J.M. Jasinski, R.N. Rosenfeld, F.K. Meyer and J.I. Brauman, J. Am. Chem. Soc. 104 (1982) 652.

79. J.T. Moseley, P.C. Cosby, J.-B. Ozenne and J. Durup, J. Chem. Phys. 70 (1979) 1474.

80. D.C. McGilvery and J.D. Morrison, J. Chem. Phys. 67 (1977) 368.

81. R.C. Dunbar and H.H. Teng, J. Am. Chem. Soc. 100 (1978) 2279.

82. J.T. Moseley, P.C. Cosby and J.R. Peterson, J. Chem. Phys. 65 (1976) 2512.

264 Robert Dunbar

83. R.G. Orth and R.C. Dunbar, J. Am. Chem. Soc. 104 (1982) 5617.

84. C.M. Rynard and J.I. Brauman, Inorg. Chem. 19 (1980) 3544.

85. J.P. Honovich and R.C. Dunbar, J. Phys. Chem. 85 (1981) 1558.

86. B.S. Freiser and J.L. Beauchamp, J. Am. Chem. Soc. 99 (1977) 3214.

87. R.C. Dunbar, J. Chem. Phys. 68 (1978) 3125.

88. R.C. Dunbar, Anal. Chem. 48 (1976) 723.

89. R.C. Dunbar, J. Phys. Chem. 83 (1979) 2376.

90. R.C. Dunbar, H.H. Teng and E.W. Fu, J. Am. Chem. Soc. 101 (1979) 6506.

91. R.C. Benz and R.C. Dunbar, J. Am. Chem. Soc. 101 (1979) 6363.

92. R.N. Rosenfeld, J.M. Jasinski and J.I. Brauman, J. Chem. Phys. 71 (1979) 1030.

93. C.A. Wight and J.L. Beauchamp, J. Am. Chem. Soc. 103 (1981) 6499.

94. J.P. Honovich and R.C. Dunbar, J. Phys. Chem. In press.

95. D.S. Bomse and J.L. Beauchamp, Chem. Phys. Lett. 77 (1981) 25.

96. R.C. Dunbar and E.W. Fu, J. Am. Chem. Soc. 95 (1973) 2716.

97. R.C. Benz, P.C. Claspy and R.C. Dunbar, J. Am. Chem. Soc. 103 (1981) 1799.

98. R.C. Dunbar and G.B. Fitzgerald, To be published.

99. P.N.T. van Velzen, W.J. van der Hart, J. van der Greef, N.M.M. Nibbering and M.L. Gross, J. Am. Chem. Soc. 104 (1982) 1298.

100. E.W. Fu, P.P. Dymerski and R.C. Dunbar, J. Am. Chem. Soc. 98 (1976) 337.

101. L.N. Morgenthaler and J.R. Eyler, Int. J. Mass Spectrom. Ion Phys. 37 (1981) 153.

102. J.P. Honovich and R.C. Dunbar, Int. J. Mass Spectrom. Ion Phys. 42 (1982) 33.

103. J.D. Hays, Ph.D. Thesis, Case Western Reserve Univ. (1980).

104. R.G. Orth and R.C. Dunbar, J. Am. Chem. Soc. 100 (1978) 5949.

105. R.G. Orth and R.C. Dunbar, J. Am. Chem. Soc. 104 (1982) 5617.

106. R. Gooden and J.I. Brauman, J. Am. Chem. Soc. 104 (1982) 1483.

SUBJECT INDEX

AUTHOR INDEX